Biomechanics and Motor Control

Defining Central Concepts

Mark L. Latash and
Vladimir M. Zatsiorsky
Department of Kinesiology, The Pennsylvania
State University, PA, USA

ELSEVIER

AMSTERDAM • BOSTON • HEIDELBERG • LONDON
NEW YORK • OXFORD • PARIS • SAN DIEGO
SAN FRANCISCO • SINGAPORE • SYDNEY • TOKYO
Academic Press is an imprint of Elsevier

Academic Press is an imprint of Elsevier
125 London Wall, London EC2Y 5AS, UK
525 B Street, Suite 1800, San Diego, CA 92101-4495, USA
225 Wyman Street, Waltham, MA 02451, USA
The Boulevard, Langford Lane, Kidlington, Oxford OX5 1GB, UK

ISBN: 978-0-12-800384-8

British Library Cataloguing-in-Publication Data
A catalogue record for this book is available from the British Library

Library of Congress Cataloging-in-Publication Data
A catalog record for this book is available from the Library of Congress

For information on all Academic Press publications
visit our website at http://store.elsevier.com/

  Working together
to grow libraries in
developing countries

www.elsevier.com • www.bookaid.org

Publisher: Nikki Levy
Acquisition Editor: Nikki Levy
Editorial Project Manager: Barbara Makinster
Production Project Manager: Caroline Johnson
Designer: Matthew Limbert

Typeset by TNQ Books and Journals
www.tnq.co.in

Dedication

To our wives, children, and grandchildren – the main source
of happiness in life.

Contents

Preface

Biomechanics of human motion and motor control are young fields of science. While early works in biomechanics can be traced back to the middle of the nineteenth century (or even earlier, to studies of Borelli in the seventeenth century!), the first journal on biomechanics, the *Journal of Biomechanics*, has been published since 1968, the first international research seminar took place in 1969, and the International Society of Biomechanics was founded in 1973 at the Third International Conference of Biomechanics (with approximately 100 participants). Motor control as an established field of science is even younger. While many consider Nikolai Bernstein (1896–1966) the father of the field of motor control, the journal *Motor Control* started only in 1997, the first conference—Progress in Motor Control—was held at about the same time (1996), and the International Society of Motor Control was established only in 2001.

Both biomechanics and motor control have developed rapidly. Currently, these fields are represented in many conferences, and many universities worldwide offer undergraduate and graduate programs in biomechanics and motor control. This rapid growth is showing the importance of studies of biological movements for progress in such established fields of science as biology, psychology, and physics, as well as in applied fields such as medicine, physical therapy, robotics, and engineering.

New scientific fields explore new topics and work with new concepts. Scientists are compelled to name them. When the field is not completely mature, terms are often used with imprecise or varying meanings. It is also tempting to adapt terms from more established fields of science (e.g., from physics and mathematics) and apply them to new objects of study, frequently with no appreciation for the fact that those terms have been defined only for a limited, well-defined set of objects or phenomena. As a result, these established terms lose their initial meaning and become part of jargon. This is currently the case in the biomechanics and motor control literature. Lack of exactness and broad use of jargon are slowing down progress in these fields. Inventing new terms, that is, renaming the same phenomena or processes without bringing a new well-defined meaning, can make the situation even worse.

The main purpose of this book is to try to clarify the meaning of some of the most frequently used terms in biomechanics and motor control. The present situation can barely be called acceptable. Consider, as an example, the title of a (nonexistent) paper: "The contribution of *reflexes* to *muscle tone*, *joint stiffness*, and *joint torque* in *postural* tasks." As the reader will see in the ensuing chapters, all the main words in this title are "hints": they are either undefined or defined differently by different researchers.

There are two contrasting views on the importance of establishing precise terminology in new fields of science. One of the leading mathematicians of the twentieth century, Israel M. Gelfand (1913–2009; a winner of all the major prizes in mathematics and a member of numerous national academies) was seriously interested in motor control. Israel Gelfand once said: "The worst method to describe a complex problem is to do this with hints." A contrasting quotation (from one of the prominent scientists in the field—we will not name him): "We should stop arguing about terms; this is a waste of time. We should work." The authors of this book consider themselves students of Israel Gelfand and share his opinion—arguing about terms is one of the very important steps in the development of science. Using undefined or ambiguously defined terms (jargon) is worse than a waste of time; it leads to misunderstanding and sometimes creates factions in the scientific community where it becomes more important to use the "correct words" than to understand what they mean.

A cavalier attitude to terminology may lead to major confusion. Consider, as an example, published data on *muscle viscosity*. In the literature review on muscle viscosity (Zatsiorsky 1997) it was found that this term had been used with at least 11 different meanings, 10 of which disagree with the definition of viscosity in the International System of Units (SI). Diverse experimental approaches applied in similar situations resulted in sharply dissimilar viscosity values (the difference was, sometimes, thousand-fold). Even the units of measurement were different. This is an appalling situation.

Biomechanics mainly operate with terms borrowed from classical (Newtonian) mechanics. By themselves, these terms are precisely defined and impeccable. However, their use in biomechanics needs caution. In some cases, new definitions are necessary. For instance, such a common term as *joint moment* does not exist in classical mechanics. It is essentially jargon. Skeptics are encouraged to peruse the mechanics textbooks; you will not find this term there. In other cases, application of notions from classical mechanics needs some refinement. For instance, the classical mechanical concept of *stiffness* cannot be (and should not be!) applied to the joints within the human body. The term *stiffness* describes resistance of deformable bodies to imposed deformation; however, the joints are not bodies and joint angles can be changed without external forces. In other words, if for deformable bodies, for instance, linear springs, there exist one-to-one relations between the applied force and the spring length, there is no such a relation between the joint angle and joint moment, or muscle length and muscle force. In the mechanics of deformable bodies, stiffness is represented by the derivative of the force–deformation relation. However, to call any joint moment–joint angle derivative, as some do, *joint stiffness*, would make this term a misnomer. In some situations adding an adjective to the main term, for example, using the term *apparent stiffness*, can be an acceptable solution.

As compared to biomechanics, the motor control terminology is at a disadvantage—in contrast to biomechanics, it cannot be based on strictly defined concepts and terms of classical mechanics. Some of the terms used in motor control, if not invented specifically for this field, are borrowed from such fields as medicine and physiology. Not all of them are precisely defined and understood by all users in the same way. Examples are such basic and commonly used terms as *reflex*, *synergy*, and *muscle tone*.

Motor control, as a field of science, aims at discovery of laws of nature describing the interactions between the central nervous system, the body, and the environment during the production of voluntary and involuntary movements. This definition makes motor control a subfield of natural science or, simply put, physics. At the contemporary level of science, relevant neural processes cannot be directly recorded. In fact, the situation is even worse. Even if one had an opportunity to get information about activity of all neurons within the human body, it is not at all obvious what to do with these hypothetical recordings of such activity. The logic of the functioning of the central nervous system cannot be deduced from knowledge about functioning of all its elements; this was well understood by Nikolai Bernstein and his students. This makes motor control something like "physics of unobservables"—laws of nature are expected to exist, but relevant variables are not directly accessible for measurement.

To overcome these obstacles, scientists introduce various models and hypotheses that can be only in part experimentally confirmed (or disproved). As a result, the motor control scientists work with unknown variables, and these unknowns should be somehow named. It is a challenging task to find a proper term for something that we do not know. A delicate balance should be maintained; the term should be as precise as possible, and, at the same time, it should not induce a false impression that we really know what is happening within the brain and the body.

The target audience of this book is researchers and students at all levels. We believe that using exact terminology has to start from the undergraduate level; hence, we tried to make the contents of the book accessible to students with only minimal background knowledge. While the book is not a textbook, it can be used as additional reading in such courses as Biomechanics, Motor Control, Neuroscience, Physiology, Physical Therapy, etc.

Individual chapters in the book were selected based on personal views and preferences of the authors. We tried to cover a broad range, from relatively clear concepts (such as *joint torque*) to very vague ones (such as *motor program* and *synergy*). There are many other concepts that deserve dedicated chapters. But some of the frequently used concepts are covered in the existing chapters (e.g., *internal models* are covered in the chapter on *motor programs*, similarly to how these notions are presented in Wikipedia); others have been covered in recent reviews (such as *normal movement*, Latash and Anson 1996, 2006); and with respect to others, the authors do not feel competent enough (e.g., *complexity*). We hope that our colleagues will join this enterprise and write comprehensive reviews or books covering important notions that are not covered in this book.

The book consists of four parts. Part 1 covers biomechanical concepts. It includes the chapters on *Joint torques*, *Stiffness and stiffness-like measures*, *Viscosity, damping and impedance*, and *Mechanical work and energy*. Part 2 deals with basic neurophysiological concepts used in the field of motor control such as *Muscle tone*, *Reflex*, *Preprogrammed reactions*, *Efferent copy*, and *Central pattern generator*. Part 3 concentrates on some of the central motor control concepts, which are specific to the field and have been used and discussed extensively in the recent motor control literature. They include *Redundancy and abundance*, *Synergy*, *Equilibrium-point hypothesis*, and *Motor program*. Part 4 includes two chapters with examples from the field of motor behavior, *Posture* and *Prehension*. Only two behaviors have been selected based on the personal experience of the authors; they cover two ends of the spectrum of human movements, from whole-body actions to precise manipulations. The book ends with the detailed Glossary, in which all the important terms are defined.

Mark L. Latash
Vladimir M. Zatsiorsky

Acknowledgments

The book reflects the personal views of the authors developed over decades of work in the fields of biomechanics and motor control. During that time, we have been strongly influenced by many of our colleagues and students. We would like to thank all our colleagues/friends (too many to be named!) who helped us develop our views, participated in numerous exciting research projects, and provided frank critique of our own mistakes. Our graduate and postdoctoral students have played a very important role not only by performing studies cited in the book but also by asking questions and engaging in discussions that forced us to select words carefully to achieve maximal exactness and avoid embarrassment. Many of our colleagues are unaware of the importance of their influence on our current understanding of the fields of biomechanics and motor control. We are very grateful to all researchers who performed first-class studies (many of which are cited in the book) leading to the current state of the field of movement science.

Part One

Biomechanical Concepts

Joint Torque

1

Concept of *joint torques*—or *joint moments* as many prefer to call them—is one of the fundamental concepts in the biomechanics of human motion and motor control. A computer search in Google Scholar for the expression *joint torques* yielded 194,000 research papers. Even if we discard the returns that are due to the possible "search noise," the number of publications in which the above concepts were used or mentioned is huge. The authors themselves were surprised with these enormous figures.

Such popularity should suggest that the term is well and uniformly understood and its use does not involve any ambiguity. It is not the case, however. In classical mechanics the concept of joint torques (moments) is not defined and is not used. Peruse university textbooks on mechanics. You will not find these terms there. One of the authors vividly remembers a conference on mechanics attended mainly by the university professors of mechanics where a biomechanist presented his data. He was soon interrupted with a question: "Colleague, you are using the term 'joint moment' which is unknown to us. Please explain what exactly you have in mind."

1.1 Elements of history

An idea that muscles generate moments of force at joints was understood already by G. Borelli (1681). Joints were represented as levers with a fulcrum at the joint center and two forces, a muscle force and external force acting on a limb, respectively. The concept of levers in the analysis of muscle action was also used by W. and E. Weber (1836). Only static tasks have been considered.

Determining joint moments during human movements is a sophisticated task (usually called the *inverse problem of dynamics*). It requires:

1. Knowledge of the mass-inertial characteristics of the human body segments, such as their mass, location of the centers of mass, and moments of inertia (German scientists Harless (1860) and Braune and Fisher (1892) were the first to perform such measurements on cadavers).
2. Recording the movements with high precision that allows computing the linear and angular accelerations of the body links. This was firstly achieved by Braune and Fisher, and the study was published in several volumes in 1895—1904. It took the authors almost 10 years to digitize and analyze by hand the obtained stroboscopic photographs of two steps of free walking and one step of walking with a load. Later, in 1920, N. A. Bernstein (English edition, 1967) improved the method, both the filming and the digitizing techniques. It took then "only" about 1 month to analyze one walking step. With contemporary techniques it can be done in seconds.

Biomechanics and Motor Control. http://dx.doi.org/10.1016/B978-0-12-800384-8.00001-6

3. Solving the inverse problem mathematically and performing all the computations. For simple two-link planar cases (such as a human leg moving in a plane), this was first done by Elftman (1939, 1940). The computations were done by hand. With the development of modern computers the opportunity arose to study more complex (but still planar) movements (Plagenhoef, 1971). For the entire body moving in three dimensions the first successful attempts of computing the joint torques during walking and running in main human joints in 3D were reported only in the mid-1970s (Zatsiorsky and Aleshinsky, 1976; Aleshinsky and Zatsiorsky, 1978).

Existence of the *interactive* forces and torques, i.e., the joint torques and forces induced by motion in other joints, and their importance, was well recognized by N. A. Bernstein (1967) for whom it was one of the main motivations for developing his ideas on the motor control.

1.2   What are the joint torques/moments?

Consider first what classical mechanics tell us. This is really an elementary material.

1.2.1   *Return to basics: moment of a force and moments of a couple*

In mechanics two basic concepts are introduced, moment of force and moment of couple.

Moment of force. According to Newton's second law ($F = ma$) a force acting on an unconstrained body induces a linear acceleration of the body in the direction of the force. Also, any force that does not intersect a certain point generates a turning effect about this point, a *moment of force*. The moment acts on a body to which the force is applied.

The moment of force $\mathbf{M}_O$ about a point O is defined as a cross product of vectors $\mathbf{r}$ and $\mathbf{F}$, where $\mathbf{r}$ is the position vector from O to the point of force application and $\mathbf{F}$ is the force vector.

$$\mathbf{M}_O = \mathbf{r} \times \mathbf{F} \tag{1.1}$$

The line of action of $\mathbf{M}_O$ is perpendicular to the plane containing vectors $\mathbf{r}$ and $\mathbf{F}$. The line is along the axis about which the body tends to rotate at O when subjected to the force $\mathbf{F}$. The magnitude of moment is $M_o = F(r\sin\theta) = Fd$, where θ is the angle between the vectors $\mathbf{r}$ and $\mathbf{F}$ and d is a shortest distance from O to the line of action of $\mathbf{F}$, the *moment arm*. The moment arm is in the plane containing O and $\mathbf{F}$. The direction of the moment vector $\mathbf{M}_O$ follows the right-hand rule in rotating from $\mathbf{r}$ to $\mathbf{F}$: when the fingers curl in the direction of the induced rotation the vector is pointing in the direction of the thumb.

Oftentimes the object of interest is not a moment of force about a point $\mathbf{M}_O$ but the moment about an axis $O-O$, for instance a flexion—extension axis at a joint. Such a moment $\mathbf{M}_{OO}$ equals a component (or projection) of the moment $\mathbf{M}_O$ along the axis $O-O$. The moment magnitude can be determined as a mixed triple product of three vectors: the unit vector along the axis of rotation $\mathbf{U}_{OO}$, the position vector from an arbitrary point on the axis to any point on the line of force action $\mathbf{r}$, and the force vector $\mathbf{F}$.

$$M_{OO} = \mathbf{U_{OO}} \cdot (\mathbf{r} \times \mathbf{F}), \tag{1.2}$$

where $\cdot$ is a symbol of a dot (inner) product of vectors.

The moment (of a nonzero magnitude) can be determined with respect to any point that is not intersected by the line of force action. The choice of the point is up to a researcher. If a body is constrained in its linear motion—it can only rotate like body links can in majority of joints—the moments are commonly determined with respect to a joint center or axis of rotation.

Moment of couple. The term a *force couple*, or simply a *couple*, is used to designate two parallel, equal, and opposite forces. Force couples exert only rotation effects. The measure of this effect is called the *moment of a couple*. The moment of a couple does not depend on the place of the couple application. In computations couples can be moved to any location. Because of that the moments of the couples are often called *free moments* (Figure 1.1).

Here is the proof of the above statement. Let $\mathbf{r}_A$ and $\mathbf{r}_B$ be the position vectors of the points A and B, respectively, and $\mathbf{r}$ is the position vector of B with respect to A ($\mathbf{r} = \mathbf{r}_A - \mathbf{r}_B$). The vector $\mathbf{r}$ is in the plane of the couple but need not be perpendicular to the forces $\mathbf{F}$ and $-\mathbf{F}$. The combined moment of the two forces about O is

$$\mathbf{M}_O = \mathbf{r}_A \times \mathbf{F} + \mathbf{r}_B \times (-\mathbf{F}) = (\mathbf{r}_A - \mathbf{r}_B) \times \mathbf{F} = \mathbf{r} \times \mathbf{F} \tag{1.3}$$

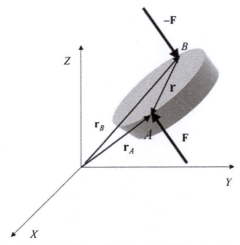

Figure 1.1 Two equal and opposite forces $\mathbf{F}$ and $-\mathbf{F}$ a distance apart constitute a couple. The magnitude of their combined moment does not depend on the distance to any point and, hence, the couple can be translated to any location in the parallel plane or in the same plane.

The product $\mathbf{r} \times \mathbf{F}$ is independent of the vectors $\mathbf{r}_A$ and $\mathbf{r}_B$, i.e., it is independent of the choice of the origin O of the coordinate reference. Hence, the moment of couple $\mathbf{M}_c = \mathbf{r} \times \mathbf{F}$ does not depend on the position of O and has the same magnitude for all moment centers.

The main differences between forces and force couples are:

1. Forces induce effects both in translation and rotation; couples generate only rotation effects.
2. Rotation action of a force depends on the point of its application (the larger the distance from the line of force action to the rotation center, the larger is the moment of force about this center). Rotation effect of a couple does not depend on the place of its application.

1.2.2 Defining joint torques

Consider a revolute joint connecting two body links, the joint allowing only rotation of the adjacent segments. The moments of force acting on the segments are computed with respect to the joint center. Assume that these moments are equal and opposite. This happens for instance in engineering when a revolute joint is served by a *torque actuator*; e.g., an electric motor that converts electric energy into a rotation motion. In such a case, by virtue of Newton's third law, the two moments of force acting on the adjacent segments are equal and opposite. The moments can be collectively referred to as a joint moment (or joint torque). Hence, the term designates not one but two equal and opposite moments acting on the adjacent body segments.

The human joints are powered, however, not by torque actuators but by muscles that are *linear actuators*. For the human body we cannot declare that the existence of the joint torques, i.e., two equal and opposite moments of force acting on the adjacent body segments, is a straightforward upshot of the Newton's third law. It is true that when a muscle, or a muscle—tendon complex, pulls on a bone an equal and opposite force is acting on the muscle. It is also true that the force is transmitted along the muscle—tendon complex. However, what effect is produced at another end of the muscle is an open question. The answer depends on where and how exactly the muscle is connected to body tissues. It can be connected to an adjacent body segment (single-joint muscles) or to a nonadjacent segment (as two-joint and multijoint muscles), to several bones (for instance, the extrinsic muscles of the hand such as the flexor digitorum profundus fan out into four tendons, that attach to different fingers), the muscles can curve and wrap around other tissues, etc. One specific case is presented at the knee joint where the quadriceps force is transmitted via the patella that acts as a first-class lever with the fulcrum at the patellofemoral contact. The location of the contact changes with the joint flexion. Therefore, at various joint angles the same force of the quadriceps is transmitted to the ligamentum patellae and then to the muscle insertion as force of different magnitude. The ratio "patellar force/quadriceps force" reaches its maximal value of 1.27 at 30° knee flexion and minimum of 0.7 at 90° and 120° knee flexion (Huberti et al., 1984). Hence the forces at the origin and insertion of the muscle could be different.

We limit our discussion to two main cases, single-joint and two-joint muscles. We consider a planar case with ideal hinge joints and only one muscle.

1.2.3 Joint torques at joints served by single-joint muscles

Consider a one-joint muscle crossing a frictionless revolute joint. The muscle develops collinear forces at the points of its effective origin, F^{mus}, and insertion, $-F^{mus}$. The forces, F^{mus} and $-F^{mus}$, are equal in magnitude but pull in opposite directions and act on the different links. Their moment arms, d, about the joint center are the same. Hence, the forces produce moments of force of the same magnitude and opposite in direction; a clockwise moment is applied to one link and a counterclockwise moment to another. According to the definition provided above, two equal and opposite moments of force about a common axis of joint rotation applied to two adjacent segments can be referred to as the joint moment (or joint torque). There is no problem here.

1.2.4 Joint torques at joints served by two-joint muscles

For the joints served by two-joint muscles, the notion of joint torque—as defined above—cannot be immediately applied. Consider a two-joint muscle spanning hinge joints J_1 and J_2 (Figure 1.2).

In the presented example the forces, F^{mus} and $-F^{mus}$, are equal in magnitude and point in opposite directions. Their moment arms about the joint center 1 (d_1) are the same. Hence the moments of force about joint center 1, M_1 and $-M_3$, are equal and opposite but they act on the nonadjacent segments, S_1 and S_3. Therefore, the moments do not satisfy the definition for the joint torques given above ("two equal and opposite moments acting on the *adjacent* body segments") and they cannot be collectively called "joint torque."

When a single-joint muscle acts at a joint, J_1, the moments of force that this muscle exerts on the adjacent segments, S_1 and S_2, induce turning effects on the both segments at this joint. If the segments are allowed to move they will rotate—due to the existing joint moment—toward each other. In contrast, when a two-joint muscle acts on the nonadjacent segments, S_1 and S_3, the moments of force are still equal and opposite, but, if there is no joint friction and $d_1 = d_3$, segment S_2 stays put and only segments S_1 and S_3 will rotate. We can imagine the situation when joint J_2 is "frozen." In such a case segments S_2 and S_3 behave like a single body and the moment of force, $-M_3$, acts on both segments. The moments of force, M_1 and $-M_3$, act on the adjacent bodies, S_1 and S_{2+3}, and can be collectively referred as the "joint torque at J_1."

In research practice, the existence of the joint torques is almost always assumed. This assumption is based on the following consideration.

The forces acting on S_1 and S_3 (Figure 1.2(B)) can be equivalently represented by a resultant force acting at the joint contact point and a force couple acting on the segment (Figure 1.2(C)). The forces acting on S_2 do so through joint centers J_1 and J_2 and, therefore, do not generate moments about the joint they are passing through. Thus, biarticular muscles do not immediately create moments of force about the joints to an intermediate segment. However, the two forces acting on joint centers of S_2 are equal in magnitude, parallel, and opposite in direction. Consequently, they can be represented by a force couple 2 with the moment arm d_2. In general, moment arms d_1, d_2, and d_3 may be different. Thus, force couples 1, 2, and 3 may be different too.

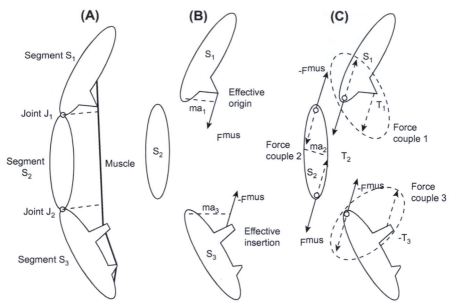

Figure 1.2 Forces and couples acting in a three-link chain served by one two-joint muscle. The muscle does not exert a moment(s) of force on an intermediate segment. The segment rotation is due to force couple 2, two equal and opposite forces acting at the joint centers J_1 and J_2. (A) A three-segment system (S_1, S_2, and S_3) with one biarticular muscle crossing two frictionless revolute joints J_1 and J_2. The muscle, shown by a bold line, attaches to segments S_1 and S_3 and spans segment S_2. The dashed lines show the moment arms, d_1 and d_2, the shortest distances from the joint centers 1 and 2 to the line of force action, respectively. Note that the muscle does not directly exert force—and hence a moment of force—on segment S_2. (B) Muscle force acting on segments S_1 and S_3 and the different moment arms. (C) An equivalent representation of forces produced by the biarticular muscle. Forces acting on S_1 and S_2 are equivalently represented by a force acting at the joint center and a force couple. Two equal and opposite parallel forces act on segment S_2 at the joint centers. They may cause the segment to rotate.
Reprinted by permission from Zatsiorsky and Latash (1993), © Human Kinetics.

According to the previous definition, the term *joint torque* collectively refers to two equal in magnitude moments of force acting on adjacent segments about the same joint rotation axis. Because the magnitudes of the moments acting on the adjacent segments served by a biarticular muscle are different, if the definition is strictly followed, these moments cannot be collectively referred to as joint torque.

The relation between torques T_1, T_2, and T_3 is determined by the relation between moment arms d_1, d_2, and d_3 (the same muscle force F^{mus} causes all three moments). Since all three moment arms are perpendicular to the line of muscle action, they are running in parallel. Therefore, the moment arm of force couple 2 equals the difference between d_1 and d_3: $d_2 = d_1 - d_3$. Torque acting on S_2 about J_1 is:

$$T_2 = F^{mus} d_2 = F^{mus}(d_1 - d_3) = T_1 + T_3 \tag{1.4}$$

since $T_3 = -F^{mus}d_3$. Note that torque T_2 is generated by the forces acting at the joint centers.

These forces form a force couple. In a particular case when d_1 equals d_3, d_2 is zero, and the resultant moment of forces applied to the intermediate segment S_2 at the joint centers J_1 and J_2 is zero.

When existence of the joint torques is assumed, the logic of calculations is as follows:

1. Torque T_3 acting on (distal) segment S_3 about J_2 is determined.
2. It is assumed that a torque $-T_3$ is acting on an intermediate segment S_2 about J_2 (which is not true).
3. It is assumed that torque T_3 contributes also to torque T_1 about J_1; then, T_1 is determined as an algebraic sum $T_1 = -(T_3 + T_2)$.

As a result, the moments of forces acting on the intermediate segment S_2 are estimated as: $-T_3$ in J_2 and $-T_1$ ($=T_3 + T_2$) in J_1. According to Eqn (1.4) the algebraic sum of these two moments is exactly T_2. Therefore, the addition of $-T_3$ to the moment acting on S_2 in J_2 and subsequent subtraction of the same value when the moment is calculated in J_1 does not change the external effects as the torque systems T_2 and $(T_2 + T_3 - T_3)$ are equivalent. Thus, when concern is given only to the external effects of the forces, e.g., in static analysis of kinematic chains consisting of rigid links, joint torques may be introduced. They should, however, be considered "equivalent" rather than "actual" joint torques.

The difference between actual joint torques (produced by single-joint muscles) and equivalent torques (calculated for the system served by two-joint muscles) may be important in some situations.

1.2.5 On the delimitations of the joint torque model

All models simplify the situation that the model addresses. Something is inevitably lost. For instance, if human body segments are considered solid bodies, their deformation cannot be studied. So what exactly is lost when the joint torques model is used and muscles are replaced with torque actuators, essentially motors, located at the joint centers?

Evidently, all the effects associated with activity of individual muscles are neglected. A most evident example is co-contraction of antagonists. If a movement analysis is limited to the joint torques, we cannot know whether antagonists are active or not.

Muscles commonly produce moments of force not only in the desired direction (*primary moments*) but also in other directions (*secondary moments*; Mansour and Pereira, 1987; Li et al., 1998a,b). To counterbalance the secondary moments, which are not necessary for the intended purpose, additional muscles should be activated. Consider, for example, a forceful arm supination with the elbow flexed at a right angle, as in driving a screw with a screwdriver. During the supination effort, the triceps, even though it is not a supinator, is also active. A simple demonstration proves this: perform an attempted forceful supination against a resistance while placing the second hand on the biceps and triceps of the working arm. Both the biceps and the triceps spring into action simultaneously. The explanation is simple: when the biceps acts as a supinator, it also produces a flexion moment (secondary moment). The flexion moment is counterbalanced by the extension moment exerted by the triceps. Such effects cannot be understood in the framework of the joint torque model.

Also, the difference between activity of single-joint and two-joint (multijoint) muscles is lost. The difference is mainly in their effects on (1) the forces acting on the involved body segments and (2) the mechanical energy expenditure. For instance, in a three-link system with two one-joint muscles (Figure 1.3(A)), when $T_1 = T_3$, a bending stress is acting on the intermediate link. For a similar system with one two-joint muscle, the bending stress is zero—a compressive force is only acting on S_2. When the actual forces and moments acting at the joints are replaced by joint torques, this distinction is lost.

(A)   **(B)**

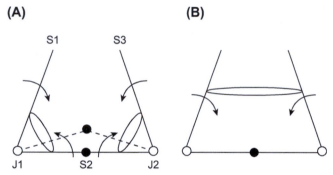

Figure 1.3 A three-link system served by two one-joint muscles (A) or one two-joint muscle (B). An additional locked pin joint (filled circle) is located in the middle of segment S_2. If the locked joint is released, (A) left and right parts of S_2 will rotate in the direction shown by the arrows (upward) and (B) the unlocked joint will be in a state of unstable equilibrium. Adapted by permission from Zatsiorsky and Latash (1993), © Human Kinetics.

Even more striking differences between joints served by one-joint muscles or multi-joint muscles occur when attempts at calculation of total mechanical work (or total mechanical energy expenditure) for several joints are made. The reason is that two-joint muscles can transfer mechanical energy from one body segment to another one to which they are attached. In some cases the length of the muscle stays put and the muscle does not produce any mechanical work, and they act as ropes or cords (so-called "tendon action of two-joint muscles"). Hence, if the joints are served by single-joint muscles only and if a muscle is forcibly stretched, the energy expended for the stretching can either be temporarily stored as elastic energy or dissipated into heat. The energy cannot be transferred to a neighboring joint. In contrast, in the chains served by two-joint muscles the mechanical energy can be transferred from one joint to another. Consider a simplified example (we will return to this example later in the text for a more detailed analysis), in Figure 1.4.

Consider a slow horizontal arm extension with a load in the hand (Figure 1.4). The mass of the body parts, as well as the work to change the kinetic energy of the load, is neglected. During the movement represented by the broken line, the muscles of

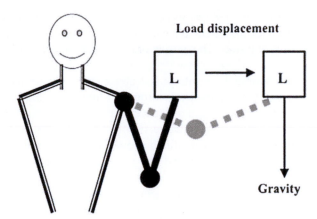

Figure 1.4 A slow horizontal arm extension with a load in the hand. The work done on the load is zero but the work of joint torques is not.
Reprinted by permission from Zatsiorsky and Gregor (2000), © Human Kinetics.

the shoulder joint perform positive (concentric) work—they generate an abduction moment and elevate the arm. The elbow joint extends while producing a flexion moment against the weight of the load. The external load does the work on the elbow flexors, forcibly stretching them. The flexors, however, actively resist the stretching; they are spending energy for that. Therefore, it can be said that the flexors of the elbow produce negative (eccentric) work. The work of the force exerted by the hand on the load is zero. The direction of the gravity force is at a right angle to the direction of the load displacement and, hence, the potential energy of the load does not change. The total work done *on* the system (arm plus load) is zero. What is the total amount of work done *by* the subject? How should we sum positive work/power at the shoulder joint with the negative work/power at the elbow joint?

The problem is whether the negative work at the elbow joint cancels the positive work at the shoulder joint (or, in other words, the positive work at the shoulder compensates for the negative work at the elbow). The correct answer depends on the information that was not provided in the preceding text.

If the joints are served by only one-joint muscles, the joint torque model is valid; the system is operated by the "actual" joint torques and the mechanical energy expended at one joint is lost; it does not return to other joints. Hence, if the joint powers at the shoulder and elbow joint are P_1 and $-P_2$, respectively, the total power can be obtained as the sum of their absolute values:

$$P_{tot} = |P_1| + |-P_2| = 2P_1 \qquad (1.5)$$

If the body does not spend energy for resisting at the elbow joint (e.g., the resistance is due to friction) $P_2 = 0$, and the equation becomes $P_{tot} = |P_1|$.

When one two-joint muscle serves both joints, total produced power equals zero:

$$P_{tot} = P_1 + (-P_2) = 0 \qquad (1.6)$$

Negative power from decelerating joint 2 is used to increase the mechanical energy at joint 1. In this example, the length of the muscle is kept constant and the muscle does not produce mechanical power. The muscle only transfers the power from one link to another.

According to the terminology introduced above, the joint torques produced by two-joint muscles are the "equivalent" torques, while those that are due to one-joint muscles are the "actual" torques. Hence, Eqn (1.5) describes the total power supplied by the actual joint torques while Eqn (1.6) represents the situation when the torques are equivalent. In real life the joints are commonly served by both single-joint and multijoint muscles and the situation is more complex than it is described above.

In general, an expression "a torque/moment at joint X," if not defined explicitly, can be misleading. The expression "a moment of force Y exerted on segment Z around a joint axis X" does not lead to confusion.

1.3   Joint moments in statics and dynamics

In spite of the above criticism about the "joint torque model," the model is indispensable in movement analysis. The problem is not that the concept is bad; the problem is that the concept should be properly understood.

Some models of motor control are based on an assumption that the central nervous system (CNS) plans and immediately controls joint torques. This is a debatable idea, especially when "equivalent" joint torques that are due to two-joint muscles are involved. Since equivalent joint torques are just abstract concepts, a proper understanding of the expression "the CNS controls joint torques" is important. Consider as an example another abstract concept—the center of mass (CoM) of a body. It is well known that the sum of external forces acting on a rigid body equals the mass of the body times the acceleration of the body's CoM. Does it mean that a central controller, in order to impart a required acceleration to the body, controls the force at the CoM? Yes and no—in a roundabout way "yes" but actually "no." The CoM is an imaginable point, a fictitious particle that possesses some very important features. The CoM can be outside the body, like in a bagel. One cannot actually apply a force to the CoM of a bagel since it is somewhere in the air. Hence, the expression "to control a force at the CoM" should not be taken or thought of literally. A similar situation occurs with joint torques. The CNS cannot immediately control joint torques because they are just abstract concepts—like force acting at the CoM of a bagel.

1.3.1   Joint torques in statics. Motor overdeterminacy and motor redundancy

To manipulate objects as well as to move one's own body people exert forces on the environment. Biomechanics provide tools for determining: (1) the force exerted on the environment from the known values of the joint torques (*direct problem of statics*) and (2) the joint torques from the known values of the endpoint force (*inverse problem of statics*).

In statics the relation between the force $\mathbf{F}$ exerted at the end of a kinematic chain, such as an arm or a leg, and the joint torques is represented by the equation

$$\mathbf{T} = \mathbf{J}^T \mathbf{F} \tag{1.7}$$

where $\mathbf{T}$ is the vector of the joint torques ($\mathbf{T} = T_1, T_2 \ldots T_n)^T$ and $\mathbf{J}^T$ is the transpose of the Jacobian matrix that relates infinitesimal joint displacement $d\alpha$ to infinitesimal end effector displacement $d\mathbf{P}$. Equation (1.7) describes a solution for the inverse problem of statics. In three-dimensional (3D) space, the dimensionality of force vector $\mathbf{F}$, called a *generalized external force*, is six: three force components, acting along the axes X_1, X_2, and X_3, and three moment components M_1, M_2, and M_3 about these axes. A generalized external force $\mathbf{F}$ is often called simply *contact force* or *end effector force*. The dimensionality of vector $\mathbf{T}$ equals the number of degrees of freedom (DOF) of the chain, N. In three dimensions the Jacobian is a $6 \times N$ matrix. In a plane, $\mathbf{F}$ is a 3×1 vector and the Jacobian is a $3 \times N$ matrix. The joints are assumed frictionless and gravity is neglected.

According to Eqn (1.7), for a given arm posture the joint torques are uniquely defined by the force vector $\mathbf{F}$, i.e., an individual joint torque T_i cannot be changed without breaking the chain equilibrium. If $N > 6$, or in a planar case $N > 3$, the task is said to be *overdetermined*. An example can be a pressing with an arm or a foot against an external object. For a given force vector $\mathbf{F}$, all the involved joint torques should satisfy Eqn (1.7). There is no freedom for the performer here. For instance, when a subject exerts a given force with her fingertip, the six joint torques (at the DIP, PIP, MCP, wrist, elbow, and shoulder joints) should satisfy the equation "joint torque = endpoint force × joint moment arm," where the moment arms are the shortest distances from the joint center to the line of fingertip force action. The torques should be exerted simultaneously and in synchrony. We should admit that—with our current knowledge—we do know how the central controller does this.

The overdeterminacy is an opposite side of the well-known problem of *motor redundancy*, also called *motor abundance* (Latash, 2012; see Chapter 10). The problem of motor redundancy arises when the system has more degrees of freedom that are absolutely necessary for performing a motor task. Therefore the task can be performed in various ways and the problem for a researcher is to figure out why the central controller prefers some solutions over the others. In overdetermined tasks the system still may have a large number of degrees of freedom but there is no freedom in solution for the central controller.

The skeletal system is often modeled as a combination of serial and/or parallel chains. Overdeterminacy can occur for the serial chains in statics and parallel chains in kinematics, e.g., when several fingers act on a grasped object. Motor redundancy can occur for (1) serial chains in kinematics, and (2) parallel chains in statics. Both these events happen when the number of control variables exceeds the number of the task constraints (see Zatsiorsky, 2002, Chapter 2).

When a motor task, e.g., an instruction given by the researcher, does not prescribe all components of vector $\mathbf{F}$, the performer has freedom to perform the task in different ways. For instance, the performer is asked to exert a force of a given magnitude in a prescribed direction, but nothing is said about the moment, e.g., grasp moment, production. The performer may produce a moment at his/her will and change the joint torques correspondingly. This is an example of an *underspecified task*. In such tasks the performer's freedom is limited to selection of the nonprescribed components of vector **F**. When all components of vector **F** are specified, the task is not redundant and Eqn (1.7) is strictly obeyed.

To clarify the geometric meaning of the transpose Jacobian presented in Eqn (1.5), we consider a simple planar two-link chain. For such a chain (see Figure 1.5 below) the Jacobian is:

$$
\mathbf{J} = \begin{bmatrix} -l_1 S_1 - l_2 S_{12} & -l_2 S_{12} \\ l_1 C_1 + l_2 C_{12} & l_2 C_{12} \end{bmatrix}
$$

(1.8)

where the subscripts 1 and 2 refer to the angles α_1 and α_2, and correspondingly, the subscript 12 refers to the sum of the two angles, $(\alpha_1 + \alpha_2)$, and the symbols S and C designate the sine and cosine functions, respectfully. Equation (1.7) assumes the following form:

$$
\begin{bmatrix} T_1 \\ T_2 \end{bmatrix} = \begin{bmatrix} -l_1 S_1 - l_2 S_{12} & l_1 C_1 + l_2 C_{12} \\ -l_2 S_{12} & l_2 C_{12} \end{bmatrix} \begin{bmatrix} F_X \\ F_Y \end{bmatrix}
$$

(1.9)

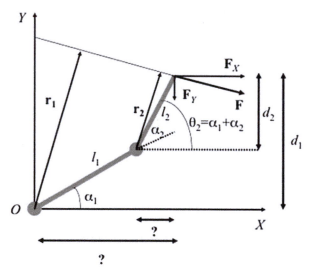

Figure 1.5 The correspondence between the moment arms of the external force **F** and the rows of the transpose Jacobian. The equation can be analyzed both in the vector form and in the scalar form, in projections on the coordinate axes (using the transpose Jacobian). Greek letters α designate the joint angles and the letters θ designate the angles in the external reference frame O-XY. The numbers 1 and 2 refer to the first and second joint/link. Symbols d_1 and d_2 in the figure illustrate the magnitude of the moment arms of the horizontal force component F_X about joints 1 and 2, respectfully. The thick horizontal arrows with the question marks below the figure illustrate the moment arms of the vertical force component F_Y. The readers are invited to determine their values.

The first row of the transpose Jacobian in Eqn (1.9) represents the coefficients of the equation used to determine the torque at joint 1 $T_1 = (-l_1S_1 - l_2S_{12})F_X + (l_1C_1 + l_2C_{12})(-F_Y)$. Similarly, the second row refers to the torque at the second joint.

A comment on the sense of the coefficients in the equations: in the presented example the horizontal force component $\mathbf{F}_X$ is in a positive direction and the vertical component $\mathbf{F}_Y$ is in a negative direction. Hence, both force components produce moments of force at joints 1 and 2 in the negative direction, i.e., clockwise.

The joint torques $\mathbf{T}_1$ and $\mathbf{T}_2$ can also be computed by using the cross product of vectors $\mathbf{r}_i$ and $\mathbf{F}$ ($\mathbf{T} = \mathbf{r}_i \times \mathbf{F}$), where both $\mathbf{T}$ and $\mathbf{r}_i$ are the 2×1 vectors. The torques have the magnitude $T_1 = Fr_1$ and $T_2 = Fr_2$ where r_1 and r_2 are the perpendicular distances from the corresponding joint to the line of $\mathbf{F}$.

1.3.2 Control of external contact forces: from the joint torques to the external force

This section deals with the static exertion of an intended contact force on the environment. We adopt a joint torque model and—because we are mainly interested in key principles—limit analysis to planar tasks.

The question under discussion is: what joint torques should be produced to exert a desired endpoint force? As already mentioned above, this question represents the direct problem of statics. If the position of a kinematic chain, i.e., an arm or leg, is specified, biomechanics offer at least two ways of analysis. The task can be analyzed either in projections on the coordinate axes, i.e., in the scalar form or with a vector method.

Scalar method (*Jacobian method*). This method naturally leads to relying on Eqn (1.7) and its by-products. If the kinematic chain is not a singular position, i.e., is not completely extended, Eqn (1.7) can be inverted and endpoint force, i.e., two force components along the coordinate axes and the moment, determined:

$$\mathbf{F} = \left[\mathbf{J}^T\right]^{-1}\mathbf{T} \tag{1.10}$$

where for a planar chain $\mathbf{F}$ is a 3×1 endpoint force vector, $[\mathbf{J}^T]^{-1}$ is the inverse of the chain transpose Jacobian, $\mathbf{T}$ is an $N \times 1$ vector of joint torques, where N is the number of joints.

For a two-link chain the inverse of the transposed Jacobian is:

$$[\mathbf{J}^T]^{-1} = \frac{1}{l_1 l_2 S_2} \begin{bmatrix} l_2 C_{12} & -l_1 C_1 - l_2 C_{12} \\ l_2 S_{12} & -l_1 S_1 - l_2 S_{12} \end{bmatrix} \tag{1.11}$$

For a three-link planar chain, for instance for an arm model that includes an upper arm, forearm, and hand and describes a human arm grasping a handle, the entire equation is:

$$\mathbf{F} = \left[J^T\right]^{-1}\mathbf{T} = \begin{bmatrix} \dfrac{C_{12}}{l_1 S_2} & \dfrac{-l_2 C_{12} - l_1 C_1}{l_1 l_2 S_2} & \dfrac{C_1}{l_2 S_2} \\[2ex] \dfrac{S_{12}}{l_1 S_2} & \dfrac{-l_2 S_{12} - l_1 S_1}{l_1 l_2 S_2} & \dfrac{S_1}{l_2 S_2} \\[2ex] \dfrac{l_3 S_3}{l_1 S_2} & \dfrac{-l_2 l_3 S_3 - l_1 l_3 S_{23}}{l_1 l_2 S_2} & \dfrac{l_3 S_{23} + l_2 S_2}{l_2 S_2} \end{bmatrix} \begin{bmatrix} T_1 \\ T_2 \\ T_3 \end{bmatrix} = \begin{bmatrix} F_X \\ F_Y \\ M \end{bmatrix}$$

(1.12)

where $\mathbf{F}$ is a 3×1 endpoint force vector that includes two force components and the grasping moment (the rotation moment exerted on the handle), T_1, T_2, and T_3 are the torques at the shoulder, elbow, and wrist joints, respectively, and other symbols have been defined previously. As seen from Eqn (1.12), the endpoint force components are determined as additive functions of all three joint torques. Each endpoint force component equals a dot product of a corresponding row of matrix $[\mathbf{J}^T]^{-1}$ and the joint torque vector. For instance, the grasp (endpoint) moment can be determined from the equation

$$M = \frac{l_3 S_3}{l_1 S_2} T_1 + \left[\frac{-l_2 l_3 S_3 - l_1 l_3 S_{23}}{l_1 l_2 S_2}\right] T_2 + \frac{l_3 S_{23} + l_2 S_2}{l_2 S_2} T_3$$

(1.13)

Such equations are convenient for computing but they do not allow simple graphical representation and they are difficult to comprehend. This can be done, however, when a vector approach is used.

Vector method (*geometric method*). The method is based on the postulate that the joints under consideration are ideal rotational joints (hinges). "Ideal" in this context means that the joint movements are frictionless and do not involve any deformation of the joint structures, such as for joint cartilage. Also, no linear translation in the joints has place. Under such assumptions, the old adage of mechanical engineers is valid: "hinge joints transmit only forces; they do not transmit moments." Having this motto in mind, let us consider a two-link chain—which can be seen as a highly simplified arm model—that exerts an endpoint force on the environment (Figure 1.6).

The endpoint force is a vector sum of the two forces: (1) due to the shoulder joint torque—along the pointing axis, and (2) due to the elbow joint torque—along the radial axis. Force (1) is transmitted along the second link (the forearm-hand segment). This force does not generate a moment at the elbow joint; force (2) does not generate a moment at the shoulder joint. With the described approach the individual joint torques are converted into the endpoint force components that are summed up vectorially.

When the number of the links at the chain exceeds two, the force exerted on the environment still can be resolved into the components associated with the individual

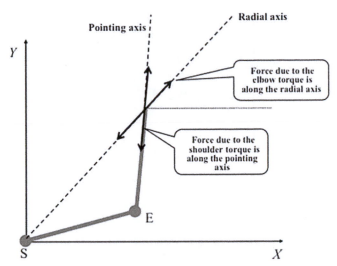

Figure 1.6 The pointing and radial axes of a two-link arm and the endpoint forces that are generated by shoulder (S) and elbow (E) torques. Both axes intersect at the endpoint of the chain. The pointing axis intersects also the elbow joint center while the radial axis intersects the shoulder joint center. Flexion torques are in counterclockwise direction, and extension torques are in the clockwise direction. A flexion torque at the E (S) joint generates endpoint force along the pointing (radial) axis toward the S (E) joint, and the extension torque generates the force in opposite direction. The actual endpoint force (not shown) can be considered a vector sum of the above two forces.

joint torques. However, the components are usually not concurrent at the endpoint and they cannot be reduced to merely one resultant force. Instead, the overall effect on the environment can be represented by a resultant force and a couple (in 3D case, by a force and a wrench). Consider a planar three-link chain in a nonsingular configuration (Figure 1.7).

An external force is exerted on the end link of the chain at a point P. It is not necessarily for P to be at the endpoint of the distal link (unlike ballet dancers who can stand on their toes, most people stand on the entire plantar surface of the foot). To find the contributing forces, we introduce lines passing through the joint centers, L_{23}, L_{13}, and L_{12}, where the subscripts refer to the corresponding joint centers. Because a force that intersects a joint center does not produce a moment of force at this joint, the line of force action that is solely due to the torque at joint 1 must intersect joint centers 2 and 3. The same is valid for other joints. The following rule exists: individually applied joint torques, T_1, T_2, and T_3, cause the end effector to apply forces to the environment along the lines L_{23}, L_{13}, and L_{12}, respectively.

The forces $\mathbf{F}_1$ and $\mathbf{F}_2$ are along the lines L_{23} and L_{13}. In the two-link chains, these lines would be along the radial and pointing directions, respectively. Forces $\mathbf{F}_1$ and $\mathbf{F}_2$ are concurrent at joint 3 but not at the end point. Force $\mathbf{F}_3$ is—rather contraintuitively—along the proximal link. The three forces, $\mathbf{F}_1$, $\mathbf{F}_2$, and $\mathbf{F}_3$, are coplanar and may be reduced to a single resultant force $\mathbf{F}$ and a couple C applied to the end link of the chain.

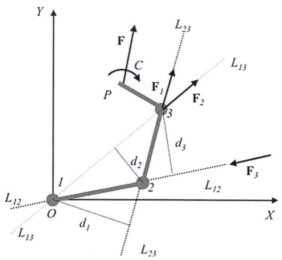

Figure 1.7 Static analysis of a planar three-link chain. The torque actuators at joints 1, 2, and 3 produce the joint torques that contribute to the end effector force **F**. The torque T_1 acting at joint 1 develops a contributing force $\mathbf{F}_1$ along the line L_{23}. The magnitude of $\mathbf{F}_1$ is equal to the ratio T_1/d_1 where d_1 is the moment arm. The magnitudes of the contributing forces from the other joints can be computed in a similar way as the quotients $F_2 = T_2/d_2$ and $F_3 = T_3/d_3$. These forces are acting along the lines L_{13} and L_{12}, correspondingly. Forces $\mathbf{F}_1$ and $\mathbf{F}_2$ are shown with their tails at joint 3. Force $\mathbf{F}_3$ is shown along the line of its action. Note that with this representation the force is along the proximal link. A couple C, represented in the figure by a curved arrow, is also exerted on the environment.

The end link transmits the force-couple system to the environment. Consequently, the three-link systems allow for not only exerting push–pull forces on the environment but also for producing rotational effects. In particular, both a force and a couple can be exerted on working tools.

In motor control literature, some researchers discussed whether the central controller plans the movements and force generation on the environment in the internal or external coordinates. In static tasks, the first approach corresponds to the direct problem of statics (from the joint torques to the endpoint force) and the second to the inverse problem (from the endpoint force to the joint torques). While the present authors are not sure whether any of these two approaches is valid, it is worth mentioning that computationally the inverse problem of statics allows for much simpler solutions than the direct problem. Computation of the products $T_i = Fr_i$ (the symbols are explained in the caption to Figure 1.5) is evidently simpler than using either vector method or a scalar (Jacobian) method to compute the endpoint force from the known joint torques.

1.3.3 Joint torques in dynamics

Let us start with a simple illustration. A subject is sitting at the table with his or her upper arm horizontal and supported by the table. The elbow joint is flexed 90°.

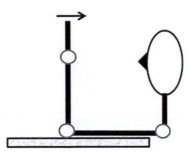

Figure 1.8 An example. A subject performs a fast elbow flexion. Consider two scenarios: (1) the muscles of the wrist joint are completely relaxed and (2) the muscles are co-contracted, such that the forearm and hand behave as a single solid body.

The forearm is oriented vertically and the wrist joint is at 180°, i.e., the hand is in extension of the forearm and vertically oriented. The subject performs a fast elbow flexion movement (Figure 1.8).

Various scenarios of the wrist/hand behavior are possible. Consider two of them:

1. The wrist joint is completely relaxed; no resistance to the wrist joint movements is provided. By definition, the joint torque at the wrist is zero. As a result of the elbow flexion, the hand location changes. The hand translation (its acceleration and deceleration) is due to the joint force. According to the model (assuming ideal rotational joints), the force is acting at the joint center and therefore is not exerting a moment about it. Besides the handle location its orientation also changes: the hand is rotated in the direction opposite to the forearm rotation, and the hand "flaps." A take home message from this example is that body links can rotate at a joint even when muscles crossing the joint are relaxed and joint torques (as they are defined above, in Section 1.2.3) are zero. Such movements can be seen in the above-knee amputees walking with a knee prosthesis. The users can rotate the shank by applying a force—not a moment, there is no actuator in the prosthesis—at the knee.
2. The wrist angle is statically fixed and remains at 180°. The hand continues to be in extension of the forearm. As a consequence of the elbow flexion, the hand location and orientation change. This indicates that both a force and a moment acted at the wrist joint on the hand.

In the latter example, when a force and a moment act on the hand equal and opposite force and moment act on the forearm. The same is valid for the elbow joint and forearm—upper arm system. If the arm were not supported, these forces and moments will propagate to the shoulder joint, trunk, and further downward. One of the authors remembers as one student asked him: does it mean that when I am talking the forces to accelerate and decelerate my chin propagate to my feet? The answer is definitely "yes." With contemporary sensitive force plates, these forces—for a standing person—can be recorded.

The joint torques and forces induced by the motion in other joints are called *inter-active* forces and torques. The term *reaction forces* (an old term) was also in use. The following simple experiment demonstrates their existence. Starting with an arm extended straight down at the side of the body, flex the elbow vigorously. If an upper arm is unsupported, the shoulder extension will be observed. The shoulder extension occurs despite the fact that none of the elbow flexors is also a shoulder extensor. Thus, the shoulder extension was performed due to the activity of the elbow flexors.

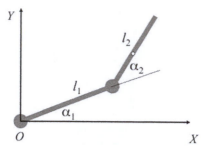

Figure 1.9 A two-link planar chain. Length of the links is l_1 and l_2, correspondingly. α_1 and α_2 are the joint angles. Also the following symbols not shown in the figure will be used: T_1 and T_2 are the joint torques; m_1 and m_2 are the masses of the links; I_1 and I_2 are the moments of inertia of the links with respect to their centers of mass; l_{c1} and l_{c2} are the distances of the link center of mass to the corresponding joint center; C and S stand for a cosine and sine function, correspondingly; subscripts 1, 2, and 12 refer to the α_1 and α_2, and α_{1+2}, correspondingly; $\dot{\alpha}$ and $\ddot{\alpha}$ stand for angular velocity and acceleration; and g is acceleration due to gravity.

The existence of the reactive forces was a motivation for Nikolai A. Bernstein to develop his theories of motor control. He recorded the effects of these forces in several human movements, such as walking, running, and playing piano.

We limit our discussion to the forces and moments acting during movement of a simple planar two-link chain (Figure 1.9) and their interaction effects. The mass-inertial characteristic of the links (their mass, locations of the CoM, and moments of inertia), position of the links, their velocity, and acceleration are supposed to be known. The goal is to find the forces and moments that caused the observed motion. We are going to write the dynamic equations of motion in the so-called *closed form*, i.e., with all the variables explicitly presented, and then analyze them.

The equations are:

$$T_1 = \underbrace{\left[I_1 + m_1 l_{c1}^2 + I_2 + m_2 \left(l_1^2 + l_{c2}^2 + 2 l_1 l_{c2} C_2 \right) \right]}_{\text{Moment of inertia of the chain about joint 1}} \ddot{\alpha}_1 + \underbrace{\left[I_2 + m_2 \left(l_{c2}^2 + l_1 l_{c2} C_2 \right) \right]}_{\substack{\text{Coupling inertia;} \\ \text{inertial effect of angular} \\ \text{acceleration of joint 2 on joint 1}}} \ddot{\alpha}_2$$

$$\underbrace{- \left(m_2 l_1 l_{c2} S_2 \right)}_{\substack{\text{Centripetal} \\ \text{coupling} \\ \text{coefficient}}} \dot{\alpha}_2^2 \; \underbrace{- \left(2 m_2 l_1 l_{c2} S_2 \right)}_{\substack{\text{Coriolis' coupling} \\ \text{coefficient}}} \dot{\alpha}_1 \dot{\alpha}_2 + \underbrace{\left[m_1 g l_{c1} C_1 + m_2 g (l_1 C_1 + l_{c2} C_{12}) \right]}_{\text{Gravity term, } G_1}$$

$$T_2 = \underbrace{\left(I_2 + m_2 l_{c2}^2 \right)}_{\substack{\text{Moment of} \\ \text{inertia of link 2} \\ \text{about joint 2}}} \ddot{\alpha}_2 + \underbrace{\left[I_2 + m_2 \left(l_{c2}^2 + l_1 l_{c2} C_2 \right) \right]}_{\substack{\text{Coupling inertia;} \\ \text{inertial effect of angular} \\ \text{acceleration of joint 1 on joint 2}}} \ddot{\alpha}_1 + \underbrace{\left(m_2 l_1 l_{c2} S_2 \right)}_{\substack{\text{Centripetal} \\ \text{coupling} \\ \text{coefficient}}} \dot{\alpha}_1^2 + \underbrace{m_2 g l_{c2} C_{12}}_{\text{Gravity term, } G_2}$$

$$(1.14)$$

Even for a simple planar two-link chain, the equations of motion are complex. When a chain has a larger number of links (>3) and moves in 3D, the closed-form equations of motion are becoming very lengthy and complex. With contemporary computers they still can be solved but usually cannot be grasped in their entirety (at least by these authors). The complexity of these equations was completely understood by Bernstein who questioned the capability of the central controller to "solve" or memorize them and looked for other ways of controlling human and animal movements.

As follows from Eqn (1.14), the *coupling inertia coefficients* determining the inertial effect of joint acceleration (1) at joint 2 ($\ddot{\alpha}_2$) on joint 1 (on T_1) and (2) at joint 1 ($\ddot{\alpha}_1$) on joint 2 (on T_2) are equal (for readers interested in mathematical proof of this statement, see Zatsiorsky, 2002, pp. 377−381). The same is valid for the so-called *centripetal coupling coefficients* at the terms that determine the dynamical effects associated with the joint angular velocity, i.e., with the effect of $\dot{\alpha}_2^2$ on T_1 and effect of $\dot{\alpha}_1^2$ on T_2.

The symmetric effect of the movements at one joint on another joint—velocity and acceleration at joint A affects the torque at joint B in the same way as velocity and acceleration at joint B affects the torque at joint A—raises a question about the meaning of the motor control theory on the existence of leading, or dominant, joints. It is a fact of mechanics that interjoint effects are symmetric, and if the velocity and acceleration at two joints are the same, their interaction effects are similar. However, if joint A moves faster—at larger velocity and acceleration—than joint B, the interaction effect of A on B will be larger than the opposite effect. Another option is that while the effects are equal in absolute values, e.g., in Nm, they may have different impacts on the large and small joints, for instance on the shoulder and wrist torques. The interaction torques of equal magnitude may affect a large joint to a smaller extent than they affect a small joint. Therefore, a clarification of what exactly is understood under a "dominant joint" is required.

On the whole, the concept of joint torques is an indispensable tool in biomechanics and motor control. To apply this tool in research, its biomechanical background should be well understood.

1.4   The bottom line

Moment of force and *moment of couple* are the fundamental concepts of classical mechanics. They describe the rotational effect on a body of a force or a force couple, i.e., two equal and opposite forces acting in opposite directions. The concept of joint torques in classical mechanics is not defined and is essentially jargon used in biomechanics as well as in some branches of mechanical engineering, in particular in robotics.

In biomechanics, the term joint torque (or a joint moment) refers to two equal and opposite moments of force acting on the adjacent body links. In animals such moments of force are generated by single-joint muscles; in technical devices, by rotational actuators, such as electric motors. When a single-joint muscle exerts forces on the adjacent segments these forces are equal, act in opposite directions, and act at the same distance from the joint center of rotation, i.e., have the same moment arms; hence, they generate on the adjacent body links equal and opposite moments of force. These two moments of force can be collectively called a joint moment.

In contrast, two-joint muscles are not attached to the intermediate body link and hence do not immediately exert a force and a moment on it. Strictly speaking, the conditions for the joint moments are not satisfied in this case. It can be shown, however, that the rotational effects of the joint forces acting at the joint centers on the intermediate body link (to which the two-joint muscle is not attached) equal the rotational effects of the muscle force acting on the adjacent body segments. Hence, existence of the joint moments (torques) can be assumed. Such *equivalent* joint torques can be used for solving many tasks of statics and dynamics. Caution should be exercised, however, when the task is to determine the mechanical loads experienced by the body segments and when determining the performed mechanical work. The chains with the *actual joint torques* (generated by single-joint muscles) and *equivalent joint torques* (due to two-joint muscles) should be analyzed differently. Note that a concept of joint torques, especially the equivalent joint torques, involves a high level of abstraction. The central controller has no tools to immediately control joint torques—it controls muscle forces whose mechanical action can be expressed via the joint torques.

In statics, biomechanics provide tools for determining: (1) the force exerted on environment from the known values of the joint torques (*direct problem of statics*) and (2) the joint torques from the known values of the endpoint force (*inverse problem of statics*).

For serial kinematic chains, such as an arm or a leg, the relation between the force $\mathbf{F}$ exerted at the end of a chain and the joint torques is represented by the equation

$$\mathbf{T} = \mathbf{J}^T \mathbf{F} \quad (1.7)$$

where $\mathbf{T}$ is the vector of the joint torques ($\mathbf{T} = T_1, T_2...T_n)^T$ and $\mathbf{J}^T$ is the transpose of the Jacobian matrix that relates infinitesimal joint displacement $d\alpha$ to infinitesimal end effector displacement $d\mathbf{P}$. The equation describes a solution for the inverse problem of statics. In 3D space, the dimensionality of force vector $\mathbf{F}$, called a *generalized external force*, is six: three force components, acting along the axes X_1, X_2, and X_3, and three moment components M_1, M_2, and M_3 about these axes. A generalized external force $\mathbf{F}$ is often called simply *contact force* or *end effector force*. The dimensionality of vector $\mathbf{T}$ equals the number of DOF of the chain, N. In three dimensions the Jacobian is a $6 \times N$ matrix. According to Eqn (1.7), for a given arm posture the joint torques are uniquely defined by the force vector $\mathbf{F}$, i.e., an individual joint torque T_i cannot be changed without breaking the chain equilibrium. If $N > 6$, or in a planar case $N > 3$, the task is said to be *overdetermined*.

The direct problem of statics deals with the question: what joint torques should be produced to exert a desired endpoint force? With the Jacobian method, the solution is described by the equation

$$\mathbf{F} = [\mathbf{J}^T]^{-1}\mathbf{T} \quad (1.10)$$

where $[\mathbf{J}^T]^{-1}$ is the inverse of the chain transpose Jacobian and other symbols are explained above. Equation (1.10) is solvable only if the transpose Jacobian $\mathbf{J}^T$ is invertible, i.e., if (1) the **J** is a square matrix and (2) the chain is not in a singular configuration, i.e., not completely extended.

For planar kinematic chains with only two or three links, the geometric method (vector method) can be used. For two-link chains, with this approach the endpoint force is treated as a vector sum of the two forces: (1) due to the shoulder joint torque— along the pointing axis (i.e., along the forearm—hand link), and (2) due to the elbow joint torque—along the radial axis from the shoulder joint center to the point of force application. When the number of the links at the chain exceeds two, the force exerted on the environment still can be resolved into the components associated with the individual joint torques. However, the components are usually not concurrent at the endpoint, and they cannot be reduced to merely one resultant force.

In dynamics, accelerated movements of one body segment occur when certain forces and moments act on it. According to Newton's third law, the equal and opposite forces and moments act on the adjacent segments (so-called *interactive* or *reactive* forces and moments). The forces and moments propagate further to other body segments and joints. This makes the movement mechanics complex even for simple planar two-link chains, see Eqn (1.14) as an example. Referring to Eqn (1.14), note that the *coupling inertia coefficients* determining the inertial effect of joint acceleration (1) at joint 2 ($\ddot{\alpha}_2$) on joint 1 (on T_1) and (2) at joint 1 ($\ddot{\alpha}_1$) on joint 2 (on T_2) are equal. The same is valid for the *centripetal coupling coefficients* at the terms that determine the dynamical effects associated with the joint angular velocity, i.e., with the effect of $\dot{\alpha}_2^2$ on T_1 and effect of $\dot{\alpha}_1^2$ on T_2. In general the effects of the movements at one joint on another joint are symmetric—velocity and acceleration at joint A affects the torque at joint B in the same way that velocity and acceleration at joint B affects the torque at joint A.

When a chain has a larger number of links (>3) and moves in 3D, the closed-form equations of motion—i.e., the equations with all the variables explicitly presented— are becoming very lengthy and complex. With current computer power these equations still can be solved but usually cannot be grasped in their entirety. The complexity of these equations was completely understood by Bernstein who questioned the capability of the central controller to "solve" or memorize them and looked for other ways of controlling human and animal movements.

References

Aleshinsky, S.Y., Zatsiorsky, V.M., 1978. Human locomotion in space analyzed biomechanically through a multi-link chain model. Journal of Biomechanics 11, 101–108.

Bernstein, N.A., 1967. The Co-ordination and Regulation of Movement. Pergamon Press, Oxford.

Borelli, G.A., 1681. De Motu Animalium. A Bernabò, Rome.

Braune, W., Fischer, O., 1892. Bestimmung der Tragheitsmomente des menschlichen Körpers und seiner Glieder. S. Hirzel, Leipzig [English translation: Maquet, P., Furlong, R., 1988. Determination of the moments of inertia of the human body and its limbs. Springer-Verlag: Berlin, New York.].

Braune, W., Fischer, O., 1895—1904. Der Gang des Menschen. B.G. Teubner, Leipzig [English translation by Maquet, P., Furlong, R., 1987. The human gait. Springer-Verlag: Berlin, New York. The chapters of the book were originally published separately. Chapter 1 appeared in 1895 under the names of Braune, W. and Fischer, O.; Braune, W., died immediately after the initial experiments. The data analysis was conducted by Fisher, O., only. Chapters 2—6 were signed by Fischer only.].

Elftman, H., 1939. Forces and energy changes in the leg during walking. American Journal of Physiology 125, 339—356.

Elftman, H., 1940. The work done by muscles in running. American Journal of Physiology 129, 672—684.

Harless, E., 1860. Die statische Momente der menschlichen Gliedermassen. In: Abhandlungen Der Mathematische-Physikalischen Klasse Der Königlich-Baverischen Akademie Der Wissenschaften, München, vol. 8, pp. 69—97.

Huberti, H.H., Hayes, W.C., Stone, J., Shybut, G.T., 1984. Force ratios in the quadriceps tendon and ligamentum patellae. Journal of Orthopaedic Research 2, 49—54.

Latash, M.L., 2012. The bliss (not the problem) of motor abundance (not redundancy). Experimental Brain Research 217, 1—5.

Li, Z.M., Latash, M.L., Zatsiorsky, V.M., 1998a. Force sharing among fingers as a model of the redundancy problem. Experimental Brain Research 119 (3), 276—286.

Li, Z.M., Latash, M.L., Zatsiorsky, V.M., 1998b. Motor redundancy during maximal voluntary contraction in four-finger tasks. Experimental Brain Research 122 (1), 71—78.

Mansour, J., Pereira, J., 1987. Quantitative functional anatomy of the lower limb with application to human gait. Journal of Biomechanics 20, 51—58.

Plagenhoef, S., 1971. Patterns of Human Motion. A Cinematographic Analysis. Prentice Hall, Inc, Englewood Cliffs, NJ.

Weber, W., Weber, E., 1992. (first German edition in 1836). Mechanics of the Human Walking Apparatus. Springer-Verlag: Berlin, Heidelberg, New York (Translated from German by Maquet, P., and Furlong, R.).

Zatsiorsky, V.M., 2002. Kinetics of Human Motion. Human Kinetics, Champaign, IL.

Zatsiorsky, V.M., Aleshinsky, S.Yu., 1976. Simulation of the human locomotion in space. In: Komi, P.V. (Ed.), Biomechanics V-B. University Park Press, Baltimore, London, Tokyo, pp. 387—394.

Zatsiorsky, V.M., Gregor, R.J., 2000. Mechanical power and work in human movement. In: Sparrow, W.A. (Ed.), Energetics of Human Activity. Human Kinetics, Champaign, IL, pp. 195—227.

Zatsiorsky, V.M., Latash, M.L., 1993. What is a joint torque for joints spanned by multiarticular muscles? Journal of Applied Biomechanics 9, 333—336.

Stiffness and Stiffness-like Measures

2

"Stiffness" (of muscles, joints, body limbs, etc.) is one of the most broadly used terms in human biomechanics and motor control literature. Regrettably, the term is also frequently ill-used, that is, used incorrectly, without a precise understanding of its meaning. The origin of the confusion is in the application of the concept developed for relatively simple deformable bodies to much more complex biological objects such as muscles, joints, or kinematic chains that may not deform but move and show changes in the configuration. As a result, a *homonymy* occurs—the same term *stiffness* is used for designating different properties. This may lead to communicative conflicts and wrong understanding.

2.1 Elements of history

The concept of stiffness (with the meaning explained below in Eqns (2.1a) and (2.1b)) is known in mechanics from the seventeenth century. British physicist Robert Hooke (1635—1703), who studied deformation of springs under external loads, found (in 1660) that "the extension is proportional to the force." Since then, this statement has been known as Hooke's law.

The stiffness of the passive muscles was first studied by M. Blix (1893). One end of the frog muscle gastrocnemius was fixed and various loads were attached to the other end. Due to the suspended load, muscle length increased. Blix himself did not determine stiffness, that is, did not divide the change in load by the change of muscle length, but the presented data easily allowed doing this. Application of the concept of stiffness to passive muscles as well as to other passive tissues such as tendons and ligaments can be technically convoluted but still conceptually unambiguous. In contrast, using this concept to active muscles and joints is associated with numerous conceptual difficulties and in many situations is questionable (discussed later in this chapter). The measurement results may represent diverse biological mechanisms and may be directly noncomparable.

In studying biomechanical properties of individual muscles substantial progress was achieved in 1920 when A. V. Hill and his coworkers (Gasser and Hill, 1924; Hill, 1925, 1950) suggested a three-component lumped-parameters model, consisting of (1) series elastic component, (2) parallel elastic component, and (3) contractile components (see Subsection 2.3.3). Since then, various methods of the measurement of the stiffness of the individual muscle components have been developed.

The term *joint stiffness* was used by clinicians for centuries to designate sensation of difficulty and pain in moving a joint as well as increased resistance at the joints seen in some patients, for instance, those with rheumatoid arthritis. The joint resistance to motion is a complex phenomenon. It depends on the mechanical characteristics of

Biomechanics and Motor Control. http://dx.doi.org/10.1016/B978-0-12-800384-8.00002-8

the joint motion, for instance, it can or cannot depend on the joint motion amplitude or joint angular velocity; it can arise from different joint structures, for example, muscles, ligaments, joint capsule, the skin, etc.; and it may have different physiological mechanisms, for example, stretch reflexes; and arise due to various clinical syndromes, from joint swelling to neural disorders.

Quantitative studies of joint stiffness started in the 1960s. The study by Johns and Wright (1962) may serve as a representative example. The authors studied resistance in the passive joints (joint stiffness) and separated it into (1) elastic stiffness that depends on the magnitude of the joint displacement and does not depend on time, (2) viscous stiffness, which is a function of rotational joint velocity, (3) plastic resistance, which is due to yielding properties of the tissues and depends on time, (4) inertia resistance that is proportional to the acceleration, and (5) friction resistance, which does not depend on amplitude, velocity, or acceleration but depends on the (unknown) force normal to the joint surfaces contact.

It is suggested below that the term *stiffness*—with or without a grammatical modifier—should be used only with the first meaning, that is, as the resistance that only depends on the displacement (but not on the velocity or acceleration). Unfortunately, studies on joint stiffness do not form a progressing line of research where subsequent studies are based on the preceding ones. Instead, overall the individual studies very often look unconnected to each other. This is at least in part due to a nonunified terminology—various authors used the term *joint stiffness* with different meanings.

In the studies on stiffness of kinematic chains, two lines of research can be seen. They focus on arm and leg stiffness, correspondingly. The studies on arm stiffness were pioneered by Mussa-Ivaldi et al. (1985) and Flash (1987). The starting point for the research was accepting that for planar and spatial movements the stiffness cannot be represented by a scalar (a number). More complex representations (matrices, vector, and scalar fields, etc.) are necessary. The authors introduced necessary mathematical tools, investigated the endpoint arm apparent stiffness, and related it to the joint stiffness. The studies concentrated only on small arm displacements, which allowed the authors to assume that during the perturbation the chain Jacobian did not change. In contrast, the studies on the leg stiffness concentrated on the stiffness in only one direction, usually in the direction "along the leg." The studies mainly deal with such activities as running and hopping where movement mechanics can be realistically modeled as movement of a mass—spring system (a pogo-stick model, "springs in the legs"). A report by Alexander and Bennet-Clark (1977) was one of the first—if not the first—in this direction. The driving point behind the research was estimating the elastic potential energy stored during landing and returned back to the system during the takeoff.

2.2 The concept of stiffness

2.2.1 The definition

In a nutshell, stiffness is a clear-cut concept. When elastic bodies—that is, the bodies that deform under force application and return to their normal shape after force removal—are being deformed, they resist the deformation. Stiffness is a measure of such a resistance.

For extension forces, similar to those that act upon tendon ends, stiffness is defined as the amount of force necessary to extend the object by one unit of length; its dimensionality in the SI system is N/m. When subjected to small forces, many deformable bodies behave linearly. That is, elongation (Δl) is linearly proportional to the force: $\Delta l = cF$, and the proportionality coefficient c is called *compliance*. Compliance is measured in m/N. The expression $\Delta l = cF$ is known as Hooke's law. For the bodies following Hooke's law, stiffness equals the quotient

$$S = \frac{F}{\Delta l} \tag{2.1a}$$

If the relation between the applied force and the deformation is nonlinear, the derivative

$$S = dF/dl \tag{2.1b}$$

is employed. For rotational movements, for instance for a single joint, the corresponding expression would be

$$S = dT/d\alpha \tag{2.1c}$$

where T is moment of force (*joint torque*, see Chapter 1) and α is joint angle.

According to Eqn (2.1c) the term *joint stiffness* refers to the resistance that depends on the magnitude of angular displacement, specifically to the amount of joint torque increase per unit of angular change. Such understanding of the term is common in biomechanics, as well as in mechanics in general. To avoid confusion, the term should not be used if the resistance does not depend on the extent of angular displacement. Unfortunately, in clinical practice this requirement is often neglected and the term is used for describing phenomena only loosely related to resistance to deformation as described in Eqn (2.1). For instance, the term may refer to pain on moving a joint (at any motion amplitude; a common symptom in osteoarthritic patients) and/or decreased range of joint motion. The expression "stiff muscles" may refer to inability to relax the muscles. In this book, the term *stiffness* is used in accordance with the definitions provided above.

As material objects are different in size, to eliminate the size effects, such variables as *stress* and *strain* are used. *Stress* is the amount of force per unit of area, N/m^2. *Strain* is a relative elongation, $\Delta l/l_0$, where l_0 is the initial length of the spring and Δl is its elongation. Strain is a dimensionless quantity. The mathematical relations between stresses and strains are called *constitutive equations*.

The simplicity of the notion of stiffness defined via Eqn (2.1) encourages its broad use. Indeed, there is nothing wrong in dividing the recorded values of force by the values of displacement. The problems are not with the computation itself but with interpretation of this quotient.

2.2.2   Passive and active objects—stiffness and apparent stiffness

The notion of stiffness has been introduced in classical mechanics for passive bodies. In the absence of external forces, passive bodies maintain constant shape; in particular, they maintain constant length. Under an external force passive bodies deform. For a given passive object, there exists a one-to-one relation between the applied force and deformation. If one variable, for example, force, is known, the matching value of the second variable, that is, length, is set, and vice versa. Passive elements of the musculoskeletal system include tendons, ligaments, fasciae, cartilage, bones, skin, and relaxed (not activated) muscles. For these objects, the notion of stiffness can be applied without conceptual difficulties.

In contrast to passive bodies, length of an active muscle as well as joint angle can change without a change in external forces. Therefore, there is no one-to-one relation between muscle force and the matching muscle length, or between joint torque and the matching joint angle. People can exert different forces at a given joint position, and they can exert the same force at various joint positions. Also, joints are not bodies; rather they are connections between adjacent segments of the human body. Hence, the situation is rather different from what was assumed in classical mechanics when the concept of stiffness was introduced to study passive deformable bodies.

There are at least two important differences between the properties of passive and active objects:

1. Because in active objects torque (force) and angle (length) can be changed independently, the unique force—length relations for active objects do not exist. For instance, a force—length relation for a muscle changes with its activation level. The measurements make sense only if the level of activation and its time course are specified, which is very hard to achieve given that muscle activation level depends on the activity of peripheral receptors sensitive to both muscle force and length (see Chapter 6). Hence, instead of a single force—length relation that is typical of passive objects, there are families of such relations for active objects. As a result, force—length combinations measured before and after an external force application may belong to different relations from such a family, and their direct comparison makes little sense.
2. In passive elastic objects, or at least in objects with ideal properties, the force—length relations, and hence the object stiffness values, depend neither on the time after the extension cessation nor on the immediately preceding history of the object behavior. For such an object, for example, for an ideal spring, the derivative $S = dF/dl$ does not depend on whether the spring was extended in small increments or in one large tug, whether the spring arrived at the current length as a result of the spring extension or contraction, etc. In short, there are no history effects there. In contrast, for active objects the history effects are strong.

For passive objects, derivatives $S = dF/dl$ or $dT/d\alpha$ characterize the stiffness of the object. For active objects, similar derivatives can also be computed. They formally satisfy Eqns (2.1b) and (2.1c). Some authors call these derivatives stiffness. However, in our opinion, such use of the term should be discouraged. The term should be reserved only for a deflection from an equilibrium position (*incremental stiffness*). Consider an example. When subjects produce maximal voluntary contractions at various joint angles, the force values change. The corresponding curves are called *joint strength curves*. The derivatives $dT/d\alpha$ or dF/da of these curves can be computed but they do not represent joint stiffness (for passive objects these derivatives would indeed represent stiffness).

Let us discuss the concept of *joint stiffness* in more detail. Humans can react in different ways to an external load applied to a kinematic chain, for example, the arm. The reaction depends on the *motor task*, that is, on the instruction given to the subjects, for instance, "resist" or "do not resist" the perturbation, and the subject's will to follow the instruction. Even under the same instruction, the resistive force depends on many factors such as the background force, the amplitude and speed of the induced change in the kinematic chain configuration, the time elapsed after the load application, the co-contraction of agonist—antagonist muscle pairs, etc. If not all of the details of the task and measurement procedure are specified, the measured "stiffness" values at the same joint can be very different. In active objects, the "stiffness" (i.e., the resistive force per unit of deflection) is always motor task specific and time dependent. Its mechanisms, for example, disruption of actin—myosin bonds in active muscles, reflex control (described in Chapter 6) or preprogrammed reactions (Chapter 7), are of biological nature and are completely different from the mechanisms of deformation of passive mechanical objects. Some of these mechanisms act instantaneously while others are characterized by time delays, which make values of "stiffness" computed in such experiments dependent on the time between the force application and measured displacements. It is unfortunate that the same term *stiffness* is used in the literature to designate the property of passive objects such as tendons and ligaments and behavior of the active objects such as muscles and joints.

The term *stiffness* should be reserved only for describing a property of passive objects. To describe stiffness-like parameters computed based on experiments with active objects the term *apparent stiffness* has been suggested (Latash and Zatsiorsky, 1993). The apparent stiffness of active objects may look like stiffness of passive objects but it has different—diverse and more complex—mechanisms. Using the two terms—*stiffness* for the passive objects and *apparent stiffness* for the active biological objects—decreases possible confusion and improves clarity of the texts.

Note that stiffness analysis per se describes only steady-state responses, from an equilibrium state to another equilibrium state. It does not describe the transient response, that is, the behavior during the transition from one state or position to another.

2.2.3 Velocity and acceleration effects—dynamic stiffness

Equation (2.1) does not include time, velocity, and/or acceleration. It essentially assumes that the measurements are performed at equilibria. The equation becomes inapplicable if the measurements are performed while the object is still moving.

Although stiffness is measured as a force/displacement ratio, not every force/displacement ratio or a derivative dF/dx refers to stiffness, even for passive objects. Consider, for instance, the movement of a material particle on a horizontal surface without friction. The equation of motion is $F = ma$. The derivative

$$dF/dx = m\ddot{x}/\dot{x} \tag{2.2}$$

can easily be calculated (x stands for displacement, $\dddot{x}$ with three dots for jerk, and $\dot{x}$ with one dot for velocity). However, this expression is, as a rule, unusable and does not represent stiffness in any meaningful way.

The joints are occasionally modeled as viscoelastic hinges where resistance to external load changes is provided by elastic and damping forces. The elastic forces depend on the amount of joint angular displacement, and the damping forces depend on the joint angular velocity. For the purposes of our discussion, we substitute the joint angular motion for the rectilinear deformation. If a model includes, along with inertia, a damping element and a spring, the equation of motion is

$$F(t) = m\ddot{x} + b\dot{x} + k(x - x_0) \tag{2.3}$$

where m stands for mass, b is a damping coefficient, k is an elastic, or spring, constant, $\ddot{x} = d^2x/dt^2$, and $\dot{x} = dx/dt$. All the three coefficients—m, b, and k—are scalars. Even when m, b, and k are not time dependent, the derivative dF/dx is a rather complex function. Let us differentiate both sides of Eqn (2.3) by t and then divide both sides by dx/dt:

$$\frac{dF}{dx} = m\frac{\dddot{x}}{\dot{x}} + b\frac{\ddot{x}}{\dot{x}} + k \tag{2.4}$$

The right side of the equation is quite different from the spring constant k. The magnitude of dF/dx depends not only on the spring constant k and the magnitude of the deviation from an equilibrium position, x, but also on the velocity, acceleration, and jerk at the instant of measurement. It only equals k when measurements are performed at equilibria and the velocity, acceleration, and jerk are equal to zero. Therefore, using the term *stiffness* for the derivative dF/dx when an object of interest is moving is misleading. This name is reserved for the spring constant k.

The dF/dx derivative for a moving body can be called *dynamic stiffness*. As follows from Eqn (2.4), dynamic stiffness values depend on velocity, acceleration, and jerk. Dynamic stiffness has little in common with stiffness as it is understood in classical mechanics. Whether computation of dynamic stiffness makes sense or not depends on the situation at hand.

2.3   Elastic properties of muscles and tendons

2.3.1   Elastic properties of tendons and relaxed muscles

Determining elastic properties of tendons and passive muscles could be technically difficult, but conceptually this is a straightforward procedure—the force is applied and the deformation measured, or vice versa.

2.3.1.1   Tendons (reviewed in Wang (2006))

In everyday life, tendons are exposed to large tensile forces. When a tendon is subjected to an external stretching force, the tendon length increases. When the force is removed, the tendon returns to its original length. If the force exceeds a certain critical magnitude, the tendon will rupture. The tendon elastic behavior is usually characterized by its *elastic modulus* that equals the slope of the stress—strain curve, $d\sigma_T/d\varepsilon_T$ where σ_T is the tendon stress (N/m^2), that is, the ratio of tendon force to its cross-sectional area, and ε_T is the strain, the ratio of the tendon elongation to the tendon length at rest (dimensionless). A typical stress—strain curve for a tendon is presented in Figure 2.1. The curve has three regions: (1) an initial toe region where the slope increases, (2) a linear region where the slope is constant, and (3) a failure region.

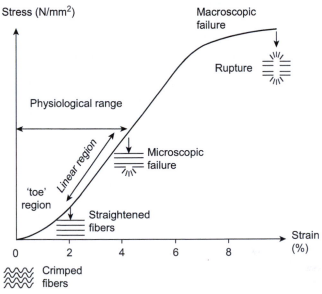

Figure 2.1 Stress—strain curve of a tendon, schematics.
Adapted with permission from J.H-C. Wang (2006), © Elsevier.

In the toe region, the tendon can be strained to approximately 2% (range 1.5—4%). Under these low-strain conditions, collagen fibers straighten and lose their crimp pattern but the fiber bundles themselves are not stretched. The crimped collagen fibers are lengthened until they become straight, similar to stretching a helical spring until it becomes a straight wire. The modulus of elasticity increases with strain until it reaches a constant value at the start of the linear region. In some joint configurations, the tendon may be *slack*—it does not resist force when elongated. For such cases, elongation is defined relative to the slack length, the length at which the tendon begins to resist external force. A more detailed description of the mechanical behavior of tendons—both elastic and viscoelastic (time dependent)—can be found in Zatsiorsky and Prilutsky (2012).

2.3.1.2 *Relaxed muscles (reviewed in Gajdosik (2001))*

When a relaxed skeletal muscle is stretched beyond its equilibrium length, that is, the length without mechanical loads (Figure 2.2), it provides resistance to the stretch.

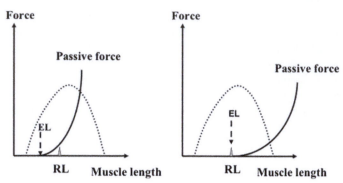

Figure 2.2 Force—length properties of the muscles with different amount of intramuscular connective tissues (schematics). RL—*rest length*, the length of the muscle in the body during a natural posture (for humans this would be the anatomical posture—standing upright on a horizontal surface with arms hanging straight down at the sides of the body, head erect). EL—*equilibrium length* (also called *initial length*), the length of the passive muscle without mechanical load. At the EL and below it the passive force is zero. The dotted curved line is the active force—length relation. A solid curved line—the passive force—deformation relation. Such relations are recorded by fixing one end of the muscle and applying incremental loads to its free end. The load (force) is then plotted versus deformation. For an active muscle, an actual force recorded at the muscle end (not shown in the picture) is equal to the sum of the active and passive forces. Left panel—a muscle with large amount of connective tissue. Note: (1) the passive force—length is shifted to the left (to shorter muscle length), (2) there is a large difference between the EL and RL, and (3) the passive force—length curve exhibits high stiffness. Right panel—a muscle with small amount of intramuscular connective tissue. As compared with the left panel: (1) the passive force—length curve is shifted to the right (to longer muscle length), (2) the EL and RL are closer to each other, and (3) the passive force—length curve is less steep, exhibiting smaller stiffness.

The resistance does not require metabolic energy and, hence, is called "passive." The muscle force—deformation curve is not linear. With increased stretching the muscles become stiffer, that is, they demonstrate toe-in mechanical response to lengthening. The behavior of the passive muscle in extension is often compared with the behavior of a knitted stocking—the passive muscle elasticity is mainly due to the web of connective tissues within the muscle. During small stretches the web deforms, its threads become progressively taut, and during large stretches the threads themselves may also deform.

When muscle fibers are stretched, resistance to extension is provided by three main structural elements: (1) connective tissues within and around the muscle belly (parallel elastic components); (2) stable cross-links between the actin and myosin filaments existing even in passive muscles—the crossbridges resist the stretch a short distance from the resting position before the contacts break and restore at other binding sites; and (3) noncontractile proteins, mainly titin. Actin and myosin filaments slide with respect to each other without visible length changes (this claim was challenged in several papers, Goldman and Huxley (1994), Takezawa et al. (1998)).

At lengths smaller than the equilibrium length, passive muscles are flaccid. During joint movement the relaxed antagonist muscles, if flaccid, do not provide much resistance to movement (e.g., triceps brachii during elbow flexion or biceps brachii during elbow extension).

At lengths larger than the equilibrium length, muscles exhibit resistance to extension. According to Hill's three-component muscle model (described later in this chapter) this resistance is provided by the *parallel elastic component* of the muscle. Force—length properties of the passive muscles depend on the amount of connective tissue in the muscle (Figure 2.2). As compared with the arm muscles, the leg muscles contain larger amount of connective tissue.

In a living body muscles and tendons are connected in series; they form *muscle—tendon units* (MTU). In an MTU the muscle and tendon experience approximately the same force (small differences are possible due to the lateral force transmission via shear force). Under the force, both muscle and tendon can deform. As a result, changes in muscle—tendon complex length could be larger than the change in the muscle length itself. They can also be smaller, if the muscle belly shortens while the tendon is extended (discussed in Subsection 2.3.4).

Passive resistance in joints results from the interaction of two components, one of which depends on joint angle and displacement (*elastic resistance*), and the other that is independent of these factors, for example, joint friction. The elastic resistance in the middle range of joint motion is usually small and in the majority of cases can be neglected. The elastic resistance increases exponentially as the joint motion approaches its maximal limit. In our opinion, the term *stiffness* should be reserved only for the elastic component of the joint resistance, that is, the component that depends on the magnitude of joint displacement. The use of such terms as *viscous stiffness*—which some authors apply to describe joint resistance that depends on angular velocity—is not desirable.

2.3.2   Reaction of active muscles to stretch

Under a constant muscle stimulation (no reflex contribution), response of an active muscle (muscle belly) to stretch depends not only on the stretch magnitude, but also on the speed of stretching and history. During and after a muscle stretch, the muscle force is not constant at the same muscle length. Hence, the ratio "muscle force/muscle deformation" defined above as the "stiffness" is also not constant. This makes application of the concept of stiffness doubtful. Let us consider the basic facts first.

Figure 2.3 illustrates typical changes in the muscle force occurring under a muscle stretch from one constant length to another with a moderate speed. The force—time curve has a rather complex pattern. We will not discuss here the mechanisms of these changes—they involve both crossbridge and noncrossbridge contributions—and mention only the force pattern itself.

At the beginning of the stretch the fast force rise occurs. It originates from the deformation of the engaged crossbridges. Muscle resistance to the stretch during this phase of muscle elongation is called *short-range stiffness*. The force rise during this period is a linear function of muscle elongation (the elastic response). After a break point, the second phase starts, in which force continues to increase but with a progressively

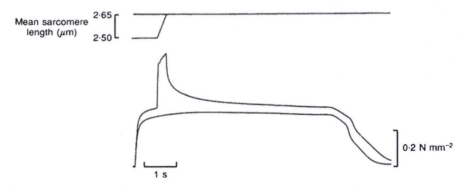

Figure 2.3 Force and displacement records from a single frog muscle fiber during tetanic stimulation. Comparison of stretch from 2.50 to 2.65 μm sarcomere length with isometric tetanus at 2.65 μm. Both the dynamic force enhancement and the residual force enhancement are seen.
Reprinted with permission from Edman and Tsuchiya (1996), © Wiley.

slower rate until the force reaches its peak at the end of the stretch. The force enhancement during the second phase increases with the stretch velocity (*velocity-dependent response*). These two phases constitute the *dynamic force enhancement.*

When the stretch is completed and the muscle length is kept constant at a new level, muscle force starts decreasing and reaches a value, which is still larger than the force of isometric action at the same muscle length. This *residual force enhancement* after a stretch lasts as long as the muscle is active. If one wants to determine the muscle stiffness, for example, the ratio "muscle force/muscle length change," the result will strongly depend on the time instant when the force is measured.

When stretch velocity is high (>20 muscle length/s), the so-called *give effect*—a sudden reduction in muscle force—may occur. The effect is due to the forced detachments of the crossbridges.

The behavior of human muscles in vivo is similar to that of isolated muscles or even isolated fibers whose response to stretch is not influenced by such serial elastic structures as tendons or aponeuroses. Although reflexes affect muscle responses to stretch in vivo conditions, many features of the force response to stretch described for isolated muscles are observed in healthy humans. In particular, both the dynamic force enhancement and the residual force enhancement phases are seen in muscles with and without stretch reflex. However, the intact muscle responds to stretch with higher resistance than the muscle without stretch reflex (Figure 2.4). In addition, with reflexes the response is more linear (Nichols and Houk, 1976).

The short-range stiffness allows instantaneous resistance to sudden small-magnitude external perturbations that occur, for example, as a result of unexpected contacts with external objects. An external perturbation leads to changes in joint angles and, in turn, to muscle stretch. Owing to *short-range stiffness* the stretched muscles resist length changes before the fastest reflexes (*monosynaptic stretch reflexes*, see Chapter 6) start to operate in about 20–40 ms. Thus *short-range stiffness* can be viewed as the first line of defense against unexpected postural perturbations. *Stretch*

Soleus

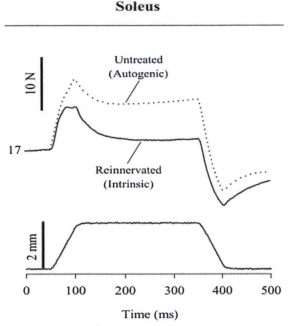

Figure 2.4 Comparisons of responses to a constant velocity stretch (2 mm in 50 ms) of an intact cat soleus and the muscle without stretch reflex. Upper panel: Dotted line—a force response of the intact muscle; solid line—response of the muscle without stretch reflex. The number 17 at the force traces corresponds to the muscle initial force (in N). The muscle was activated physiologically by eliciting the withdrawal reflex of the contralateral hind limb (cross-extension reflex). Stretch reflex of ipsilateral soleus and gastrocnemius was removed by self-reinnervating soleus and gastrocnemius muscles. Self-reinnervation was performed nine months before the experiments by cutting the nerves innervating the muscles and suturing the cut nerve stumps together. Axons of motoneurons grew back and reinnervated the muscles over several months, but stretch reflex did not recover. Bottom panel: Imposed length changes.
Reprinted with permission from Huyghues-Despointes et al. (2003), © American Physiological Society.

reflexes, the neural excitation of muscles in response to stretch, provide the second line of defense, which allows the muscles to resist stretch beyond the span of short-range stiffness (see Chapter 6). The response to perturbation within the time shorter than the latency of the monosynaptic stretch reflex is called by some authors a *preflex* (Loeb et al., 1999).

2.3.3 Elastic properties of muscle components

Muscles are complex organs. They include multiple elements such as muscle fibers, aponeuroses, connective tissues, blood vessels, etc. Their architecture can be rather complicated, for example, in pennate and convergent muscles or in curved and wrapping muscles. In some cases, the muscle behavior is similar to the behavior of the fiber-reinforced composites. In such composites, the fibers made from a material with high tensile strength are embedded in another material (called *matrix*), which glues the fibers together and transfers external stresses. Various muscle elements possess different mechanical properties.

A simplifying approach is to construct a lumped-parameter model, that is, to assign the muscle mechanical properties (e.g., stiffness, damping, mechanical inertia, etc.) to a limited number of muscle "components." The most popular is the Hill three-component model, according to which the muscle can be viewed as consisting of the following:

1. Contractile component (CC) whose function is to generate force. This component is damped, that is, it provides internal resistance to length changes.
2. Serial elastic components (SEC) connected in series with the CC.
3. Parallel elastic components (PEC) that provide resistance to muscle deformation (extension) beyond its equilibrium length (EL; see Figure 2.2).

While these elements can be associated with certain morphological structures—for instance, SEC is mainly associated with the muscle tendon and PEC with the muscle perimysium—the model is not intended to realistically replicate muscle architecture or morphology. The model is intended to represent only the muscle behavior at a gross phenomenological level.

The model can be represented by either one of two structures presented in Figure 2.5. Both structures have similar mechanical properties.

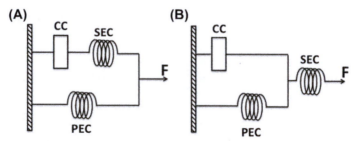

Figure 2.5 Element arrangements in the Hill-type models. Both arrangements have similar mechanical properties. The actual arrangement cannot be inferred from mechanical experiments.

In the model, the contractile component possesses the properties of force generation, the force—length, and force—velocity relations. It also possesses damping properties (a nonlinear increase of resistance with the velocity). The action of the contractile component depends on its current length and velocity as well as on the activation level.

The elastic elements in the model, PEC and SEC, represent mainly the properties of the connective tissues (albeit it was shown that crossbridges also possess elastic properties and some mechanical properties of structural proteins that constitute PEC, that is, titin, are activation dependent). The model predicts the following phenomena of muscle behavior:

1. With both muscle ends fixed, muscle activation results in CC shortening and SEC lengthening. The muscle—tendon unit as a whole exerts force.
2. At lengths longer than the rest length, muscle force is the sum of the passive force (caused by PEC resistance to forceful stretching) and the active force generated by the CC (as illustrated in Figure 2.2).

3. When a muscle, fixed at both ends, is activated and then allowed to shorten against a smaller load (quick release), the muscle length–time history consists of two parts: (1) a rapid length decrease due to SEC shortening and (2) slow length changes due to CC contraction (Figure 2.6).

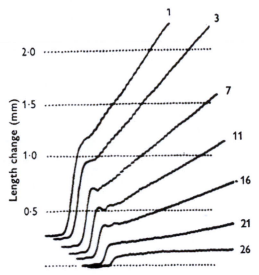

Figure 2.6 Experimental records of muscle shortening, which occur when the load on the muscle is suddenly reduced (quick release) from the full isometric tension to the value (in g of weight) shown alongside each trace (frog sartorius, $L_o = 31$ mm, weight 76 mg, temperature 2 °C; muscle is released 0.7 s after the start of a 1-s tetanus (30 shocks/s); time interval between dots 1 ms).
Modified with permission from Jewell and Wilkie (1958), © Wiley.

The first phase (1) has high speed that does not depend on the load, that is, the slope of the shortening trace is essentially constant. However, the magnitude of the shortening depends on the load. The magnitude is larger for smaller loads. By plotting the amount of muscle length change during the initial shortening versus the corresponding load, the force–length relation for the SEC is obtained. The slope of this relation represents the *stiffness of SEC*.

Besides the quick-release method described above, to determine the SEC properties the so-called *controlled-release method* is also used. In this method, an activated muscle is suddenly released and allowed to shorten over a given distance, say 2 mm. Then, at the new length, the muscle exerts isometric tension. The starting value of the newly developed force is, however, smaller than the value prior to the release. By comparing the drop in the muscle tension (ΔF) with the amount of the length change (Δl), the stiffness of the SEC, or its compliance, can be determined. The quick-release and controlled-release methods are illustrated in Figure 2.7. The methods yield similar results.

The stiffness of SEC, as well as its compliance, does not depend on the initial muscle length, or joint angle, but it strongly depends on the values of the exerted force, or joint moment (Figure 2.8).

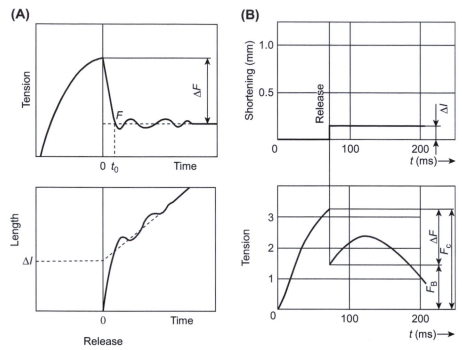

Figure 2.7 Quick-release (A) and controlled-release (B) methods, schematics. An active muscle exerting an isometric tension against a stop is suddenly released and allowed to shorten. In the quick-release method, the muscle shortens against a constant load; in the controlled-release method, it shortens over a given distance. When the SEC properties are the object of interest, the changes in the muscle tension (ΔF) are compared with the changes in the muscle length (Δl). The figure is adapted from Zatsiorsky et al. (1981), FiS Publishers.

2.3.4 Elastic properties of muscle–tendon units

In the muscle–tendon units the muscle belly and the tendon are connected in series. In an ideal case (no lateral force transmission), the muscle and tendon experience the same force. If an MTU is extended by a certain amount Δl_{MTU}, the total length changes depend in an obvious way on the changes in both muscle length Δl_M and tendon length Δl_T. Note that muscle length may change in both directions, that is, the muscle either can shorten or be forcibly stretched.

$$\Delta l_{MTU} = \Delta l_T \pm \Delta l_M \tag{2.5}$$

This allows for different combinations of tendon length increase and changes in the muscle belly length. The most evident case is static force production. As by definition the MTU length stays put, the tendon subjected to a force increase should elongate and the muscle itself contracts. Hence, strictly speaking, static force production by an MTU is not identical to static force production recorded at the muscle belly ends.

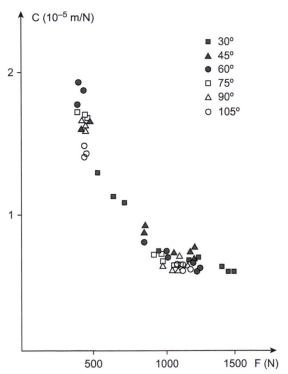

Figure 2.8 The relation between the estimates of SEC compliance (C) and the muscle tension (F) for a group of elbow flexor muscles obtained with the quick-release method. The subjects exerted elbow flexion torques of different magnitudes at different joint angles, from 30° to 105°. Then the trigger was suddenly released. The lumped muscle force was estimated for an "equivalent elbow flexor" assuming that its moment arm equals the moment arm of the biceps brachii. The angular values in the figure are for the initial angles before release. Note that the compliance (and stiffness) values do not depend on the joint angle, and hence on the muscle length. Adapted from Cnockaert et al. (1978), © University Park Press.

Application of such methods as sonomicrometry (in animals) and ultrasonography (in humans) allows for recording muscle fascicle length as well as the MTU length in vivo. The general conclusion is that the changes in the MTU length do not represent well the changes in the muscle belly or muscle fascicle length. When an active MTU is forcibly stretched, the muscle belly length may change at a different speed than the MTU speed or even change in the opposite direction (Cronin et al., 2013). In such a situation, interpreting the "MTU stiffness" requires thoughtful consideration.

Elastic bodies, when stretched, accumulate potential elastic energy. The amount of energy is proportional to the force and the amount of deformation. When subjected to the same force, more compliant (less stiff) bodies accumulate more energy. Hence, when a tendon length changes more under stretch than a muscle belly length the tendon can accumulate a larger amount of elastic potential energy. Animals who are fast runners, for example, horses or antelopes, have short, strong inextensible muscles and lengthy compliant tendons. Such tendons work as springs; they allow for storing and recoil a large amount of mechanical energy at each step (Alexander and Bennet-Clark, 1977).

2.4 Apparent stiffness of joints and kinematic chains

We limit our discussion to the joints with one degree of freedom (DoF) and planar kinematic chains, concentrating mainly on the biomechanical aspects of the problem.

2.4.1 One-DoF joints

Passive resistance in joints is due to two components, one that depends on joint angle and displacement, and the other that is independent of these factors, for example, joint friction. The first component is called *elastic resistance*. The elastic resistance in the middle range of joint motion in healthy people is small and is usually neglected. However, as the joint approaches its maximal limit the resistance, that is, the joint stiffness, increases exponentially. Increased resistance at the joints—for example, as seen in patients suffering from spasticity—may be of a neural origin, for example, due to hyperactive stretch reflexes, or to altered mechanical properties of the muscles and tendons (nonneural origin). Separating the contribution of these mechanisms is clinically important.

When muscles are active, the term *apparent stiffness* is more appropriate. In addition to the factors discussed in Subsection 2.2.2, the changes in the muscle moment arms with the joint angle are important.

Consider an ideal hinge joint served by only one muscle. Let us perform a mental experiment. Perturb the joint angular position by a small amount and record both the force (moment) and angular displacement. Then the changes in the joint torque (ΔT) and joint angle ($\Delta \alpha$) are determined and the ratio $\Delta T/\Delta \alpha$ is computed. The ratio estimates the joint apparent stiffness. As there is only one muscle in the model, one may assume that the $\Delta T/\Delta \alpha$ ratio indirectly estimates the apparent stiffness of the muscle. There is, however, a caveat in this reasoning.

For a hypothetical hinge joint served by only one muscle, the equation is $T = F(\alpha) \, r(\alpha)$, where T is joint torque (the moment of muscle force), $F(\alpha)$ is the force produced by the muscle at joint angle α, and $r(\alpha)$ is the muscle moment arm at this angle. In the limit, the $\Delta T/\Delta \alpha$ ratio converges to the derivative $dT/d\alpha$ that equals

$$\frac{dT}{d\alpha} = \frac{dF(\alpha)}{d\alpha} r(\alpha) + \frac{dr(\alpha)}{d\alpha} F(\alpha) \tag{2.6}$$

The derivative has two terms. The first term includes the derivative of the force–length curve of the muscle (the change in the muscle force due to the infinitesimal joint displacement), and the second term contains the derivative of moment arm with respect to the joint angle. The second term can be large and should not be neglected. In an extreme case, the muscle force can stay put—it does not change with the joint perturbation—while the joint moment changes. It is also possible that the muscle force decreases while the joint moment increases (if the moment arm increases).

For multijoint muscles, a change in the joint angle can correspond to different changes in the muscle length. Whether the muscle is shortening or lengthening depends on the changes in other joints.

Also, as already discussed before, changes in the MTU length associated with joint angle change may not represent changes in the length of the muscle fascicles.

2.4.2 Planar kinematic chains

For illustration purposes, we limit the discussion to two-link kinematic chains. We assume that the endpoint of the chain is perturbed. The external force causes a deflection of the endpoint as well as displacements at the joints. We will consider only small displacements. Because the deflections are small, the force–displacement relations are approximately linear. In experiments of such a type, several preventive measures are usually taken. To prevent the early occurrence of voluntary reactions the subjects are given such instructions as: first, to concentrate on perceiving the direction of the displacement; second, to say "one-two" aloud; and third, to move rapidly in the direction opposite to the imposed displacement. The measurements of the endpoint apparent stiffness are made at zero hand velocity, between the end of the hand movement and the expected beginning of the voluntary reaction.

In kinematic chains, the direction of the endpoint deflection is generally not coincident with that of the line of force. More often than not, the force direction differs from the direction of deflection (Figure 2.9).

The force–displacement relation cannot be reduced to a simple ratio as it was for unidimensional objects discussed previously, and the endpoint stiffness is not a scalar quantity. The relation is described by a matrix equation where *endpoint stiffness* is represented by 2×2 matrix $[S]$

$$\begin{bmatrix} dF_x \\ dF_y \end{bmatrix} = -\begin{bmatrix} S_{XX} & S_{XY} \\ S_{YX} & S_{YY} \end{bmatrix} \begin{bmatrix} dX \\ dY \end{bmatrix} \qquad (2.7a)$$

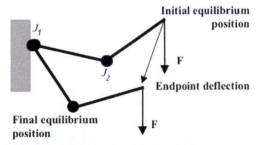

Figure 2.9 For kinematic chains, the direction of the endpoint deflection is as a rule different from the force direction. A vertical force **F** is applied to the endpoint of a two-link chain connected to a wall by a hinge joint J_1. The angular changes (deflections) occur in both joints, J_1 and J_2. Reprinted by permission from Zatsiorsky (2002), © Human Kinetics.

or

$$dF = -[S]dP \qquad (2.7b)$$

The subscripts under the elements of the stiffness matrix correspond to the force direction and the displacement direction, respectively. For instance, the element S_{XY} characterizes the force change in X direction in response to the unit displacement of the endpoint in Y direction. It is conceivable that all four elements change independently. If this happened, this would signify that the endpoint stiffness is characterized by four scalar quantities.

If matrix $[S]$ was not symmetric, $S_{XY} \neq S_{YX}$, the endpoint after a perturbation would not return to its initial position but rather would show a circular movement, which would be larger for larger differences $S_{XY} - S_{YX}$ (called *curl*). It has been experimentally established that for the human arm the curl is close to zero. Because $S_{XY} \approx S_{YX}$, the apparent stiffness of the arm is characterized by three quantities, S_{XX}, S_{YY}, and S_{XY}.

The stiffness of the endpoint of a kinematic chain can be represented by a stiffness ellipse (Figure 2.10). The perimeter of the ellipse is the locus of the force vectors for a

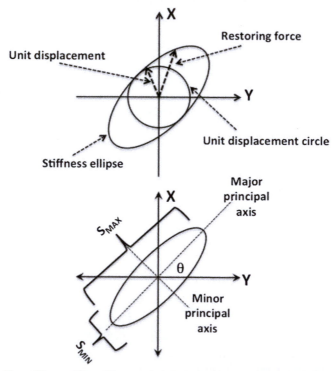

Figure 2.10 The stiffness ellipse. Upper panel: Unit displacement circle and the restoring force (stiffness ellipse). Bottom panel: Principal axes of the ellipse. The angle θ between the major principal axis and the X-axis defines the *apparent stiffness orientation*. The shape of the ellipse is given by the ratio S_{max}/S_{min}. The size of the stiffness ellipse is represented by its area, $\pi S_{max} S_{min}/4$.

unit displacement in various directions, $0 < \varphi < 2\pi$. The perimeter can be obtained by multiplying the endpoint displacement of unit amplitude by the symmetric $[S]$.

The stiffness ellipse is characterized by three main variables: (1) *Ellipse orientation* is defined by the angle between the major axis and the X-axis of the fixed reference system. (2) *Ellipse shape* is given by the ratio of the major to minor axes (the ratio of the maximal to minimal stiffness sometimes refers to an *eccentricity* of the ellipse). (3) *Size* or *magnitude* of the ellipse is characterized by the ellipse area. The area is proportional to the determinant of the matrix $[S]$, that is, to the expression $det[S] = S_{XX}S_{YY} - S_{XY}S_{YX}$.

In general, the deflection and the restoring force are not collinear except along the major and minor axes of the ellipse. The major axis is oriented along the direction of maximal apparent stiffness and the minor axis is along the direction of minimal apparent stiffness. The major and minor axes are orthogonal to each other. When an endpoint force is acting along a principal axis, the endpoint deflects in the same principal direction and the displacement takes an extreme value. In any other direction, the force and the deflection are not collinear and the deflection is not extreme.

The endpoint apparent stiffness is a function of the apparent stiffness of the individual joints and the chain Jacobian, that is, the function of the arm posture:

$$[S] = \left(\mathbf{J}^T\right)^{-1}[K]\mathbf{J}^{-1} \tag{2.8}$$

where $[S]$ is the matrix of the endpoint apparent stiffness, $[K]$ is the matrix of the joint apparent stiffness, $\mathbf{J}$ is the chain Jacobian, and superscripts T and -1 designate the matrix transpose and the matrix inverse, respectively. Both $[K]$ and $[S]$ depend on the chain Jacobian, that is, on the chain configuration. In experiments, when $[K]$ and $[S]$ are determined, the Jacobian is usually assumed constant. This assumption is only valid if the limb deflection is small. Therefore, Eqn (2.8) cannot be used to study large-range arm and leg movements without making necessary corrections to the equations (the Jacobians become variable).

In general, the matrix of the joint stiffness $[K]$ is not diagonal. The diagonal elements of the matrix, called *direct terms*, relate changes in the torque at a given joint to the angular deflection at this joint, for example, a change in the elbow torque that is due to a change in elbow angle. Off-diagonal terms, called *cross-coupling terms*, relate changes in the torque at one joint to the angular displacements at another joint, for example, a change in the elbow torque due to the shoulder joint angular displacement. In humans and animals, the cross-coupling terms are due to two factors: (1) the biarticular muscles that span both joints and (2) the heterogenic reflexes that occur when stretching one muscle changes the activity of other muscles. Mechanically, cross-coupling means that every joint deflection affects torques at all joints. Without coupling between joints, control of the endpoint stiffness would deteriorate (Hogan, 1985).

Equations (2.7) and (2.8) assume that the chain Jacobian is constant. Therefore, the equations are valid only for small changes of the chain position (small position perturbations) when variations in the chain Jacobian can be neglected.

2.4.3   Leg stiffness (in the rebound takeoffs)

Determining "leg stiffness" in such activities as running and hopping is a popular technique. The underlying assumption is that in these activities the human leg behaves similarly to a pogo stick or as a mass—spring system, that is, a mass connected in series with an ideal spring ("springs in the legs"). When such a system rebounds, the body, which initially moves downward, decreases its speed to zero and then moves upward. During the entire period of contact, the body accelerates upward with the maximal value of acceleration at the bottom point of the body movement, that is, at the instant of maximal spring deformation and zero velocity. In such a system, there exists a linear relation between the body acceleration—and hence the force acting on the body—and the displacement. This allows easy computation of the force/deformation ratio. Usually, the ground reaction force (F) and the change in the leg length (Δl) are measured and then either the ratio $S = \Delta F/\Delta l$ or the derivative $S = dF/dl$ are determined. Some variants of the computations exist, for instance, the forces and displacements are recorded in the vertical direction only and not along the leg length ("vertical stiffness").

It is evident that applicability of such an approach depends on whether in the studied movement the leg behavior can be modeled as a mass—spring system or not. For instance, if a standing person slowly bends the legs—acceleration is close to zero—and assumes a deep squat position, the leg length changes but the ground reaction force stays put ($\Delta F = 0$). Hence, computation of the stiffness ratio $\Delta F/\Delta l$ does not make sense.

The force—leg length relation can be far from linear even in simple activities. For instance, during landing to a rest position, as gymnasts do during dismounts, the ground reaction force initially rises and then decreases to magnitudes corresponding to the weight of the performer (Figure 2.11). Using the leg stiffness concept in such activities would require introducing two stiffness values: one positive, for the initial phase of landing after the ground contact when the force increases and the leg length decreases; and one negative, for the second phase of landing when the legs continue to bend but the force decreases. Expression of "negative stiffness" makes no physical sense and is an oxymoron.

In short, determining "leg stiffness" in the described way has little in common with the measures described in the previous sections of the chapter. We agree with Pearson and McMahon (2013) that to avoid misunderstanding the "leg stiffness" under discussion should be termed differently, for example, it could be referred as "quasi-stiffness" (an old term suggested in Latash and Zatsiorsky (1993)) or "rebound leg stiffness."

2.5   The bottom line

The concept of stiffness was developed in mechanics to describe behavior of elastic deformable bodies under application of external force. After removal of the external force, elastic bodies return to their initial size (during force application the bodies store

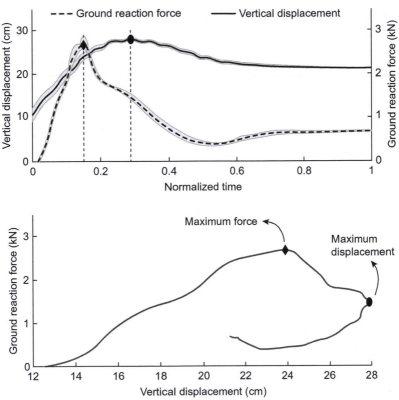

Figure 2.11 Upper panel: Ground reaction force and the COM displacement during landing after a jump. A peak of the ground reaction force occurs earlier than the maximal COM displacement. Bottom panel: Force versus displacement relation. For the segment from "Maximal force" to "Maximum displacement" the relation is negative. The data are for a representative trial.
The figure is courtesy of Dr S. Ambike (Penn State University).

elastic potential energy and return it back to the system after the force removal). For elastic bodies, stiffness was defined as amount of force, or force change, per unit of the induced deformation, *dF/dx*. The bodies studied in classical mechanics are "passive" in the sense that they maintain the same size, for example, length, if external forces are absent or constant. For the passive bodies, there exist one-to-one relations between the force and body geometry. Such relations do not exist for the active muscles and intact joints. Force (joint moment) can change without the change in the muscle length (joint angle), and vice versa. This alone makes application of the concept of stiffness to the active objects problematic.

Other complexities associated with using the term *stiffness* for describing behavior of active biological objects (muscles, joints, kinematic chains) include but are not limited to (1) the time dependence of the response of the biological objects to the external perturbation (by definition, "stiffness" characterizes only time-independent effects); (2) complex multielement structure of the biological objects; and (3) profound differences between the mechanisms of functioning of the inanimate and living objects. Neglecting these intricacies and using the term *stiffness* indiscriminately may result in misunderstanding and wrong conclusions. Hence, caution in using the term is advised. Adding some modifiers may help.

In particular, depending upon the physical nature of the system and method of measurement, the following terms for the derivative *dF/dx* have been suggested (Latash and Zatsiorsky, 1993):

1. *Stiffness.* The measurements are performed at equilibria. Resistance to the external force is provided by elastic forces (passive structures), and potential energy is being stored.
2. *Apparent stiffness.* The measurements are performed at equilibria. The physical nature of the resistive forces is being disregarded.
3. *Quasi-stiffness (q).* The measurements are performed not at equilibria.

This chapter discusses:

1. stiffness (elastic properties) of tendons and passive muscles, Subsection 2.3.1;
2. response of active muscles to stretch (Subsections 2.3.2−2.3.4). This response is time and velocity dependent, depends on various physiological mechanisms, and is different for individual elements comprising the muscles and muscle−tendon complexes. If the measurements are performed at equilibria, the term *apparent stiffness* is recommended. The readers should recognize that depending on the measurement procedure, for instance, on the instant of measurement, several (many) apparent stiffness values can be obtained;
3. apparent stiffness of joints and kinematic chains (Subsections 2.4.1 and 2.4.2); and
4. quasi-stiffness of legs (as measured in rebound activities, such as hopping), Subsection 2.4.3.

References

Alexander, R.M., Bennet-Clark, H.C., 1977. Storage of elastic strain energy in muscle and other tissues. Nature 265, 114−117.

Blix, M., 1893. Die Lange und die Spannung des Muskels. Scandinavian Archives of Physiology 4, 399−409.

Cnockaert, J.C., Pertuzon, E., Goubel, F., Lestienne, F., 1978. Series-elastic component in normal human muscle. In: Asmussen, E., Jorgensen, K. (Eds.), Biomechanics VI-A. International Series on Biomechanics, vol. 2A. University Park Press, Baltimore, pp. 73−78.

Cronin, N.J., Prilutsky, B.I., Lichtwark, G.A., Maas, H., 2013. Does ankle joint power reflect type of muscle action of soleus and gastrocnemius during walking in cats and humans? Journal of Biomechanics 46, 1383−1386.

Edman, K.A., Tsuchiya, T., 1996. Strain of passive elements during force enhancement by stretch in frog fibers. Journal of Physiology 490, 191−205.

Flash, T., 1987. The control of hand equilibrium trajectories in multi-joint arm movements. Biological Cybernetics 57, 257−274.

Gajdosik, R.L., 2001. Passive extensibility of skeletal muscle: review of the literature with clinical implications. Clinical Biomechanics 16, 87−101.

Gasser, H.S., Hill, A.V., 1924. The dynamics of muscle contraction. Proceedings of the Royal Society of London, Series B 96, 398−437.

Goldman, Y.E., Huxley, A.F., 1994. Actin compliance: are you pulling my chain? Biophysical Journal 67, 2131−2133.

Hill, A.V., 1925. Length of muscle, and the heat and tension developed in an isometric contraction. The Journal of Physiology 60, 233−263.

Hill, A.V., 1950. The series elastic component of muscle. Proceedings of the Royal Society of London, Series B 137, 273−280.

Hogan, N., 1985. The mechanics of multi-joint posture and movement control. Biological Cybernetics 52, 315−331.

Huyghues-Despointes, C.M., Cope, T.C., Nichols, T.R., 2003. Intrinsic properties and reflex compensation in reinnervated triceps surae muscles of the cat: effect of activation level. Journal of Neurophysiology 90, 1537−1546.

Jewell, B.R., Wilkie, D.R., 1958. An analysis of the mechanical components in frog's striated muscle. Journal of Physiology 143, 515−540.

Johns, R.J., Wright, V., 1962. Relative importance of various tissues in joint stiffness. Journal of Applied Physiology 17, 824−828.

Latash, M.L., Zatsiorsky, V.M., 1993. Joint stiffness: myth or reality? Human Movement Science 12, 653−692.

Loeb, G.E., Brown, I.E., Cheng, E.J., 1999. A hierarchical foundation for models of sensori-motor control. Experimental Brain Research 126, 1−18.

Mussa-Ivaldi, F.A., Hogan, N., Bizzi, E., 1985. Neural, mechanical, and geometric factors subserving arm posture in humans. Journal of Neuroscience 5, 2732−2743.

Nichols, T.R., Houk, J.C., 1976. Improvement in linearity and regulation of stiffness from actions of stretch reflex. Journal of Neurophysiology 39, 119−142.

Pearson, S.J., McMahon, J., 2013. Authors' reply to Morin and colleagues: "Lower limb mechanical properties: significant references omitted". Sports Medicine 43, 155−156.

Takezawa, Y., Sugimoto, Y., Wakabayashi, K., 1998. Extensibility of the actin and myosin filaments in various states of skeletal muscle as studied by X-ray diffraction. Advances in Experimental Medicine and Biology 453, 309−316, discussion 317.

Wang, J.H.-C., 2006. Mechanobiology of tendon. Journal of Biomechanics 39, 1563−1582.

Zatsiorsky, V.M., 2002. Kinetics of Human Motion. Human Kinetics, Urbana, IL.

Zatsiorsky, V.M., Prilutsky, B.I., 2012. Biomechanics of Skeletal Muscles. Human Kinetics, Urbana, IL.

Zatsiorsky, V.M., Aruin, A.S., Seluyanov, V.N., 1981. Biomechanics of Musculoskeletal System. FiS Publishers, Moscow (in Russian). [The book is also available in German: Saziorski WM, Aruin AS, Selujanow WN Biomechanik des menschlichen Bewegungsapparates. Berlin, Sportverlag, 1984.].

Velocity-Dependent Resistance

3

Resistance to deformation may depend not only on its magnitude—as it was discussed in Chapter 2—but also on the rate. In liquids and gels, the latter resistance is labeled as *viscosity*. Viscosity arises due to internal friction. When neighboring particles are moving at different speeds, they experience friction forces.

Some solids, among them tendons and ligaments, can deform very slowly. Such materials are described as *viscoelastic*, possessing both elasticity and viscosity (defined in this case as reaction to rate of deformation). Besides the sensitivity to strain rate, the viscoelasticity is also manifested as (1) *creep*—an increase in length under a constant load, (2) *stress relaxation*—a decrease of stress over time under constant deformation, and (3) *hysteresis*—the difference between the loading and unloading curves in the stress—strain cycle. The hysteresis area represents the energy loss in a deformation cycle, mainly due to converting mechanical work into heat.

3.1 Viscosity in physics

In the International System of Units (SI), viscosity is defined as the resistance that a liquid or gaseous system offers to flow when it is subjected to a shear stress (Figure 3.1).

Viscosity expresses the magnitude of internal friction in a substance, as measured by the force per unit area resisting uniform flow. The governing expression (Newton's equation) is

$$f = \eta A \frac{dv}{dx} \tag{3.1}$$

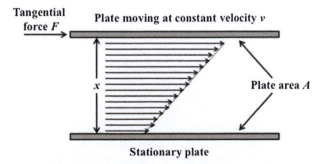

Figure 3.1 Definition of viscosity. Two plates are moving horizontally with respect to each other at velocity v at distance x. The space between the plates is filled with a viscous liquid. Force F is given by Eqn (3.1).

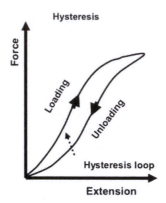

Figure 3.2 Hysteresis. The force—deformation curves are different for loading and unloading periods. The area of the hysteresis loop is representative of the energy loss and hence the object viscosity.

where *f* is the force required to maintain a velocity gradient, dv/dx, between planes of area A, and η is the viscosity coefficient. The SI metric unit for viscosity is $(N/m^2) \cdot s = Pa \cdot s$ (Pascal-second), equivalent to $(N \cdot s)/m^2$. The viscosity unit is the force per unit area required to sustain a unit velocity gradient normal to the flow direction. A high viscosity fluid does not flow as easily as a low viscosity fluid (honey does not move as easily as water).

Due to internal friction, viscoelastic materials, when moving, experience energy loss—part of mechanical energy is converted into heat. The energy loss can be estimated from the area of the *hysteresis loop* (Figure 3.2).

In summary: (1) in viscous materials the resistive force depends on the rate of deformation, (2) resistance arises due to internal friction, and (3) viscous resistance is associated with the energy loss (under repetitive deformation the object temperature increases).

3.2 Elements of history: muscle viscosity theory and its collapse

The concept of muscle viscosity was a cornerstone of the theory on muscle functioning popular in the 1920s—1930s. Both the birth and downfall of the theory was due to A.V. Hill (1886—1977, Nobel Prize in Physiology or Medicine in 1922). At that time, mechanisms of muscle contraction were not known; in particular, nothing was known about the existence of actin and myosin filaments and their interaction. Even recording of tendon forces in the contracting muscles was technically not possible (the first strain gauge measurements of tendon forces were performed only in 1947).

In the early 1920s, A.V. Hill and his coworkers (1922) found the following:

1. In experiments on frog muscles: after electrical stimulation, muscles shorten and lift suspended loads at a constant speed (i.e., in spite of the acting force the lifted objects did not accelerate, as could be expected according to Newton's second law); and
2. In experiments on humans: across elbow flexion trials with maximal effort against different inertial loads, mechanical work decreases with movement speed.

Both the constant muscle speed during shortening in a single trial and decrease of mechanical work with increasing speed across different trials were explained by the muscle viscosity (internal friction). According to the hypothesis, under standard stimulation, a muscle generates similar internal forces but these forces are not manifested externally due to the viscous resistance within the muscle. The resistance is larger for larger speed. Hence, the active muscles behave as systems consisting of elastic (spring-like) and damping elements (viscous dashpots). A mechanical analog of such a system is a spring working in a viscous medium. When a muscle progresses from rest to activity, the muscle has a shorter equilibrium length, that is, the length at which it exerts a zero force at its ends, and it accumulates mechanical potential energy proportional to the difference between the equilibrium and resting lengths. When the muscle is allowed to shorten, this energy is converted into mechanical work and heat. The percentage of the energy converted into work and heat depends on the shortening speed. The larger the speed of shortening, the larger the energy losses due to internal friction (viscosity), and the smaller mechanical work is done. However, the total amount of the stored and liberated energy (work + heat) remains constant (Figure 3.3). In other words, according to the viscoelastic theory, muscle activation adds a fixed amount of energy to the muscle.

It seemed that human experiments agreed well with this theory. The viscosity hypothesis became a dominant theory of muscle biomechanics. The theory required that the total amount of energy generated by an active muscle (work plus heat) should be constant independent of the speed of shortening (as shown in Figure 3.3). However, as early as 1924, W. O. Fenn recorded both the muscle mechanical work and heat production and demonstrated that the total energy released by the muscle was not constant when it shortened against different loads and at different speeds. (Wallace O. Fenn, 1893–1971, an American scientist; from 1922 to 1924 he worked in the laboratory of A.V. Hill in England.) When a stimulated muscle was allowed to shorten against a load, that is, when it was allowed to do mechanical work, the amount of liberated heat increased in proportion to the muscle force and the distance of shortening. The heat production (in excess of the so-called activation heat) was found to be proportional to the work done by the muscle (Figure 3.4). Hence, it was acknowledged that the total energy liberated during muscle contraction was not constant. The amount of energy is not determined solely by the muscle activation: the energy expenditure increases with the work done. This fact became known as the *Fenn effect*. In the concluding remarks of his paper, Fenn (1924, p. 395) mentioned that "…the existence of an excess heat liberation is to some extent inconsistent with the idea that a stimulated muscle is a new elastic body." Using another metaphor, muscle behavior

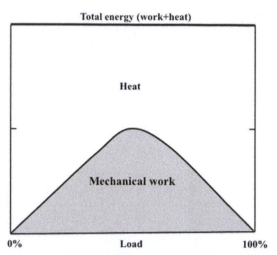

Figure 3.3 Mechanical work and heat production according to the viscoelastic theory of muscle contraction. In this schematic, the *mechanical work* and *heat* are represented as the shaded and white areas, respectively. At zero load, the muscle end is free and, when the muscle shortens, the muscle force does not do any mechanical work (because the force is zero). Hence, all the energy goes into overcoming internal friction (viscosity) and converts into heat. Note that in this case the speed of muscle shortening is maximal. When the load is maximal ($=100\%$) the muscle contracts isometrically and the mechanical work is again zero. (Remember, in the simplest case, when the force acting on the object (F) and the displacement of the point of force application (d) are in the same direction, work (W) equals the product $W = F \times d$, and if one of the factors equals zero, the product is also zero. No work is performed in this case.) Reprinted by permission from Zatsiorsky and Prilutsky (2012), © Human Kinetics.

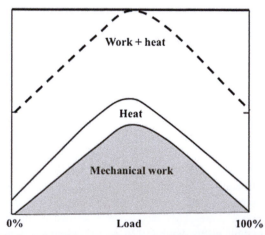

Figure 3.4 Mechanical work and heat production according to the Fenn effect. In this schematic, the mechanical work is represented as the shaded area while the heat production is represented as the area below the heat curve. The total amount of energy (work + heat) equals the sum of the two above curves (broken line). The figure is not drawn to scale.
The figure is reprinted by permission from Zatsiorsky and Prilutsky (2012), © Human Kinetics.

more closely resembles electric motors, which increase both electricity (energy) consumption and heat production when they are heavily loaded. Mechanical springs, on the other hand, possess a fixed amount of elastic potential energy when deformed.

The Fenn effect was largely overlooked until 1938, when Hill published his famous paper on the heat of shortening. In that study, Hill discovered—among other things—that during eccentric muscle action, that is, when the muscle is forcibly extended, the total energy produced by the muscle is less than in isometric contraction, a fact that is incompatible with the idea of the decisive role of muscle viscosity. This is because viscous resistance depends only on movement speed, not on the movement direction. Imagine that you move your arm in a barrel of oil. Whether you move the arm to the right or to the left, you would experience similar viscous resistance at equal speeds. For muscles, this direction independence does not hold—when active muscles shorten, they generate more energy than when they are forcibly stretched. Hill arrived at the conclusion that "the viscosity hypothesis must be dismissed" (1938, p. 193). In muscle biomechanics and physiology, this theory is not used anymore.

3.3 On muscle and joint viscosity—comments on the terminology

In the above discourse on the muscle viscoelastic theory, the term *viscosity* was used in its canonical meaning, as internal friction that depends on the movement velocity and is manifested in the heat production. It seems, however, that in muscles two characteristics, which are commonly associated with the notion of viscosity, are not related, or only weakly related, to each other:

- energy loss proportional to the velocity and
- resistive force proportional to the velocity.

There is a historic tendency in the motor control literature to use the term *viscosity* with a broader meaning, as any resistance proportional to velocity. It is done without regard to whether the resistance is due to internal friction or other mechanisms, for instance the speed-dependent component of the stretch reflex. The dimensionality of such viscosity is N/(m/s), the force per unit of velocity, or Nm/(rad/s), the force moment per unit of angular velocity (cf. with the SI system where the unit of viscosity is $(N \cdot s)/m^2$). In the author's opinion, such a use of this term should be discouraged. It can lead to terminological confusion. In the human movement science literature the phrase "muscle and joint viscosity" was used at least with 11 different meanings (Zatsiorsky, 1997).

To avoid misunderstanding, we recommend: (1) use the term *viscosity* only in accordance with the SI metric system (Eqn (3.1)) and (2) for describing the resistance proportional to velocity and measured in N/(m/s) use the term *velocity-dependent resistance*. The term is a tad lengthy but utterly precise. The coefficient in equation $f = -kv$ can be called the *damping coefficient*.

First, we consider velocity-dependent resistance (viscosity and damping) in passive biological objects and then in active muscles and joints. Differences in the behavior and parameterization of passive and active objects discussed in Section 2.1.2 in relation to the muscle and joint stiffness are also valid for the velocity-dependent resistance and will not be repeated here.

3.4 Velocity-dependent resistance of the passive objects—synovial fluid, tendons, passive muscles, and joints

3.4.1 Synovial fluid

In healthy people viscosity of the synovial fluid decreases when the joint angular velocity increases, that is, the viscosity is non-Newtonian (for a review, see Dumbleton, 1981). Such behavior does not agree with the widely held assumption in the motor control literature that viscous resistance in joints should increase with the movement velocity in a linear manner.

3.4.2 Tendons

Tendons possess viscoelastic properties. However, the hysteresis (H) dependence on the loading−unloading frequency is small. Tendons (as well as other soft biological tissues) usually have a nearly constant hysteresis over a large range of loading−unloading frequencies. For instance, when tendons are subjected to loading−unloading cycles at different rates, a 1000-fold increase of frequency changes the hysteresis loop area, that is, the energy loss, by no more than a factor of 2. Note, however, that such a large (1000-fold) increase in stretching velocity is nonphysiological. During real actions, the speed of stretch is smaller. For instance, in high-speed running, the maximal lengthening velocity of the hamstring muscle−tendon complex is estimated at only $2\ L_0/s$, where L_0 is the muscle−tendon length at rest (Thelen et al., 2005).

Several rather complex computational models of the tendon viscoelastic behavior have been suggested. The most broadly used is the *quasi-linear viscoelasticity model* (QLV model), suggested by Fung (1967, 1993).

3.4.3 Passive muscles and joints

Similarly to tendons, passive muscles and joints possess viscoelastic properties. The speed dependency of the resistance to stretch varies across stretch velocity ranges. For the individual muscle fibers, the peak resistive force is sensitive to lengthening velocities in the range between $0.1\ L_0/s$ and $2\ L_0/s$. Further increasing the velocity does not significantly change the peak resistance force (Rehorn et al., 2014).

For rhythmic perturbations performed at various frequencies, the relation between the hysteresis area and deformation frequency can be represented as

$$E = k\omega^n \tag{3.2}$$

where E is the dissipated mechanical energy (the hysteresis area), ω is the oscillation frequency, and k and n are empirical constants. If $n = 1$, the dissipated energy is proportional to velocity, and, when $n = 0$, the dissipated energy is velocity independent.

In human joints the exponent n is close to zero. For instance, for the passive knee flexion extension, $n = 0.085$ (Nordez et al., 2008). When angular velocity increased more than 20-fold, the increase in the passive torque was only 20%. For the human fingers, in the velocity range $0.85-1.29$ rad/s, the hysteresis is insensitive to joint speed (Esteki and Mansour, 1996). Similar facts were reported for the elbow joint (Boon et al., 1973). This indicates that the speed-dependent component does not play a decisive role in the resistance provided by the passive muscles to its forcible stretching. This behavior is similar to the behavior of other soft biological tissues. The intramuscular mechanisms of the small resistance dependence on the rate of muscle length change are complex and mainly unknown. Although muscles contain a large amount of water (a viscous liquid), viscous resistance (which is speed dependent across all velocity ranges) is relatively limited.

3.5 Velocity-dependent resistance of the active objects—muscles, joints, kinematic chains

3.5.1 Active muscles

When an active muscle is forcibly stretched, the resistive forces change in a complex manner illustrated in Figure 2.3, Chapter 2. The response consists of two main periods: the *dynamic force enhancement* and *residual force enhancement*, correspondingly. The dynamic phase enhancement includes two phases: a fast force rise that depends mainly on the length change (elastic response) and the subsequent slower force rise that is velocity dependent. The elastic force rise is associated with the increased strain of the crossbridges between actin and myosin while the velocity-dependent response is explained by partial detachment of some of the crossbridges and the formation of new crossbridges that are not so strongly deformed. These complex physiological transformations are quite different from the internal friction and hence cannot be labeled as viscosity.

The dynamic force enhancement depends—among other factors, such as stretch magnitude, initial muscle length, muscle fatigue, etc.—on the stretch velocity. With the increasing velocity of stretch, the force increases at a decreasing rate (Figure 3.5). Muscle force during stretch may exceed maximal isometric values. Several empirical force—velocity equations for lengthening muscles have been suggested (Otten, 1987; Krylow and Sandercock, 1997; Brown et al., 1999); however, none of them is accepted by all researchers.

During very fast stretches (>20 muscle length/s), a sudden reduction of force, called *slip* or *give*, can occur. The reduction is explained by the detachment of the crossbridges. The slip magnitude depends on stretch velocity—the higher the velocity, the greater the drop in force.

All the described reactions are reported for the muscles without reflexes; they represent the properties of the muscles themselves.

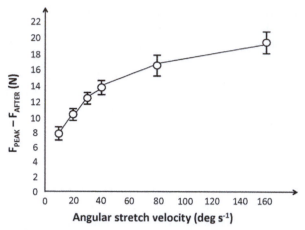

Figure 3.5 Force enhancement during stretch of human adductor pollicis muscle at different speeds. Effect of stretch velocity (angular velocity of the thumb abduction, grad/s) on the dynamic force enhancement ($F_{peak} - F_{after}$), N. The muscle was activated by percutaneous electrical stimulation of the ulnar nerve at the wrist at 80 Hz. A representative subject. Adapted by permission from De Ruiter et al. (2000), © Wiley.

3.5.2 Joints

The velocity-dependent resistance at joints in living humans and animals arises from numerous sources, first from the stretch reflex (cf. Chapter 6). Peripheral mechano-receptors located in the muscle spindles (e.g., type 1a dynamic fibers) and skin (e.g., slowly adapting type-II cutaneous mechanoreceptors) are velocity sensitive (see, e.g., Grill and Hallett, 1995). Reflex effects induced by the afferent velocity-related signals allow adjusting the motor responses to the movement speed. Such physiological responses can barely be attributed to "joint viscosity." Mechanical properties of the muscles and tendons, such as their viscosity, play a very limited role here.

To determine velocity-dependent resistance at the joints, perturbations of small (usually) amplitude are applied. The computations are commonly based on the second-order linear model:

$$M(t) = I\ddot{\alpha} + b\dot{\alpha} + k(\alpha - \alpha_0) \tag{3.3}$$

where M is joint moment (torque), α is angular coordinate, and I, b, and k are constants having, by the assumption of the author, physical meanings of inertia, viscosity/damping (resistive force proportional to the velocity), and stiffness, respectively. In the equation, Newton's notation for the time derivatives is employed. The equation is similar to Eqn (2.3) (see Chapter 2). The equation neglects physiological (control) mechanisms causing the observed outcomes, such as, for instance, stretch reflexes, and assumes that the human joint behavior is analogous to the behavior of a simple three-element mechanical system that possesses the properties of inertia, a viscous damper, and an elastic spring (stiffness).

Equation (3.3) represents, in essence, the first three terms of a Taylor series. Because any differentiable function can be expanded into a Taylor series, the coefficients I, b, and k are not obliged to have a physical meaning. They can just be mathematical abstractions. The burden of proof that the individual coefficients represent physical reality rests with the researcher. Unfortunately, this aspect of investigation is often neglected. For instance, it is repeatedly taken for granted that coefficient b stands for "viscosity." Application of this statement to active muscles, even if their activation level in the course of measurements does not change, is questionable (cf. with the Hill findings discussed above, see again Figure 2.3). Identifying b with viscosity when the reflexes are preserved has no physical meaning at all except one: b is *called* viscosity.

In experiments designed to estimate the coefficients of Eqn (3.3), the following three methods are most often used:

1. Singular torque perturbations (Ma and Zahalak, 1985)
2. Sinusoidal perturbations (Agarwal and Gottlieb, 1977; Joyce et al., 1974)
3. Randomized perturbations (Bennett et al., 1992; Hunter and Kearney, 1982; Lacquaniti et al., 1993).

The results are very sensitive to the details of the experimental measurements. For instance, when singular and sinusoidal ankle joint perturbations were applied to the same subjects, the differences in the obtained b values were almost two-fold, namely 811.6 N/(ms^{-1}) for the singular perturbation and 430.1 N/(ms^{-1}) for the vibration testing (Zatsiorsky and Aruin, 1984).

In the attempts to determine the velocity-dependent resistance (viscosity), some authors (e.g., Bennett et al., 1992) calculated it as the partial derivative of Eqn (3.3), or Eqn (2.3), with respect to the velocity. This derivative is evidently not equal to the b coefficient in the equation; it is

$$\frac{\partial F(x, \dot{x}, \ddot{x})}{\partial \dot{x}} = I\frac{\dot{x}}{\ddot{x}} + b + k\frac{\dddot{x}}{\ddot{x}} \qquad (3.4)$$

It is clear from Eqn (3.4) that the partial derivative depends on the coefficients I and k and, in addition, also on the second (acceleration) and the third (jerk) time derivatives of coordinate. The physical meaning of this derivative remains vague.

3.5.3 Kinematic chains

The matrix techniques used for estimating the apparent stiffness of kinematic chains (described in Chapter 2, section 2.3.2) can also be used for determining the velocity-dependent resistance. Consider a two-link planar kinematic chain in a static condition. An external force causes a (small) deflection of the endpoint as well as displacements at the joints. The force and displacement/velocity are not obliged to be in the same direction.

With the velocity-dependent term added, Eqn (2.7a) transforms into

$$
\begin{bmatrix} dF_x \\ dF_y \end{bmatrix} = - \left\{ \begin{bmatrix} S_{XX} & S_{XY} \\ S_{YX} & S_{YY} \end{bmatrix} \begin{bmatrix} dX \\ dY \end{bmatrix} + \begin{bmatrix} D_{XX} & D_{XY} \\ D_{YX} & D_{YY} \end{bmatrix} \begin{bmatrix} d\dot{X} \\ d\dot{Y} \end{bmatrix} \right\}
\tag{3.5a}
$$

where D stands for the damping coefficients and the symbols with overdots signify the time derivatives. In the vector-matrix notation the equation is

$$
d\mathbf{F} = - \left\{ [S]d\mathbf{P} + [D]d\dot{\mathbf{P}} \right\}
\tag{3.5b}
$$

Because the displacements are small, the chain Jacobian can be considered constant. Therefore, to determine the velocity-dependent resistance at the joints, the methods described in Chapter 2 (see Section 2.3.2, Eqn (2.8)) can be used. It seems that humans can control the displacement-related resistance at the joints (their apparent stiffness) and the speed-related resistance (damping) independently of each other (Lacquaniti et al., 1993).

3.6 More on muscle viscosity

Muscles contain approximately 70% water, which is a viscous liquid (water viscosity at 20 °C = $10.1 \times$ Pa·s). For this reason, if nothing else, viscous-like phenomena in muscles should certainly exist. However, as it was discussed above, their effects on the speed-related resistance to stretch are relatively small.

Muscles demonstrate thixotropic behavior (Hagbarth et al., 1985; Lakie and Robson, 1988). Thixotropy is the property of a substance to decrease its viscosity when it is shaken or stirred. In particular, synovial fluid is thixotropic. Also, the *ground substance (extrafibrillar matrix)* in connective tissue is thixotropic. When subjected to mechanical stress, these tissues become less viscous, they change from thick to thin. Muscles possess a property analogous to thixotropy—after long periods of rest, for instance in the morning hours, their resistance to induced motion increases. The resistance decreases after performing several motions (warming up).

3.7 Mechanical impedance

In engineering the mechanical impedance of a system undergoing simple harmonic motion is commonly defined as "the ratio of the force applied at a point to the resulting velocity at that point" (Hixson, 1961). The impedance is a function of frequency and the concept is broadly used in the studies on the effects of mechanical vibrations on the human body. For instance, Holmlund (2000) studied mechanical impedance of the human body in sitting posture in the vertical direction. Such factors as vibration level (0.5−1.4 m/s^2), frequency (2−100 Hz), body weight (57−92 kg), and relaxed and erect upper body posture were varied. The impedance was parameterized as a scalar (amount of resistance, a single number). It was concluded that the impedance increased with frequency up to a peak at about 5 Hz, after which it decreased in a complex manner.

In motor control and biomechanics studies, starting from seminal research by N. Hogan (1984, 1985), the term *mechanical impedance* is understood with a broader meaning—as total resistance of an object to motion that depends not only on its velocity but also on the displacement (stiffness) and acceleration (inertia). The relations are described by Eqn (3.3), where the three coefficients on the right side of the equations characterize the individual impedance components. Inertial resistance of the human body or a kinematic chain depends on the place of force application. Usually researchers are interested in the effect of force applied at the endpoint of a kinematic chain or in an effect of a force−couple system acting on the end link. Similar to the endpoint stiffness, described in Chapter 2, the endpoint inertia has a directional property—the force and acceleration vectors are not always collinear. The endpoint inertia can be described by an endpoint matrix (or its inverse, the *mobility tensor*) and represented by a *mobility ellipse*. The eigenvalues of the mobility ellipse correspond to the directions in which a force and acceleration induced by the force coincide. Those are the directions of the maximal and minimal inertia resistance. By changing the body position the central controller can increase inertial resistance in some directions and decrease it in other directions. A more detailed explanation of the endpoint inertia concept can be found in Zatsiorsky (2002), Section 4.5.3.

The motion mechanics are usually described with a second-order linear model (similar to Eqns (2.3) and (3.3)) and the mass (inertia), damping, and apparent stiffness matrices or scalar coefficients are estimated experimentally over a range of muscle activations, velocities, frequencies, etc. (see, e.g., Hajian and Howe, 1997). The damping and apparent stiffness coefficients depend on both neural control mechanisms and biomechanical properties of the involved muscle−tendon units, while the inertia of the kinematic chains depends on the chain configuration, for example, on the arm posture. At a given body posture the central nervous system can vary the damping and apparent stiffness by changing sensory feedback reflex gains and co-contraction of antagonistic muscles.

Some models of motor control (e.g., Hogan, 1984) are based on the hypothesis that, when movements, for example, arm movements, are performed, not only the arm kinematics but also the arm impedance is under neural control. As an example, compare simple arm extension or arm swing with the boxing punches or the tennis stroke, respectively. While kinematics of the movements can be similar, the arm impedances are different. In the boxing punches and tennis strokes, the arm should be ready to get in contact with an external object and overcome the expected resistance. At the end of a simple arm extension, the arm can be relaxed, while in the boxing punches and tennis strokes such a relaxed performance can lead to an injury.

3.8   A comment on clinical terminology

In the clinical literature the velocity-dependent resistance of a muscle/joint to passive stretch—which is partly due to the velocity-dependent stretch reflexes—is sometimes called *spasticity*. However, this term is also used with another meaning (discussed later in Chapter 5 on muscle tone).

3.9   The bottom line

Viscosity is internal friction in a substance; it is measured by the force per unit area resisting uniform flow. The SI metric unit for viscosity is $(N/m^2) \cdot s = Pa \cdot s$ (Pascal-second), equivalent to $(N \cdot s)/m^2$. Viscous resistance (internal friction) (1) depends on the rate of flow or deformation; and (2) is associated with the energy losses (under repetitive deformation the object temperature increases).

There is a historic tendency in motor control literature to use the term *viscosity* with a broader meaning, as the resistance of any source proportional to velocity. It is done without regard to whether the resistance is due to internal friction or other mechanisms, for instance, a speed-dependent component of a stretch reflex. The dimensionality of such "viscosity" is $N/(m/s)$, the force per unit of velocity, or $Nm/(rad/s)$, the force moment per unit of angular velocity (cf. with the SI system where the unit of viscosity is $(N \cdot s)/m^2$). In the author's opinion, such a use of this term should be discouraged.

To avoid misunderstanding, we recommend: (1) use the term *viscosity* only in accordance with the SI metric system (Eqn (3.1)), and (2) for describing the resistance proportional to velocity and measured in $N/(m/s)$ use the term *velocity-dependent resistance*. The coefficient in equation $f = -kv$ can be called the *damping coefficient*. If one uses the word *viscosity* to mean something different from what is accepted in the SI, one should define the term.

In passive human muscles, energy losses (damping) only slightly depend on velocity.

In active mammalian muscles, irreversible energy losses into heat are different during muscle shortening and lengthening and hence cannot be explained by viscosity of the muscle tissue (Hill, 1938). In human muscles, two characteristics that are commonly associated with the notion of viscosity are not related to each other:

- energy loss proportional to the velocity and
- resistive force proportional to the velocity.

The velocity-dependent resistance at joints in living humans and animals arises from numerous sources, first of all from stretch reflexes. To determine the velocity-dependent resistance at the joints, the matrix methods described in Chapter 2 (see Section 2.3.2, Eqn (2.8)) can be used.

The term *mechanical impedance* is understood as total resistance of an object to motion that depends not only on its velocity but also on the displacement (stiffness) and acceleration (inertia).

References

Agarwal, G.C., Gottlieb, G.L., 1977. Compliance of the human ankle joint. Journal of Biomechanical Engineering 99, 166−170.

Bennett, D.J., Hollerbach, J.M., Xu, Y., Hunter, L.W., 1992. Time-varying stiffness of human elbow joint during cyclic voluntary movement. Experimental Brain Research 88, 433−442.

Boon, K.L., Hof, A.L., Wallinga-de Jonge, W., 1973. The mechanical behavior of the passive arm. In: Cerquiglini, S., Venerando, A., Wartenweiler, J. (Eds.), Biomechanics III: Medicine and Sport, vol. 8. Karger, Basel, Switzerland, pp. 243−248.

Brown, I.E., Cheng, E.J., Loeb, G.E., 1999. Measured and modeled properties of mammalian skeletal muscle. II. The effects of stimulus frequency on force-length and force-velocity relationships. Journal of Muscle Research and Cell Motility 20, 627−643.

De Ruiter, C.J., Didden, W.J., Jones, D.A., Haan, A.D., 2000. The force-velocity relationship of human adductor pollicis muscle during stretch and the effects of fatigue. Journal of Physiology 526, 671−681.

Dumbleton, J.H., 1981. Tribology of Natural and Artificial Joints. Elsevier, Amsterdam-Oxford-New York.

Esteki, A., Mansour, J.M., 1996. An experimentally based nonlinear viscoelastic model of joint passive moment. Journal of Biomechanics 29, 443−450.

Fenn, W.O., 1924. The relation between the work performed and the energy liberated in muscular contraction. Journal of Physiology 58, 373−395.

Fung, Y.C., 1967. Elasticity of soft tissues in simple elongation. American Journal of Physiology 213, 1532−1544.

Fung, Y., 1993. Biomechanics: Mechanical Properties of Living Tissues, second ed. Springer-Verlag, New York.

Grill, S.E., Hallett, M., 1995. Velocity sensitivity of human muscle spindle afferents and slowly adapting type II cutaneous mechanoreceptors. Journal of Physiology 489, 593−602.

Hagbarth, K.E., Hagglund, J.V., et al., 1985. Thixotropic behaviour of human finger flexor muscles with accompanying changes in spindle and reflex responses to stretch. Journal of Physiology (London) 368, 323−342.

Hajian, A.Z., Howe, R.D., 1997. Identification of the mechanical impedance at the human finger tip. Journal of Biomechanical Engineering 119, 109−114.

Hill, A.V., 1922. The maximum work and mechanical efficiency of human muscles and their most economical speed. Journal of Physiology 56, 19−41.

Hill, A.V., 1938. The heat of shortening and the dynamic constants of muscle. Proceedings of the Royal Society, London, Series B 126, 136−195.

Hixson, E.L., 1961. Mechanical impedance and mobility. In: Harris, C.M., Creder, C.E. (Eds.), Shock and Vibration Handbook, first ed. McGraw-Hill, pp. 1−59 (Chapter 10).

Hogan, N., 1984. Adaptive control of mechanical impedance by coactivation of antagonistic muscles. IEEE Transactions on Automatic Control AC-29, 681−690.

Hogan, N., 1985. The mechanics of multi-joint posture and movement control. Biological Cybernetics 52, 315−331.

Holmlund, P., Lundstrom, R., Lindberg, L., 2000. Mechanical impedance of the human body in vertical direction. Applied Ergonomics 31, 415−422.

Hunter, L.W., Kearney, R.E., 1982. Dynamics of human ankle stiffness: variation with mean ankle torque. Journal of Biomechanics 15, 747−752.

Joyce, G.C., Rack, P.M.H., Ross, H.F., 1974. The forces generated in the human elbow joint in response to imposed sinusoidal movements of the forearm. Journal of Physiology 240, 351–374.

Krylow, A.M., Sandercock, T.G., 1997. Dynamic force responses of muscle involving eccentric contraction. Journal of Biomechanics 30, 27–33.

Lakie, M., Robson, L.G., 1988. Thixotropic changes in human muscle stiffness and the effects of fatigue. Quarterly Journal of Experimental Physiology 73, 487–500.

Lacquaniti, F., Carrozzo, M., Borghese, N.A., 1993. Time-varying mechanical behavior of multijoint arm in man. Journal of Neurophysiology 69, 1443–1464.

Ma, S.-P., Zahalak, G.I., 1985. The mechanical response of the active human triceps brachii muscle to very rapid stretch and shortening. Journal of Biomechanics 18, 585–598.

Nordez, A., Casari, P., Cornu, C., 2008. Effects of stretching velocity on passive resistance developed by the knee musculo-articular complex: contributions of frictional and visco-elastic behaviours. European Journal of Applied Physiology 103, 243–250.

Otten, E., 1987. A myocybernetic model of the jaw system of the rat. Journal of Neuroscience Methods 21, 287–302.

Rehorn, M.R., Schroer, A.K., Blemker, S.S., 2014. The passive properties of muscle fibers are velocity dependent. Journal of Biomechanics 47, 687–693.

Thelen, D.G., Chumanov, E.S., Hoerth, D.M., Best, T.M., Swanson, S.C., Li, L., Young, M., Heiderscheit, B.C., 2005. Hamstring muscle kinematics during treadmill sprinting. Medicine and Science in Sports and Exercise 37, 108–114.

Zatsiorsky, V.M., 2002. Kinetics of Human Motion. Human Kinetics, Champaign, IL.

Zatsiorsky, V.M., 1997. On muscle and joint viscosity. Motor Control 1, 299–309.

Zatsiorsky, V.M., Aruin, A.S., 1984. Biomechanical characteristics of the human ankle joint muscles. European Journal of Applied Physiology 52, 400–406.

Zatsiorsky, V.M., Prilutsky, B.I., 2012. Biomechanics of Skeletal Muscles. Human Kinetics, Urbana, IL.

Mechanical Work and Energy

4

Humans and animals spend energy to move. Through diverse metabolic processes, the chemically bound energy is transformed into mechanical work and heat. Examination of the mechanical work and energy in human and animal movements is an important field of research. This research direction benefits from relying on a basic law of nature: the law of conservation of energy. The less energy animals spend for ambulation, the less food they need, and the higher their chances for survival. Movement economy, that is, the ability to move expending smaller amounts of energy, was important in human evolution (Bramble and Lieberman, 2004). In particular, endurance running helped early hominids in predator pursuit to get close enough to their prey; compared to other animals, humans perform remarkably well at endurance running (Carrier, 1984). In the course of evolution, diverse mechanisms improving movement economy have been developed.

The problem of work and energy in human and animal movement is an interdisciplinary one; it is studied in biophysics, biochemistry, cell biology, physiology, motor control, and biomechanics. This chapter addresses only the biomechanical and motor control aspects of the problem.

When people move they perform (1) work on external objects called *external work* and (2) work to move the body parts and to change their mechanical energy (*internal work*). Similar terminology is applied to mechanical power, *external power* and *internal power*, respectively (Zatsiorsky and Prilutsky, 2012). For instance, in cycling the power of the force applied to the pedals is external power, and the power to move the legs themselves is internal power. Determining external power (work) is based on a straightforward application of classical mechanics and is not usually bound with conceptual difficulties. For instance, in functional testing on a bicycle ergometer the external power is determined as the product of the moment of force applied to the pedals and the angular velocity of the pedaling. In contrast, determining the power and work expended to move the legs—the internal power and work—is not a trivial problem. It does not have a definite solution and allows more than one answer. This chapter is limited to the discussion of the internal power and work, that is, the power and work expended to move body parts and the entire body.

It is crucial for the mechanical work and power determination to strictly follow the definitions of these terms in classical mechanics. Notice that in mechanics textbooks the term *work* is not defined. The basic defined term is the *work of a force*. The difference is important. Using "work" without specifying "force" may lead to ambiguity. For instance, the expression "mechanical work in walking" without specifying the forces of interest may have different meanings. The estimates of such "works" can be sharply different.

Biomechanics and Motor Control. http://dx.doi.org/10.1016/B978-0-12-800384-8.00004-1

4.1   Elements of history

The first estimation of mechanical work in human movement (work to move the general center of mass (COM) in walking) was made as early as 1836 (E. Weber and W. Weber, two of the famous Weber brothers). Substantial progress was, however, achieved only a hundred years later, in the 1930s. Two approaches, called at times Fenn's approach (*fraction approach*) and Elftman's approach (*source approach*), were developed. W.O. Fenn (1930a,b) recorded changes in the mechanical energy of the individual body segments from which the changes of the mechanical energy of the entire body and the work done on the body were computed. Elftman (1939, 1940) determined work at a joint as the integral of the product of joint moment and angular velocity over time. While both approaches yielded similar results in some simple tasks, like the sit-to-stand motion, using them for complex movements such as walking and running resulted in sharply different estimates. Application of both methods was bound with difficulties, some of which have not been overcome until now. For instance, the questions on how to account for negative muscle work, when muscles are forcibly stretched, or how to compute the total work when at some joints positive work is produced while at other joints negative work is performed are still topics of discussion and research.

In the 1970s, several seminal works on energy saving mechanisms were published. Among them were the studies on (1) the muscles' "spring action," that is, elastic energy storage and recoil in movements (Cavagna, 1970; Alexander and Bennet-Clark, 1977); (2) transfer of mechanical energy between body segments, both adjacent (Robertson and Winter, 1980) and nonadjacent (Morrison, 1970; such energy transfer was also envisioned by Elftman (1940)); and (3) mechanical energy transformation from kinetic energy to potential and vice versa as it happens in pendulum-like movements (this mechanism in human movements had already been addressed by Fenn (1930a,b) and further investigated by Cappozzo et al. (1976, 1978) and Cavagna et al. (1977)).

In the course of studies it became apparent that different methods of computation of mechanical work (especially the total work in walking and running) bring about sharply different results. For instance, Williams and Cavanagh (1983) who studied mechanical power in males running at 3.57 m/s have found that, depending on the employed computational methods, the power values ranged from 273 to 1775 W, that is, the difference was almost eight-fold. Such huge discrepancies required an explanation and understanding of "what is happening here." In 1986, S. Aleshinsky published in a single issue of the *Journal of Biomechanics* a set of five papers that considered in detail the theory of mechanical work determination in multilink systems (Aleshinsky, 1986a,b,c,d,e). These publications provided a theoretical basis for understanding the problem.

As already been mentioned in Chapter 1 (Subsection 1.2.5), computation of mechanical work in joints served by either one-joint or two-joint muscles should be done in different ways; for example, for a simple kinematic chain illustrated in Figure 1.4, either Eqn (1.5) or Eqn (1.6) should be used. In reality, human joints are usually served by both one-joint and two-joint muscles. Prilutsky and Zatsiorsky (1994) and

Prilutsky (2000) introduced a model for such actions and estimated the contribution of both the one-joint and two-joint muscles into the total work during jumping, landing, and running.

4.2 Definitions of work and power—work of a muscle

Let us start with the ground rules. Sure, everybody knows them, however, some people sometimes forget them. This was exactly the main reason behind the aforementioned eight-fold difference in the estimates of mechanical power in running found by Williams and Cavanagh (1983)—the researchers estimated the power of different forces.

4.2.1 Basic definitions

Consider a material particle on which a force $\mathbf{F}$ acts. The particle undergoes an infinitesimal displacement $d\mathbf{r}$. The *work* (or *elementary work*) done by $\mathbf{F}$ in this displacement is defined as the scalar product of the vectors $\mathbf{F}$ and $d\mathbf{r}$, $dW = \mathbf{F} \cdot d\mathbf{r}$ where both $\mathbf{F}$ and $\mathbf{r}$ are vectors. In scalar form, the equation is $dW = F\, dr \cos\theta$ where F and dr are the magnitudes of the force and displacement and θ is the angle formed by the vectors $\mathbf{F}$ and $d\mathbf{r}$. The expression $F\cos\theta$ represents the component of the force in the direction of displacement. Thus, the work of a force equals the product of the displacement of its point of action and the force component in the direction of the displacement. *Power* is the rate of doing work. The work during a finite displacement from P_1 to P_2 is obtained by integrating the equation $dW = \mathbf{F} \cdot d\mathbf{r}$ along the path traveled by the point of force application:

$$W|_{P_1}^{P_2} = \int_{P_1}^{P_2} \mathbf{F} d\mathbf{r} \tag{4.1}$$

In rotational movements, mechanical work is defined in a similar way. The elementary work produced by a moment $\mathbf{M}$ equals $dW = M d\theta$, where M is the magnitude of the moment and $d\theta$ is the small angle through which the body rotates. The above definitions should be strictly followed.

4.2.2 Some simple consequences

According to the provided definition, a force vector does work only when the object, on which it acts, moves in a direction along which the force has a nonzero component. In other cases, no work is produced. Here are several examples of workless forces:

1. Static forces. No displacement—no work. In mechanics, the expression "static work" is a misnomer.
2. A force acting perpendicularly to the displacement. Centripetal forces do not do work.

3. The ground reaction force in walking, running, and jumping, provided that the surface is not deformed. When people walk upstairs, they exert large contact forces on the ground, but the reaction forces do no work on the performer. The internal forces do.
4. Displacement of the point of force application itself (without movement of the object on which the force acts) does not produce work. For instance, if during a takeoff the point of application of the horizontal ground reaction force displaces forward but the foot does not (no slipping occurs), the work of the force is zero.

4.2.3   Work of a muscle—negative work

Work of a muscle equals the product of the muscle force and the muscle length change. When a muscle shortens (*concentric muscle action*) it performs positive work. When a muscle is forcibly stretched (*eccentric muscle action*), the external force does work on the muscle. It can be also said that the muscle does *negative work*. The expression *negative work* is not popular in classical mechanics. However, in biomechanics this term is convenient and universally accepted. The reason is that, in contrast to passive bodies, such as tendons, active muscles spend energy ("do work") to resist the extension. During eccentric muscle actions, there are two flows of energy: (1) to the muscle—external force does work on the muscle, and (2) the muscle spends energy to provide the resistance against the external force.

Presently, measuring the muscle work in human movements, especially measuring the muscle force, requires surgical intervention and has been performed to a limited extent only in several studies. If, in spite of the technical difficulties, the muscle force and the change of the muscle length are measured, then the determined muscle power (work) is the power (work) of the actual force. Such a determination is unambiguous; it satisfies the basic definition of power (and work) in classical mechanics. In contrast, when work at a joint or work done on a body segment is determined (see below), this is the work of the resultant, that is, imagined, force introduced by researchers for their convenience. Such a determination may involve conceptual difficulties with interpretation of the obtained values (explained later in this chapter).

An active skeletal muscle spends more energy than it does at rest, disregarding whether the muscle acts *isometrically* (no work is performed), *concentrically* (the muscle does positive work), or *eccentrically* (an external force does work on the muscle). In classical mechanics, energy represents the capacity to do work. In biomechanics, energy represents the capacity to do work and exert muscle force.

4.3   Work and power in human movements

To determine work of a force, both the force and the displacement of its point of application must be known. When the object of interest is the work to move body segments, such knowledge is usually not available, in particular, muscle forces are not known. Commonly, only the motion of human body segments is known. This motion occurs as a result of action of many forces produced by muscles, ligaments, bone-on-bone contacts, etc. These forces are usually not known, and their work cannot be determined. Some of the forces act in opposite directions and, when it comes to their effect on the body, cancel each other. They still produce some work, positive or negative, which we usually cannot determine. We can determine only the "total work" done by all the forces and moments, or more precisely by their resultants, acting on the body. Then one of the two options is used: the work on a body segment is estimated from the changes of the segment mechanical energy ("fraction approach") or the work done by a joint torque is computed ("source approach"). In essence, the difference in the approaches lies in the different ways of computation of the resultant forces and moments—whether they are assumed to act at the COM or at the joint center.

4.3.1 Work done on a body segment ("fraction approach")

This work equals the change in the body's mechanical energy (the *work-energy principle for a rigid body*). Consequently, the work done on a human body segment can be determined by registering changes in the mechanical energy of the segment and its three *fractions*, that is, the potential energy and the kinetic energy due to the translation and rotation, respectively.

For a rigid body in plane motion in the absence of elastic forces, the equation can be written as

$$W_{nc} = E_f - E_i = \left(mgh_f + \frac{mv_f^2}{2} + \frac{I\dot{\theta}_f^2}{2} \right) - \left(mgh_i + \frac{mv_i^2}{2} + \frac{I\dot{\theta}_i^2}{2} \right) \quad (4.2a)$$

where E stands for the total mechanical energy of the body (the sum of the potential and kinetic energy), m stands for mass, h is vertical location of the body's COM, I is moment of inertia, v is linear velocity of the COM of the body, $\dot{\theta}$ is angular velocity, g is acceleration due to gravity, and subscripts f and i refer to the final and initial values, correspondingly. The work done on the segment can also be represented as the sum of the changes in the potential energy (ΔE_P) and kinetic energy (ΔE_K) of the link, respectively:

$$W_{nc}\big|_{t_1}^{t_2} = \underbrace{mg(h_f - h_i)}_{\substack{\text{Gravitational} \\ \text{potential energy} \\ \text{change}}} + \underbrace{\left\{ \left(\frac{mv_f^2}{2} + \frac{I\dot{\theta}_f^2}{2} \right) - \left(\frac{mv_i^2}{2} + \frac{I\dot{\theta}_i^2}{2} \right) \right\}}_{\substack{\text{Change of the kinetic translational} \\ \text{and rotational energy}}}$$

$$= \Delta E_P + \Delta E_K \quad (4.2b)$$

The total work done on a rigid body equals the work of the resultant force and the force couple. The resultant force can be assumed to act at different points (researchers are free to choose any point they like). The most convenient point is, however, the

COM—a force acting at this point induces only linear acceleration of the body and does not affect the body orientation and rotation. This work can be compared to the total work performed by an imaginable rope fixed at the COM of the body and a torque generator at this place. Whether this model is acceptable or not depends on the goal of the research and the mindset of the researcher.

Work done on a body can sharply differ from work of individual forces; in particular, it can be much smaller. This happens when forces act in opposite directions and in part cancel each other, for example, as the agonist—antagonist muscles crossing a joint do. When forces cancel each other, this does not mean that the work done by these forces can also be canceled. It is quite possible that these individual works should be added, not canceled.

4.3.2 Work and power at a joint ("source approach")

By definition, *source of mechanical energy* (or simply *source*) is any force or moment acting on the system and changing its mechanical energy. In particular, joint torques and joint forces are sources of mechanical energy.

4.3.2.1 Work and power of a joint torque

In three dimensions, power of a joint torque—usually called simply *joint power*—equals the scalar product of the vectors of the joint torque ($\mathbf{T}$) and the angular velocity ($\dot{\alpha}$) at the joint, both referred to the global reference system:

$$P(\text{watts}) = \mathbf{T} \cdot \left(\dot{\mathbf{\theta}}_d - \dot{\mathbf{\theta}}_p \right) = \mathbf{T} \cdot \dot{\alpha} \tag{4.2}$$

where $\dot{\mathbf{\theta}}$ is the vector of the angular velocity of the body link rotation and subscripts d and p refer to the distal and proximal links adjacent at the joint, correspondingly (for a definition of joint torque concept, see Chapter 1). Note that joint power is a function of the joint angular velocity ($\dot{\alpha}$), that is, the velocity of one link with respect to the other. Joint power is positive when the joint torque and the joint angular velocity are in the same direction. Joint power is negative when the joint torque and joint angular velocity are in the opposite directions. When the joint links move in the same direction at the same speed, the joint power is zero. Joint work during the period from t_1 to t_2 equals the time integral of the joint power over that period. Power delivered to (or absorbed by) the adjacent distal and proximal segments is $P_{distal} = \mathbf{T} \cdot \dot{\mathbf{\theta}}_d$ and $P_{proximal} = \mathbf{T} \cdot \dot{\mathbf{\theta}}_p$, correspondingly.

Power and work at a joint are evidently consequences of the power and work done by the contributing muscles. However, computation of the joint power and work does not take into account the energy expended

1. to overcome antagonistic activity of muscles,
2. for displacement of muscles relative to the skeleton,
3. against elastic and nonelastic internal forces, for example, for ligament extension and friction, displacement of internal organs relative to the spine, etc.

Such an estimation of power and work neglects also the difference between the contribution of single-joint and two-joint muscles into the joint torques and power, that is, it neglects the contrast between "actual" joint torques (produced by single-joint muscles) and "equivalent" torques (calculated for the system served by two-joint muscles), see Chapter 1, Subsections 1.2.4 and 1.2.5. As it has already been mentioned in those sections for joints served by two-joint muscles, estimation of joint power cannot be done without knowing what is happening at a neighboring joint, at the other end of the muscle.

4.3.2.2 Work and power of a joint force

When a joint center (the point of application of the resultant joint force) moves, the joint force does work on the adjacent links. The power P_j of the joint force $\mathbf{F}_j$ equals

$$P_j = \mathbf{F}_j \mathbf{v}_j \tag{4.3}$$

where $\mathbf{v}_j$ is the velocity of the joint center. Similar to "joint torque," the expression "joint force" is just an abbreviated designation of two forces—action and reaction—acting between the two adjacent segments at a joint. While small displacements of the adjacent links could be different, the displacement of the point of their contact is the same. Otherwise, either the links impinge on each other or the contact is broken. Therefore, the power of $\mathbf{F}_j$ is equal in magnitude and opposite in sign to the power of $-\mathbf{F}_j$, and their sum is zero. As a result, the mechanical energy of one of the adjacent segments increases and the mechanical energy of the second segment decreases by the same amount, but the total mechanical energy of the entire system does not change. The joint forces redistribute the mechanical energy among the body segments without changing the total mechanical energy of the whole human body. In summary, the joint forces neither generate nor absorb the mechanical energy; they only transfer it between adjacent segments.

4.4   Energy saving mechanisms

Suppose a researcher determined energy expenditure for walking. Then consider a mental experiment. The body of the performer is dismembered at all joints into individual body segments but maintained in the same body configuration as in the intact body. By the condition of the experiment, all the body parts move along the same trajectories, at the same velocities and accelerations as they move during real walking. Kinematics of such a movement are indistinguishable from the kinematics of real walking. The only difference is that in real walking the body parts interact with each other while in the disjointed body the body parts do not interact. It has been computed that the required work for the disjointed body is approximately five-fold the work in real walking. Hence, if we take the work for moving the disjointed body parts as 100%, in real walking people spend only about 20% of that value. The other 80% is conserved due to energy-saving mechanisms.

The following energy-saving mechanisms are present in human motion:

1. Transformation of the kinetic energy of body segments (as well as of the COM of the body) into the gravitational potential energy and vice versa—the pendulum-like motion.
2. Transfer of mechanical energy between the adjacent and nonadjacent segments and compensation of energy loss at one joint by positive energy inflow at another joint (*intercompensation of sources*). In human and animal movements, the intercompensation occurs mainly due to activity of two-joint muscles: if one joint moves in a direction that corresponds to the muscle lengthening and the second joint moves in a direction that corresponds to the same muscle shortening, the total work balance can be either positive (the muscle does work), negative (the work is done on the muscle), or zero (the muscle acts isometrically, its length is not changed).
3. Transformation of the kinetic energy of body segments into potential energy of deformation of the soft tissues of the musculoskeletal system (muscles, tendons, ligaments, joint capsules) and vice versa. This mechanism is also known as *energy recuperation* or *compensation during time*.

The abstract terminology—*intercompensation of sources* and *compensation during time*—was introduced to stress that similar events occur not only in human motions but also in the inanimate world, in particular in engineering. Examples of the intercompensated sources are robots with electrical motors as torque actuators. The motors convert electrical energy into mechanical energy and vice versa, mechanical energy into electrical energy. When an external force does positive work at a joint (so, the work of the joint actuator is negative), the mechanical energy is converted into electrical current and used in another joint(s). This arrangement allows for saving energy. Recuperation of energy (*compensation during time*) is used for instance in underground transportation. To save energy, the stations are usually located higher than portions of the railroads between the stations. Gravity assists train acceleration from a station, saving energy.

4.4.1   Pendulum-like motions

Pendulum-like motion associated with energy transformation from potential energy into kinetic energy and back is common in human motion. This phenomenon habitually occurs in cyclic movements, such as walking and running, and is indicative of the *conservation of energy*. When a segment moves downward, part of its gravitational potential energy is used to accelerate the segment and is transformed into kinetic energy. When the segment moves up, the kinetic energy of the segment may be used to lift the segment (to do work against the gravity).

The presence of the pendulum-like motions in human movement is sometimes taken for granted. However, in each case, the existence of pendular motion must be tested biomechanically. For an ideal pendulum:

1. the total mechanical energy (the sum of the kinetic and potential energy) is constant over an oscillation cycle, and
2. the kinetic and potential energy change exactly out of phase (Figure 4.1).

Consider from this standpoint leg movement during the swing phases of walking. Visually, the leg movement resembles the movement of a compound pendulum.

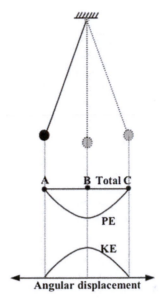

Figure 4.1 Energy changes during one swing of an ideal pendulum. The pendulum is released at point A and moves to point C. Kinetic energy (KE) of the pendulum is maximal at the bottom point B; the potential energy is minimal at this point. The gravitational potential energy (PE) is maximal at the highest positions of the pendulum, A and C. The sum of the kinetic and potential energy (total energy) is constant. The changes in the kinetic and potential energy mirror each other.

The legs are periodically raised and lowered under the influence of gravity. Therefore, the idea of pendulum-like leg movements during walking seems very natural (the idea was mentioned by Weber and Weber (1836)). However, the peak values of leg kinetic energy during the swing phase greatly exceed the peak values of its potential energy. As a result, the total mechanical energy of the leg during a stride is not constant. This is especially evident for the shank and foot. During the support period, when the foot is on the ground, both its potential and kinetic energy fractions are at a minimum—the foot assumes the lowest possible position and does not translate. Thus, the leg during walking can barely be considered a free pendulum, Figure 4.2.

In contrast, the total energy of the trunk at a certain speed of walking fluctuates only slightly, and the changes in the trunk potential and kinetic energy are out of phase. Hence, the trunk rather than the legs moves in such a way that total energy is conserved.

To estimate possible conservation of energy via pendulum-like mechanisms, the sum $(\Delta E_K + \Delta E_P)$, known as the *quasi-mechanical work*, is computed. It represents an imagined work that would be done on the segment by the resultant force and couple if there were no transformation between the gravitational potential energy and kinetic energy. The quasi-mechanical work is a pure theoretical construct. The difference between the sum of the gains of the kinetic and gravitational potential energy $(\Delta E_K + \Delta E_P)$ and the gain of the total energy of the segment, ΔE, can be used for

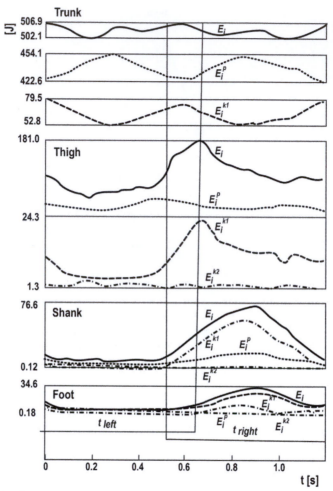

Figure 4.2 The time course of the fractions of the mechanical energy of body links during walking at 4.2 km/h. E_i is the total mechanical energy of the link, E_i^p is the potential energy, E_i^{k1} is the translational kinetic energy and E_i^{k2} is the rotational kinetic energy, t_{left} and t_{right} are the stance periods.
Reprinted by permission from Zatsiorsky et al. (1982), © Human Kinetics.

estimating the magnitude of the conserved energy. It should be done with care—a positive difference $D = [(\Delta E_K + \Delta E_P) - \Delta E] > 0$ is a necessary but not sufficient condition for energy transformation from the potential into kinetic form and vice versa. In other words, if energy transformation takes place, then $D > 0$. However, if $D > 0$, we cannot claim that this is due to energy transformation. For instance, if a person lowers and lifts a barbell, it is quite possible that $D > 0$; however, one cannot claim that the weight lifting was performed at the expense of the kinetic energy that the barbell had during its downward movement.

The *energy conservation coefficient* was defined as the ratio of D to quasi-mechanical work:

$$CC = \frac{(\Delta E_K + \Delta E_P) - \Delta E}{(\Delta E_K + \Delta E_P)} \times 100\% \qquad (4.4)$$

The coefficient is intended to characterize the proportion of the mechanical energy conserved in a given movement. A similar approach can be used to estimate conservation of energy associated with movement of the general COM of the body. In a walking cycle, the potential and kinetic energy of the COM change out of phase, and the fluctuations of the total mechanical energy are small. The highest position of the COM and hence the maximal potential energy is at mid-stance when the hip of the stance leg passes over the ankle. The COM velocity is minimal near this instant.

One of the ways to estimate the possibility of energy conservation during locomotion is to compare the maximal values of changes of the kinetic and potential energy of a link, or the COM, in a cycle. If the values are different, the energy must be supplied by external sources. The kinetic energy increases as a quadratic function of the speed while changes of the potential energy with speed are less obvious. When the speed of a body link doubles, its kinetic energy quadruples, but potential energy of the link in a cycle does not fluctuate that much. For example, we do not lift the legs much higher when we walk faster. Therefore, the conditions for energy conservation due to the transformation of the kinetic energy into the potential energy and vice versa depend on the speed of ambulation (Figure 4.3).

4.4.2 Transfer of mechanical energy

Joint torques can transfer energy between the body segments. We consider first how such transfer can occur between adjacent segments and then for nonadjacent segments.

4.4.2.1 Energy transfer between adjacent segments

Consider a two-link chain with a hinge joint (Figure 4.4). The joint is served by a single-joint muscle. Apart from a trivial case, when only one segment is rotating (and hence the second segment neither gives away nor acquires energy), there are two distinct situations. The adjacent segments may rotate either in opposite directions or in the same direction. When the adjacent links rotate in the opposite directions, energy is not transferred between the links. When they rotate in the same direction, the transfer of energy occurs.

When the adjacent segments rotate in the same direction with *equal* angular velocities ($\dot{\theta}_d = \dot{\theta}_p$), the joint angular velocity is zero. Hence, the joint power is also zero. The absolute values of power of the moments of force acting on the adjacent segments are equal, $|T\dot{\theta}_d| = |-T\dot{\theta}_p|$. One segment loses energy and the second gains energy by the same amount. As a result, "pure" energy transfer from one segment to an adjacent segment occurs. Energy is transferred between the segments while the total energy of

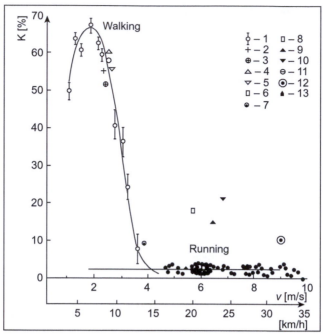

Figure 4.3 Conservation coefficient for the body's center of mass (COM) in walking and running. The numbers correspond to different subjects and methods, in particular the optical methods or computation based on the ground reaction forces. In walking, the conservation of energy associated with the COM movement sharply depends on the speed, and the energy conservation coefficient can be as high as 65%. In running, the conservation of energy due to pendulum-like mechanisms is very low at all speeds.
Adapted by permission from Zatsiorsky and Yakunin (1980), © Nauka (Science).

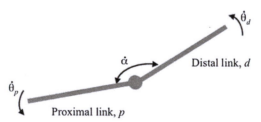

Figure 4.4 A two-link system with a hinge joint. A planar movement. $\dot{\alpha}$ is joint angular velocity, $\dot{\theta}_p$ and $\dot{\theta}_d$ are angular velocities of the proximal and distal links, respectively. The angular velocities of the links are in the external coordinates.

the system is conserved. The muscles that cross the joint act isometrically and do not do any work. They behave like nonextensible struts.

When the adjacent segments rotate in the same direction with *unequal* angular velocities $(\dot{\theta}_d \neq \dot{\theta}_p)$, the joint angular velocity is $\dot{\alpha} = \dot{\theta}_d - \dot{\theta}_p$. The power flow to/from the individual segments is $P_d = T\dot{\theta}_d$ and $P_p = -T\dot{\theta}_p$, respectively, where T is a joint torque. One segment gives away the energy to the other segment; the second segment acquires the energy. Because $\dot{\theta}_d \neq \dot{\theta}_p$, the energy gain of one segment and the energy loss of the other segment are not equal in magnitude. The difference equals the joint power, $P = T \cdot (\dot{\theta}_d - \dot{\theta}_p)$ or $P = T \cdot \dot{\alpha}$. Two flows of energy exist: (1) from a segment to a segment and (2) between the segment(s) and the joint actuators (the muscles and tendons). The joint actuators transfer the energy from one segment to the other, and, in addition, they either provide mechanical energy to the system (when the joint power is positive) or absorb the energy (when the joint power is negative). The entire situation is summarized in Figure 4.5.

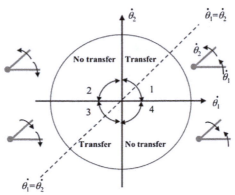

Figure 4.5 Energy transfer for a two-link system. The curved arrows indicate the direction of the link rotation. Transfer occurs when the angular velocities of the links $\dot{\theta}_i$ are in the same direction, in sectors 1 and 3. There is no energy transfer in sectors 2 and 4, where the links are rotating in opposite directions. Link i acquires energy when $T\dot{\theta}_i > 0$ and it gives away energy when $T\dot{\theta}_i < 0$.
Adapted by permission from Zatsiorsky (2002), © Human Kinetics.

4.4.2.2 Energy transfer between nonadjacent segments— energy intercompensation

When two-joint muscles cross joints, energy transfer may occur between nonadjacent segments. It happens when the segments, to which a two-joint muscle is attached, rotate in the same direction. In this case, the joint power at one of the joints is negative while the power at the second joint is positive. The mechanism is analogous to the one illustrated in Figure 4.5 but with one important difference. For a single joint and a single-joint muscle, the energy transfer is determined by the angular velocities of the body links as it was explained above. For a kinematic chain with a two-joint muscle, the energy transfer depends also on muscle moment arms at the joints. At the same movement kinematics, the muscle can perform either positive work (the muscle shortens), negative work (the muscle length increases), or no work at all (the muscle length does not change); see Figure 4.6.

In the framework of the *fraction approach*, calculation of the energy transfer between the segments in multilink kinematic chains is difficult. Within the *source*

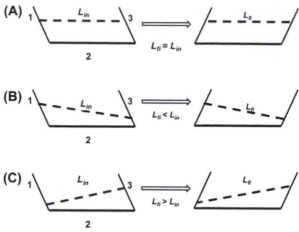

Figure 4.6 The effects of the differences in the points of origin and insertion of a two-joint muscle on its length at the same angular displacements at both joints. The solid lines are links of a three-link kinematic chain and the dashed line is a two-joint muscle attached to links 1 and 3. The angular displacements are being caused by external moments (not shown), while the muscle is generating force. L_{in} and L_{fi} are initial and final muscle lengths. (A) The muscle length does not change (isometric action). (B) Muscle length decreases (concentric action). (C) Muscle length increases (eccentric action).

approach, the power and work of the joint torques are determined and the mechanical energy of the body segments is not immediately known. The energy (power, work) transfer is considered among joints (joint actuators), not body segments. The difference is subtle, however, and energy transfer between the segments or joints is usually discussed together. The main idea of the between-joint power transfer (energy inter-compensation) was already discussed in Subsection 1.2.4. To avoid duplication, we refer the readers to this section, especially to the example considered in Figure 1.4 and Eqns (1.5) and (1.6).

At a certain combination of angular velocities, the length of a two-joint muscle may remain constant (item (A) in Figure 4.6). In this case, the muscle acts isometrically and does not do any work. This situation is known as the *tendon action of the two-joint muscles*. In this particular case, the amount of positive power generated by the muscle at one joint equals the amount of the negative power absorbed by the same muscle at the second joint.

Due to energy transfer, the joint power, calculated as the product $P_j = T_j \cdot \dot{\alpha}_j$, does not have to be equal to the sum of powers developed by each muscle crossing the joint, $P_j^M = \sum_j P^M$. The difference between the joint moment power P_j and the sum of powers across all muscles crossing the joint indicates the rate at which energy is transmitted through two-joint muscles to or from the joint. Table 4.1 illustrates possible variants of mechanical energy transfer by two-joint muscles.

Energy transfer between body segments not only allows decreasing energy expenditure but may also lead to performance improvement, for example, increasing the

Table 4.1 Mechanical energy transfer by two-joint muscles

#	Joint power, $P_j^T(t)$	Summed muscle power, $\sum_j P^M(t)$	Difference $P_j(t)$	Energy transfer		Muscles of the jth joint generate (gen) or absorb (abs) power at a rate of
				Rate	Direction	
1	>0	≥ 0	>0	$|P_j(t)|$	To joint j; from joints $(j-1)$ and/or $(j+1)$	$|P_j^T(t)| - |P_j(t)|$, gen
2	≥ 0	>0	<0	$-|P_j(t)|$	From joint j; to joints $(j-1)$ and/or $(j+1)$	$|P_j^T(t)| + |P_j(t)|$, gen
3	≤ 0	<0	>0	$|P_j(t)|$	To joint j; from joints $(j-1)$ and/or $(j+1)$	$-[|P_j^T(t)| + |P_j(t)|]$, abs
4	<0	≤ 0	<0	$-|P_j(t)|$	From joint j; to joints $(j-1)$ and/or $(j+1)$	$-[|P_j^T(t)| - |P_j(t)|]$, abs
5	≥ 0 <0	≥ 0 <0	$=0$	1. $|P_j(t)| = 0$, or 2. $|P_j(t)|$ and $-|P_j(t)|$	No energy transfer or To joint j from one of the joints and from joint j to another joint.	$|P_j^T(t)|$, gen or $-|P_j^T(t)|$, abs

Adapted by permission from Prilutsky and Zatsiorsky (1994), © Elsevier.

jump height. For instance, during the vertical jumps, timely activation of a two-joint gastrocnemius allows for the energy generated by the knee extensors to be in part transmitted to the calcaneus to increase the work done at the ankle, leading to an increase in the jump height. This mechanism can be illustrated with a physical model, a Jumping Jack (Figure 4.7).

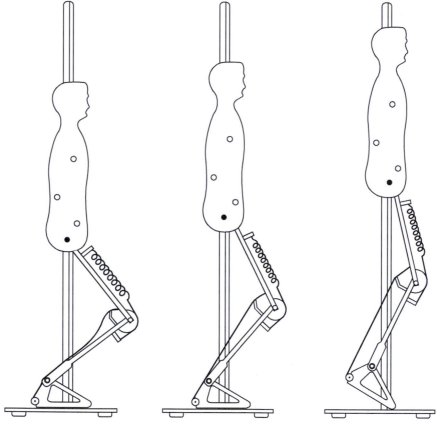

Figure 4.7 Jumping Jack, a physical model demonstrating the role of two-joint gastrocnemius in jumping performance. During the demonstration, the model is manually pushed down and then released.
Reproduced by permission from van Ingen Schenau (1989), © Elsevier.

The trunk of the model can move only in the vertical direction along a vertical rail. A one-joint knee extensor is represented by a spring that can store potential energy when the trunk is pushed downward. A wire representing the gastrocnemius has adjustable length. The length determines the knee angle at which the wire engages and starts ankle extension following model release. There is an optimal length of the wire at which the Jumping Jack achieves maximum jumping height. Without the wire or when the wire is very long, the energy from the knee extension cannot be delivered to the ankle joint and the Jumping Jack cannot jump.

The concept of source intercompensation provides a convenient framework and language for exploring energy transfer. The sources are called *intercompensated* if the energy expended by one source can be compensated for by the energy absorbed by the second source. For instance, a two-joint muscle can simultaneously produce negative power at one joint and positive power at the other joint. Hence, when two joints are served by one two-joint muscle, the power from one joint may be used in the neighboring joint and the sources of energy (the joint torques) are intercompensated. Contrary to that, if the joints are only crossed by single-joint muscles, the negative power from a joint cannot be used at the other joint. The joint torques are in this case *nonintercompensated sources* of energy.

When sources of energy are intercompensated, the total power produced by them equals the algebraic sum of the powers developed by them in individual joints:

$$P_{tot} = \sum_{s} P_s \tag{4.5}$$

where P_s is the power of individual sources of energy (forces, moments). Referring back to Figure 1.4 (the horizontal arm extension), Eqn (4.5) explains why the total power of intercompensated sources in the absence of friction can be zero when a movement is performed.

When the sources are *not intercompensated*, the total power they produce equals the sum of *absolute values* of the power developed by the sources:

$$P_{tot} = \sum_{s} |P_s| \tag{4.6}$$

The joint moments generated by one-joint muscles are not intercompensated sources of mechanical energy. If muscles, rather than joint torques, are considered the sources of energy, the sources are not intercompensated. As a rule, energy expended by one muscle cannot return to the system through the simultaneous absorption of energy by another muscle.

Equation (4.6) yields the total amount of power generated or absorbed by several nonintercompensated sources of energy. It represents the so-called *mechanical energy expenditure* (MEE), a work-like measure used in biomechanics when computing mechanical work is not feasible. By definition,

$$MEE = \int_{t_1}^{t_2} \sum_{s} |P_s| dt \tag{4.7}$$

where S stands for sources of energy, for example, joint torques, and other symbols are self-evident. In summary, to compute mechanical energy expended by the sources, summation of the absolute values is performed. In contrast, for mechanical work computations, the real values of the work done by the individual sources—which can be both positive and negative—is summated across the sources.

Returning from the general theory to real life examples, we can say that a two-joint muscle can reduce the amount of mechanical energy necessary for performing a given motion. This reduction occurs if both of the following conditions are met:

1. The two-joint muscle produces force
2. Powers developed by the muscle at the two joints have opposite signs.

In human motion, these conditions are often met. For example, during walking the gastrocnemius, rectus femoris, and hamstrings (all two-joint muscles) develop oppositely signed power at the joints they cross in several phases of the gait cycle, for example, the rectus femoris in late stance demonstrates positive power at the hip and negative power at the knee, while power produced by the muscle itself has much lower peak values.

4.4.3   Compensation during time (energy recuperation)

When muscles develop force eccentrically (in the direction opposite to the motion), the external force does positive work on the muscle. Hence, the energy from the external source goes to the muscle. The energy can be stored temporarily as elastic energy of muscle and tendon deformation and then recoiled to the system later in time ("springs in the legs"). This feature is called *energy recuperation* or *energy compensation during time*.

Consider two examples, with and without energy recuperation, respectively. Assume that a muscle during an eccentric phase behaves like a spring; it does not expend energy to resist the forcible stretching and can store elastic energy. During a forcible extension of the muscle−tendon complex by an external force, the force does the work of 10 J on the complex. This energy is absorbed by the complex. During the following recoil phase, positive work done by the muscle is 15 J. Assuming that the entire 10 J from 15 J are recuperated, the energy expended by the muscle for motion is $MEE = |-10 + 15| = 5$ J. The 5 J are delivered by metabolic processes.

In contrast to the previous example, if the muscle does not recuperate the mechanical energy and expends energy for resisting the forcible extension (performs negative work), the total amount of energy expended for the motion is $MEE = |-10| + |15| = 25$ J. All 25 J are due to energy delivered by metabolic processes.

In general, for a source of energy compensated over time, MEE equals the absolute value of the integral over time of the power developed by the source:

$$MEE = \left| \int_{t_1}^{t_2} P_s dt \right| \tag{4.8}$$

While for a nonrecuperative source,

$$MEE = \int_{t_1}^{t_2} |P_s| dt \tag{4.9}$$

Note the different location of the vertical bars in Eqns (4.8) and (4.9). When the absolute values of power are integrated over time as in Eqn (4.9) ($MEE = \int |P_s| dt$), the equation signifies that energy recuperation is prohibited. When a source is compensated during time, its MEE is expressed as $MEE = \int P_s dt$.

4.5 The bottom line

Mechanical energy of the human body parts and its change can be determined from the kinematic data and mass-inertial characteristics of the body segments. However, to determine mechanical work or mechanical energy expended for motion, additional information is necessary. The six main situations occur that correspond to affirmative or negative answers to the following questions: (1) Are two-joint muscles included in the model (intercompensation)? (2) Do muscles and tendons store and recoil elastic energy (recuperation)? (3) Does negative work of joint torques require metabolic energy from the system (is *active*) or not (is *passive*)? Unfortunately, the precise proportion of the intercompensated energy (transferred from a joint to joint) and recuperated energy (stored in elastic form and lately returned to the system) in human movement is commonly unknown. The cost of the negative work is also known only for some activities. Consequently, studies are limited to approximate models. However, these models are still useful in many applications as well as an intermediate step to a conclusive solution of the problem. The precise determination of mechanical work in human movements will become possible when new experimental methods are developed. These methods should allow measuring the individual muscle forces together with the muscle length changes.

References

Alexander, R.M., Bennet-Clark, H.C., 1977. Storage of elastic strain energy in muscle and other tissues. Nature 265, 114–117.

Aleshinsky, S.Y., 1986a. An energy "sources" and "fractions" approach to the mechanical energy expenditure problem–I. Basic concepts, description of the model, analysis of a one-link system movement. Journal of Biomechanics 19, 287–293.

Aleshinsky, S.Y., 1986b. An energy "sources" and "fractions" approach to the mechanical energy expenditure problem–II. Movement of the multi-link chain model. Journal of Biomechanics 19 (4), 295–300.

Aleshinsky, S.Y., 1986c. An energy "sources" and "fractions" approach to the mechanical energy expenditure problem–III. Mechanical energy expenditure reduction during one link motion. Journal of Biomechanics 19, 301–306.

Aleshinsky, S.Y., 1986d. An energy "sources" and "fractions" approach to the mechanical energy expenditure problem–IV. Criticism of the concept of "energy transfers within and between links". Journal of Biomechanics 19, 307–309.

Aleshinsky, S.Y., 1986e. An energy "sources" and "fractions" approach to the mechanical energy expenditure problem–V. The mechanical energy expenditure reduction during motion of the multi-link system. Journal of Biomechanics 19, 311–315.

Bramble, D.M., Lieberman, D.E., 2004. Endurance running and the evolution of Homo. Nature 432, 345–352.

Cappozzo, A., Figura, F., Marchetti, M., 1976. The interplay of muscular and external forces in human ambulation. Journal of Biomechanics 9, 35—43.

Cappozzo, A., Figura, F., Leo, T., Marchetti, M., 1978. Movements and mechanical energy changes of the upper part of the human body during walking. In: Assmussen, E., Jorgansen, K. (Eds.), Biomechanics VI-A. University Park Press, Baltimore, pp. 272—279.

Carrier, D.R., 1984. The energetic paradox of human running and hominid evolution. Current Anthropology 25, 483—495.

Cavagna, G.A., 1970. Elastic bounce of the body. Journal of Applied Physiology 29, 279—282.

Cavagna, G.A., Heglund, N.C., Taylor, C.R., 1977. Mechanical work in terrestrial locomotion: two basic mechanisms for minimizing energy expenditure. American Journal of Physiology 233, R243—R261.

Elftman, H., 1939. Forces and energy changes in the leg during walking. American Journal of Physiology 125, 339—356.

Elftman, H., 1940. The work done by muscles in running. American Journal of Physiology 129, 672—684.

Fenn, W.O., 1930a. Frictional and kinetic factors in the work of sprint running. American Journal of Physiology 92, 583—611.

Fenn, W.O., 1930b. Work against gravity and work due to velocity changes in running. American Journal of Physiology 93, 433—462.

van Ingen Schenau, G.J., 1989. From rotation to translation: constraints on multi-joint movements and the unique action of bi-articular muscles. Human Movement Science 8, 301—337.

Morrison, J.B., 1970. The mechanics of muscle function in locomotion. Journal of Biomechanics 3, 431—451.

Prilutsky, B.I., 2000. Coordination of two- and one-joint muscles: functional consequences and implications for motor control. Motor Control 4, 1—44.

Prilutsky, B.I., Zatsiorsky, V.M., 1994. Tendon action of two-joint muscles: transfer of mechanical energy between joints during jumping, landing, and running. Journal of Biomechanics 27, 25—34.

Robertson, D.G.E., Winter, D.A., 1980. Mechanical energy generation, absorbtion and transfer amongst segments during walking. Journal of Biomechanics 13, 845—854.

Williams, K.R., Cavanagh, P.R., 1983. A model for the calculation of mechanical power during distance running. Journal of Biomechanics 16, 115—128.

Weber W., and Weber E., (1836 first German edition; 1992 English translation). Mechanics of the Human Walking Apparatus. Springer-Verlag: Berlin, Heidelberg, New York. (Translated from German by Maquet P., Furlong R.).

Zatsiorsky, V.M., 2002. Kinetics of Human Motion. Human Kinetics, Champaign, IL.

Zatsiorsky, V., Aleshinsky, S., Yakunin, N., 1982. Biomechanical Basis of Endurance. FiS Publisher, Moscow (in Russian). The book is also available in German. Saziorski W.M., Aljeschinski S., Jakunin N.A., 1986. Biomechanische Grundlagen der Ausdauer. Sportverlag: Berlin.

Zatsiorsky, V.M., Prilutsky, B.I., 2012. Biomechanics of Skeletal Muscles. Human Kinetics, Champaign, IL.

Zatsiorsky, V.M., Yakunin, N.A., 1980. Mechanical work and energy in human locomotion. Human Physiology 6, 579—696 (in Russian).

Part Two

Neurophysiological Concepts

Muscle Tone

5

Muscle tone is arguably one of the most commonly used and least commonly defined notions in studies of movement, posture, and movement disorders. While most researchers imply under this expression something like "state of relaxed muscle under the spontaneous excitation by the central nervous system," methods of assessment of muscle tone commonly used in the clinical practice are likely to reflect a host of factors including properties of peripheral tissues that do not receive neural excitation. The lack of a clear and unambiguous definition for muscle tone has resulted in much misunderstanding in the literature and the emergence of various devices that claim to measure muscle tone objectively. Most commonly, these devices measure resistance of tissues to deformation applied to the surface of a body part; so, they measure apparent stiffness (see Chapter 2) of all the tissues, which is defined by numerous factors related and unrelated to the neural control of muscle state.

5.1 Elements of history

The usage of the term *muscle tone* (*muscle tonus*) dates back to the works of Galen (Bernstein and Kots, 1963), a great Roman physician and scientist of the first century A.D. This term has been used actively in the field of muscle physiology, movement science, and motor disorders since the middle of the nineteenth century. In particular, neuromuscular disorders have been traditionally classified into hypotonic and hypertonic (Watts and Köller, 2004; Fahn and Jankovic, 2007) referring to an increased or decreased muscle tone as compared to healthy persons. (Some disorders are also classified as *dystonic*, but this term seems to have only remote relation to muscle tone.) While the word *muscle* is relatively unambiguous, the word *tone* allows different interpretations. The *Oxford Dictionary* defines tone as "the normal level of firmness or slight contraction in a resting muscle." The definition from Wikipedia is "continuous and passive partial contraction of the muscles, or the muscle's resistance to passive stretch during resting state." Other current definitions emphasize the continuous nature of muscle tone, its relation to resistance to stretching, and partial contraction of the muscle caused by neural impulses.

Two great scientists of the twentieth century, Sir Charles Sherrington and Nikolai Bernstein (Bongaardt, 2001; Stuart et al., 2001), contributed significantly to the current understanding of muscle tone. Sherrington emphasized the relation of muscle tone to posture and its reflex origin. Bernstein agreed with the importance of reflexes for muscle tone and linked the notion of muscle tone to the adaptive function of the neuromotor apparatus; he assumed that the apparatus tuned its current state and the excitability of its components to tasks of active postural or movement control. The latter definition

Biomechanics and Motor Control. http://dx.doi.org/10.1016/B978-0-12-800384-8.00005-3

makes muscle tone an active contributor to movement and postural tasks, which contrasts some contemporary definitions available on the Internet that emphasize the passive nature of muscle tone.

Two main meanings of the word *tone* relevant to the notion of muscle tone are "state of alertness" and "steady state." When one speaks about a healthy person and mentions that his tonus is low, this usually implies low alertness, even sleepiness. Drinking strong coffee to elevate one's tonus is a common practice. On the other hand, classification of motor reactions to stimuli (see Chapter 6 on reflexes) into phasic and tonic, uses the word *tonic* in the meaning of long-lasting, steady-state, continuous. For example, Sir Charles Sherrington used the term *tonic stretch reflex* to mean a lasting resistance of a muscle to stretch due to changes in the amount of excitation it receives from the central nervous system. While the term *tonic stretch reflex* is used broadly in the field of motor control, the actual mechanical response of a muscle to stretch is known to be rather complex (see Zatsiorsky and Prilutsky, 2012).

5.2   Current definitions

Despite its frequent use in movement science literature, muscle tone is rarely defined explicitly. In contrast, its operational definitions are rather common. Clinicians frequently call muscle tone a feeling of resistance offered by a person's body segment when the person is asked to relax, and the clinician moves the limb relatively slowly over its range of motion (Definition #1). Studies using electromyography (EMG) sometimes equate muscle tone with baseline EMG level, again when the person tries to be completely relaxed (Definition #2). There are devices that claim to measure muscle tone by applying deformation orthogonal to the surface of the body and measuring the magnitudes (and, sometimes, frequencies) of the resistive force and deformation (Definition #3). Mechanically, such devices usually measure a property that in engineering is called *indentation hardness*. Note that none of these methods addresses the Bernstein understanding of muscle tone as a contributor to active movement and postural tasks, since the subject is always asked to relax and do nothing.

The first of the three operational definitions (feeling of resistance) is practically useful, but it has limited explanatory value because a host of factors can affect the resistance experienced by the clinician. For example, consider the well-established notion of hypotonia (low muscle tone) in persons with Down syndrome (DS) (Rarick et al., 1976; Morris et al., 1982). Assuming that a person is relaxed completely, and during the examination muscle activation shows no measurable deviations from zero, resistance to slow motion of a limb segment will be defined primarily by the mechanical properties of peripheral tissues. In a first approximation, resistance to an external motion gets contribution from inertia, apparent stiffness (resistance to stretch, see Chapter 2), and damping (resistance to velocity, see Chapter 3):

$$R_{TOT} = R_{IN} + R_{DA} + R_{ST},$$

where R stands for resistance and the subscripts refer to total (TOT), inertial (IN), apparent stiffness (ST), and damping (DA) components of resistance. When applied to a single joint, in this equation,

$$R_{IN} = I \cdot d^2\phi/dt^2; \quad R_{DA} = b \cdot d\phi/dt; \quad R_{ST} = k \cdot (\phi - \phi_0), \quad (5.1)$$

where ϕ stands for the joint angle, I—moment of inertia, b—damping coefficient, k—apparent stiffness, and ϕ_0—zero angle, at which the "joint spring" offers no resistance. Overall, Eqn (5.1) describes impedance (see Chapter 3), i.e., complex resistance that depends on the amplitude of the deformation, its speed, and acceleration. In particular, human tendons and ligaments provide for resistance of passive joints to motion due to their nonlinear viscoelastic properties, while relaxed muscle fibers show relatively low resistance to stretch (Borg and Caulfield, 1980; reviewed in Zatsiorsky and Prilutsky, 2012). As muscles and tendons are connected in series, deformation of a muscle—tendon complex in a state when the muscle is relaxed means that the muscle will be mainly deformed. If a muscle is activated, the tendon will be mainly deformed.

The following purely peripheral factors can potentially lead to lower resistance to externally applied motion in persons with DS: laxity of ligaments, lower inertia of the body segments (in particular, they are shorter in DS persons), and more compliant tendons. The first two factors may be viewed as well documented (Ulrich and Ulrich, 1993; Spanò et al., 1999). In particular, the shorter limbs of persons with DS allow expecting significantly smaller moments of inertia. In fact, a 20% shorter limb segment (assuming proportional scaling of all its dimensions) provides less than half of the inertial resistance to motion given the same angular acceleration.

The second definition is based on measurement of residual muscle activity (level of muscle activation, EMG) under the instruction "to relax completely." The ability to relax muscle completely cannot be taken for granted. In some cases, such as spasticity, people cannot relax muscles at certain joint configurations, while under emotional stress, or when their body parts move (reviewed in Dietz, 2000; Calota and Levin, 2009). Even a typical, healthy person may need explanation and time to be able to relax when another person manipulates his/her limb segments. It is common knowledge that some people are able to relax almost "completely" during massage or other manipulations while others cannot or, at least, require reminders and time to relax. Complete relaxation can only be achieved when muscle relaxants are used. This procedure is, however, dangerous since it can induce relaxation of the muscles involved in breathing, and therefore it is used only in the medical/surgical practice.

If muscles crossing a joint are not completely relaxed, they offer much larger resistance to external motion that stretches the muscles as compared to the same joint in the absence of the baseline muscle activation. Moreover, the resistance increases significantly with an increase in the level of muscle activation. This factor makes the first and second operational definitions of muscle tone nonmutually exclusive. Indeed, resistance offered by limb segments that are passively moved by the examiner about joints that are not completely relaxed is expected to increase with the baseline muscle activation.

The third operational definition of muscle tone is based on measuring apparent stiffness (see Chapter 2) of the tissues in the area of examination. This is clearly the least informative and interpretable method. For example, it is expected to show the highest readings of muscle tone in a dead person whose muscles are in the state of rigor mortis. Its readings are also expected to be affected by such factors as skin mechanical properties and amount of fat tissue in the area of examination, which are marginally relevant to issues of neuromuscular physiology.

The definition offered by Bernstein links the notion of muscle tone to active postural and movement tasks. This definition deserves attention. First, it does relate to muscle function explicitly. Second, it suggests that muscle tone in an apparently relaxed state may reflect a state of preparedness to a movement and, hence, can hardly be estimated using any measurements when the person is asked to relax and not to make any movement. We will develop this definition and try to link it to procedures that can be used to estimate muscle tone in the next sections.

5.3   Relation of muscle tone to the tonic stretch reflex

Since the classical studies of Liddell and Sherrington (1924), the notion of muscle tone has been intimately linked to muscle reflexes and the notion of tonic stretch reflex. In particular, Sherrington claimed a dramatic drop in muscle tone estimated as muscle resistance to passive stretch after the sensory fibers leading from the muscle to the spinal cord had been cut. This procedure, known as deafferentation, eliminates the autogenic reflexes (those that originate from sensory organs in the muscle). Since reflexes contribute significantly to muscle activation (see Chapter 6), which has a profound effect on the muscle resistance to stretch, the drop in muscle tone after deafferentation is a strong argument in favor of an important role of muscle activation in muscle tone. Note, however, that these estimates of muscle tone changes were based on Sherrington's definition of muscle tone as muscle resistance to passive stretch and the animal preparations were not involved in any movement or postural tasks.

When a muscle is stretched slowly by an external force, in an animal preparation with intact neural connections between the muscle and the spinal cord, at relatively low values of length the muscle shows rather weak resistance to stretch and no signs of activation (the thin, dashed curve in Figure 5.1). At some length value (λ) muscle resistance starts to increase at a much higher rate with further stretch (the thick line in Figure 5.1). This is associated with emergence of muscle activation, which increases with further stretch (reflected with the different sizes of the word *EMG*—electromyographic signals). The thick curve in Figure 5.1 is addressed as the tonic stretch reflex characteristic, and the value of muscle length, at which the first signs of muscle activation can be seen, is called the threshold of the tonic stretch reflex (λ in Figure 5.1).

Figure 5.1 suggests that, when muscle length is shorter than λ, no muscle activation can be seen, while the muscle is active when its length is longer than λ. What does this

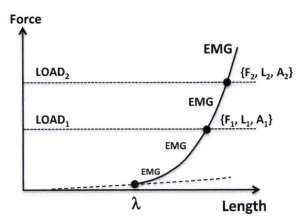

Figure 5.1 The tonic stretch reflex produces a dependence of active muscle force (and level of activation, EMG) on muscle length. Threshold of the tonic stretch reflex (λ) is the length when muscle activation starts during very slow stretch. Depending on external load (L), the system comes to an equilibrium point (black dots) defined by both λ and external load. Equilibrium points differ in muscle length, force, and activation level, {F, L, A}.

mean if one associates muscle tone with partial muscle contraction and/or action of the tonic stretch reflex? This means that, for any nonzero muscle tone, actual muscle length is always longer than λ since for shorter muscles no contraction and no visible reflex action are expected. This conclusion is hardly compatible with observations of complete muscle relaxation (no signs of activation seen in the surface EMG) and no visible changes in muscle activation with slow muscle stretch in healthy persons (Katz and Rymer, 1989; Musampa et al., 2007).

To link muscle tone with movement production, in accordance with Bernstein's definition, consider how control of a single intact muscle is organized by the central nervous system. Many experiments, both in humans and animal preparations, have shown that a descending command to a segment of the spinal cord, from where the muscle is innervated, cannot be associated with an encoded value of muscle length, muscle force, or muscle activation level (Matthews, 1959; Feldman and Orlovsky, 1972). Indeed, as suggested by Figure 5.1, all three mentioned characteristics of muscle output change along the characteristic of the tonic stretch reflex. The exact point on the characteristic where the muscle will be in equilibrium depends on the external load. Two such points are shown in Figure 5.1 for two load magnitudes (LOAD$_1$ and LOAD$_2$) corresponding to different combinations of muscle force, length, and activation level: {F$_1$, L$_1$, A$_1$} and {F$_2$, L$_2$, A$_2$}.

According to the equilibrium-point hypothesis (Feldman, 1986; for more detail see Chapter 12), descending control of a muscle can be adequately described as shifts of the threshold (λ) of the tonic stretch reflex. Nearly parallel shifts of the tonic stretch reflex force-length characteristics have been measured in both animal and human experiments in support of this assumption (Feldman, 1966; Feldman and Orlovsky, 1972). Peripheral consequences of a shift in λ depend on the external load characteristic.

Figure 5.2 illustrates that a shift of λ (from λ_1 to λ_2) can lead to muscle movement to a new length value (L_0 to L_1), a change in muscle force (from F_0 to F_2), or both (to $\{L_3, F_3\}$) depending on the external loading conditions. In isotonic conditions (load is constant), muscle movement takes place, in isometric conditions (movement is blocked), muscle force changes, and when the muscle acts against a spring-like load, both length and force change. Note that, when a muscle is in a steady state, its active force, i.e., the force in excess of the force exhibited during stretching a passive muscle, is nearly proportional to the level of activation (Enoka, 2008); therefore, in all conditions with force changes, muscle activation level changes as well.

Healthy humans can relax muscles to a level at which no signs of muscle electrical activity can be recorded, at any position within the available range of motion of the corresponding joint. On the other hand, humans are able to generate large muscle forces (joint torques) also at any position within the range of motion. These observations suggest that the range of changes in the tonic stretch reflex threshold (λ) for all healthy muscles is larger than the anatomic range of motion.

Figure 5.3 illustrates this statement by showing the available range of muscle motion with two vertical dashed lines and two tonic stretch reflex characteristics. The leftmost characteristic corresponds to a value of the threshold of the tonic stretch reflex (λ_1) that is shorter than the shortest possible muscle length value. This characteristic corresponds to relatively large active muscle force generated at a short muscle length (the black circle in Figure 5.3). The rightmost characteristic corresponds to a value of λ (λ_2) that is longer than the longest possible muscle length value. It corresponds to muscle relaxation over the whole range, from the minimal (L_{MIN}) to the maximal (L_{MAX}) muscle length values.

From Figure 5.3, it follows that, if a person moves the threshold of the tonic stretch reflex beyond the largest value in the anatomical range of muscle length (to λ_2), no

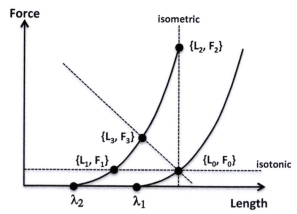

Figure 5.2 A shift in the tonic stretch reflex threshold, λ (from λ_1 to λ_2) can lead to changes in muscle length (in isotonic conditions, from $\{L_0, F_0\}$ to $\{L_1, F_1\}$, $F_0 = F_1$) or in muscle force (in isometric conditions, from $\{L_0, F_0\}$ to $\{L_2, F_2\}$, $L_0 = L_2$) or both (elastic external load, from $\{L_0, F_0\}$ to $\{L_3, F_3\}$).

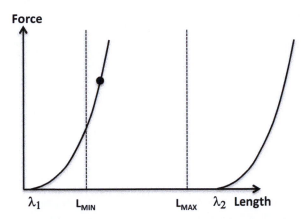

Figure 5.3 The tonic stretch reflex characteristic can be shifted beyond the biomechanical limits of muscle length (shown with vertical dashed lines at L_{MIN} and L_{MAX}). Shifting λ beyond these limits allows generating relatively large forces with short muscles (λ_1) and relaxing a long muscle (λ_2).

visible muscle activation occurs during slow joint motion. All definitions that link the notion of muscle tone to muscle activation via reflex mechanisms (e.g., Sherrington's definition and the operational definition via muscle activation level) suggest that no value of muscle tone can be assigned to this situation. Since healthy humans are able to relax muscles throughout the range of muscle length values, these definitions make the notion of muscle tone useless when applied to a healthy person since muscle tone becomes either undefined or equal to zero.

5.4   Muscle tone and ability to relax

To resolve this apparent controversy, consider the following mental experiment that represents a typical situation of measuring muscle tone in a clinical setting. A person is asked to relax, and then the examiner moves a joint of that person smoothly throughout the range of motion. Typically, this motion is produced at a moderate speed and represents a few cycles of flexion-extension. Consider first a single muscle that is being stretched during the joint motion.

Figure 5.4 illustrates a relaxed muscle. This means that its length (L_0) is lower than the current value of the tonic stretch reflex threshold. Three values of the threshold (λ) are illustrated. One of them (λ_1) is very close to L_0, the second one (λ_2) is within the biomechanical range of muscle length (L_{MIN} to L_{MAX}), while the third one (λ_3) is beyond that range ($\lambda_3 > L_{MAX}$). Imagine now that another person (the examiner) moves the joint causing an increase in the muscle length.

Muscle activation will start immediately after the motion initiation for λ_1, it will start after some time (when the muscle reached the corresponding length) for λ_2, and it will not be seen over the whole range of motion for λ_3. So, when the muscle

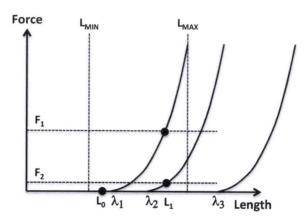

Figure 5.4 When a person is asked to relax, the threshold of the tonic stretch reflex (λ) is larger than the current muscle length (L_0). Stretching the muscle to a new length (L_1) would produce different resistive forces for a muscle with the threshold value close to L_0 (λ_1) as compared to the muscle with the threshold value farther away from L_0 (λ_2). Very small resistance (not illustrated) would be produced by a muscle with the threshold λ_3, which is beyond the biomechanical length limit (shown with vertical dashed lines, L_{MIN} to L_{MAX}).

reaches a certain intermediate length (L_1 in Figure 5.4), it would resist very weakly in the case of λ_3 (due to the weak resistance offered by the relaxed muscle, see the dashed line in Figure 5.1; not shown in this figure), stronger in the case of λ_2, and much stronger in the case on λ_1 ($F_1 > F_2$). The examiner in this mental study would probably conclude that muscle tone is the highest for the case illustrated with λ_1 (hypertonia) and the lowest in the case with λ_3 (hypotonia).

There is, however, no difference in the properties of the muscle and its reflex mechanisms across the three scenarios illustrated in Figure 5.4. The only difference is in the ability of the subject of this experiment to relax. One can say that the subjects who moved the tonic stretch reflex threshold beyond the anatomical range (to λ_3) followed the instruction "to relax completely" better than the subjects who kept λ within the range of muscle motion. Such results may be expected in a study of a person with DS (cf. Davis and Kelso, 1982) leading to a conclusion of low muscle tone, while in fact this person may have unchanged muscle and reflex properties and simply follows the instruction better compared to a person without DS. According to this theory, muscle tone is not immediately determined by the peripheral muscle state including its activation level (indeed, the muscle was equally relaxed in all three cases illustrated in Figure 5.4). Instead, muscle tone is determined by central commands setting the threshold for the tonic stretch reflex.

5.5 Factors causing "low muscle tone"

If we accept the definition "muscle tone is what a clinician calls muscle tone" (feeling of resistance offered by a person's body segment when the person is asked to relax, and the clinician moves the limb relatively slowly over its range of motion), a number of factors may contribute to the feeling of low resistance, which is interpreted as low muscle tone. Some of these factors have been already mentioned. These include the inertial resistance offered by the segment and the resistance offered by stretched tissues that are unaffected by signals from the central nervous system. In particular, laxity of ligaments and short limb segments may contribute to the feeling of low muscle tone.

However, there is another factor mediated by the central nervous system. As described in the previous section, when a person tries to relax, the actual distance from the current muscle length to the threshold for muscle activation via the tonic stretch reflex loop may vary. As shown experimentally (Powers et al., 1988), when a healthy person is asked to relax, stretching a relaxed muscle can lead to its activation, even when the stretch is produced at a relatively low velocity comparable to velocities produced during clinical examinations. This means that a typical healthy person does not move the threshold of the tonic stretch reflex beyond the biomechanical range, and this is revealed in nonzero muscle activation when the threshold is reached during muscle stretch produced by an external force.

Why do people not relax "completely"? This may be conditioned by everyday experience when muscle activation may become urgently needed at unpredictable times. Assuming that the threshold of the tonic stretch reflex can be changed only at a limited speed (cf. Feldman, 1986; Latash et al., 1991), a quick reaction to an unexpected event requires the distance between the current length of a relaxed muscle and its threshold length to be small. When the threshold is far away from the current muscle length, shifting it to cause muscle activation over a relatively large distance may add substantially to reaction time. During clinical examinations or experiments in a typical laboratory, no unexpected events happen and this constraint can be lifted. However, this may require time and experience (cf. the earlier distinction between people who can relax "fully" and those who cannot.)

This analysis suggests that what clinicians call "normal muscle tone" is a reflection of several factors including the relatively small distance from the initial muscle length at the start of the examination to the threshold of the tonic stretch reflex. A feeling of low muscle tone can happen when the examined person is very good at following the instruction "to relax completely" and no joint motion leads to muscle activation (as in the case of λ_3 in Figure 5.4). In other words, a completely typical, healthy person may be diagnosed as hypotonic if this person is very good at relaxing his or her muscles.

Theoretically, low muscle tone may also get contribution from low gain of the tonic stretch reflex. Consider Figure 5.5. In this illustration two tonic stretch reflex characteristics, i.e., their force—length relations, are shown, which differ in their slopes while the thresholds are identical (λ). If a muscle is stretched over a fixed length (for example, from the initial length L_0 to L_1), the case illustrated with the thick curve (lower slope) will produce lower resistive force (F_2) as compared to the case illustrated by the thin curve (larger slope, F_1). While this is in theory possible, no studies have documented reliable changes in the tonic stretch reflex gain across persons with "normal" and low muscle tone. In fact, similar tonic stretch reflex gains (similar slopes of the force—length characteristics) were reported in a study that compared persons

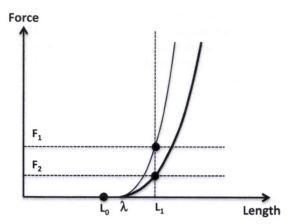

Figure 5.5 A hypothetical illustration of two muscles with different gains of the tonic stretch reflex. The difference in the gains is reflected in the different slopes of the force—length curves. The muscle with the larger gain (thin line) would produce higher resistive force ($F_1 > F_2$) to passive stretch from L_0 to L_1 as compared to the muscle with the lower gain (thick line).

with and without DS (Kelso and Holt, 1980), while persons with DS are most commonly addressed as hypotonic.

5.6 Factor causing "high muscle tone"

Just like in the case of low muscle tone, peripheral factors such as mechanical properties of segments and connective tissues may contribute to the feeling of resistance experienced by the clinician examiner. Changes in properties of peripheral tissues, such as an increase in the stiffness of connective tissues about joints of patients who suffer from spasticity, have been documented (Dietz and Berger, 1983). Nevertheless, cases of high muscle tone have been primarily linked to problems within the central nervous system, from the cortex (e.g., in stroke) to the spinal cord (spinal cord injury and other disorders).

Several factors mediated by the central nervous system may contribute to the feeling of high resistance during passive motion of a body segment. These may be classified into three groups: (1) higher gain of the tonic stretch reflex; (2) problems with control of the threshold of the tonic stretch reflex; and (3) suprathreshold activation of a motoneuronal pool that makes it insensitive to feedback from the tonic stretch reflex.

Several studies tried to quantify gain of the tonic stretch reflex in conditions of spasticity, which is typically associated with perception of high muscle tone. Most of these studies (Dietz and Berger, 1983; Levin and Feldman, 1994) failed to detect a significant increase in the gain of the tonic stretch reflex (the slope of the dependence of active muscle force on muscle length). One study (Latash et al., 1990; Latash, 1993) measured the apparent stiffness characteristic of the elbow joint of a person with spasticity before and after a single intrathecal injection—injection into the space between the spinal cord and protective meninges—of a drug that effectively suppressed spasticity (baclofen, a precursor of an inhibitory neuromediator, gamma-amino-butyric acid). In that study, a significant drop in the slope of the dependence of active joint torque on joint displacement was observed, suggesting a change in the gain of the tonic stretch reflex of muscles crossing the joint. This, however, remains so far a single report suggesting an association between high muscle tone and tonic stretch reflex gain.

Recently, spasticity has been discussed as a problem of control of the tonic stretch reflex threshold (Jobin and Levin, 2000). Impaired ability to shift the threshold of the tonic stretch reflex within its normal range can lead to both limitations in voluntary control of a spastic muscle and its involuntary activation leading to an impression of high muscle tone. Figure 5.6 illustrates this idea for a muscle with a particular biomechanical range of motion (shown by two vertical dashed lines). In a healthy muscle, λ can be shifted far beyond the biomechanical length limits (see also Figure 5.4); this allows both activating the muscle to high levels when it is short and relaxing the muscle when it is long. Imagine now that the range of λ shifts is constrained to a narrow interval limited by λ_S and λ_L (as in Figure 5.6). In this case, involuntary muscle activation would be seen at all length values $L > \lambda_L$, and it would be impossible to activate the muscle at length values $L < \lambda_S$.

This scheme unites high muscle tone and impaired voluntary control into a single scheme, which is against the traditional attitude to spasticity as a combination of relatively independent "positive" and "negative" signs (Hughlings Jackson, 1889; Landau, 1974). On the other hand, several studies have documented positive effects of the mentioned treatment with intrathecal baclofen on both spastic muscle contractions (contributing to high muscle tone) and residual voluntary control of the affected

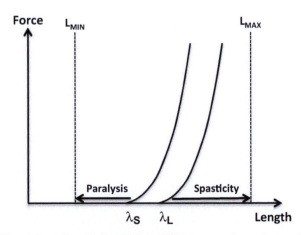

Figure 5.6 An illustration of a muscle with lost ability to move the tonic stretch reflex threshold (λ) over its full range. If the range of λ shifts is limited by λ_S and λ_L, it becomes impossible to voluntarily activate the muscle when its length is shorter than λ_S. It also becomes impossible to relax the muscle when its length is longer than λ_L. The biomechanical range of muscle length changes in shown with vertical dashed lines, L_{MIN} to L_{MAX}.

extremities (Latash et al., 1990; Latash and Penn, 1996). These observations are readily compatible with the scheme in Figure 5.6.

Factors that lead to the lack of voluntary control of λ over its whole range are unknown. It is possible, for example, that joint motion can lead by itself to a shift in the tonic stretch reflex characteristic toward the current muscle length. It is also possible that neural processes apparently unrelated to joint motion and action of the tonic stretch reflex contribute to λ shifts. These may include reflex effects from cutaneous and subcutaneous receptors leading, in particular, to involuntary muscle contractions caused by a touch to or a scratch of a skin area.

While tonic stretch reflex is one of the main mechanisms of muscle activation, it is possible to activate muscles in the absence of this reflex. For example, monkeys after complete deafferentation (cutting all the afferent fibers originating in the limb) can perform accurate movements by the deafferented extremity (Polit and Bizzi, 1979). Humans in a rare case of large-fiber peripheral neuropathy ("deafferented humans," e.g., Sainburg et al., 1995) can also produce voluntary movements, which, however, require continuous visual control. These observations suggest that alpha-motoneuronal pools can be activated by descending signals alone, i.e., without contribution from peripheral reflexes, resulting in a state of ongoing contraction, which could be associated with high muscle tone.

5.7   The bottom line

Given the widespread use of the notion of muscle tone in clinics, its practical utility in everyday clinical practice seems beyond doubt. In contrast, the current understanding of the peripheral and neural mechanisms contributing to muscle tone is fragmented at best. The following questions seem central for improving the current situation:

Is muscle tone a property of a relaxed muscle (no signs of activation) or of a partially activated muscle? While the available definitions differ on this issue, given the clinical practice, we believe that the definition refers to a muscle "relaxed to the maximum of the person's ability."

Is muscle tone related to properties of the tonic stretch reflex? It seems that the answer is a definite "yes." Muscle tone reflects the control of the threshold of the tonic stretch reflex and, possibly, its gain.

Is muscle tone related to active movement and posture control? The answer is "yes." So-called "normal muscle tone" in a relaxed muscle is likely to reflect its state of preparedness for future action.

How can "low muscle tone" and "high muscle tone" be interpreted? Unfortunately, there is no single answer to this question. There are too many potential causes for the feeling of reduced or increased resistance during a typical clinical examination of muscle tone.

Is muscle tone a useful concept in biomechanics and motor control? The answer seems to be "no." First, mechanisms of muscle tone and of its changes are basically unknown. Second, its most common method of measurement is subjective.

References

Bernstein, N.A., Kots, Y.M., 1963. Tone. In: Grand Medical Encyclopaedia, vol. 32. Moscow: State Encyclopaedia, pp. 418–422.

Bongaardt, R., 2001. How Bernstein conquered movement. In: Latash, M.L., Zatsiorsky, V.M. (Eds.), Classics in Movement Science. Human Kinetics, Urbana, IL, pp. 59–84.

Borg, T.K., Caulfield, J.B., 1980. Morphology of connective tissue in skeletal muscle. Tissue and Cell 12, 197–207.

Calota, A., Levin, M.F., 2009. Tonic stretch reflex threshold as a measure of spasticity: implications for clinical practice. Top Stroke Rehabilitation 16, 177–188.

Davis, W.E., Kelso, J.A.S., 1982. Analysis of "invariant characteristics" in the motor control of Down's syndrome and normal subjects. Journal of Motor Behavior 14, 194–211.

Dietz, V., 2000. Spastic movement disorder. Spinal Cord 38, 389–393.

Dietz, V., Berger, W., 1983. Normal and impaired regulation of muscle stiffness in gait: a new hypothesis about muscle hypertonia. Experimental Neurology 79, 680–687.

Enoka, R.M., 2008. Neuromechanics of Human Movement, fourth ed. Human Kinetics, Urbana, IL.

Fahn, S., Jankovic, J., 2007. Principles and Practice of Movement Disorders. Churchill Livingstone Elsevier, Philadelphia, PA, USA.

Feldman, A.G., 1966. Functional tuning of the nervous system with control of movement or maintenance of a steady posture. II. Controllable parameters of the muscle. Biophysics 11, 565–578.

Feldman, A.G., 1986. Once more on the equilibrium-point hypothesis (λ-model) for motor control. Journal of Motor Behavior 18, 17–54.

Feldman, A.G., Orlovsky, G.N., 1972. The influence of different descending systems on the tonic stretch reflex in the cat. Experimental Neurology 37, 481–494.

Hughlings Jackson, J., August 17, 1889. On the comparative study of disease of the nervous system. British Medical Journal 355–362.

Jobin, A., Levin, M.F., 2000. Regulation of stretch reflex threshold in elbow flexors in children with cerebral palsy: a new measure of spasticity. Developmental Medicine and Child Neurology 42, 531–540.

Katz, R.T., Rymer, W.Z., 1989. Spastic hypertonia: mechanisms and measurement. Archives of Physical Medicine and Rehabilitation 70, 144–155.

Kelso, J.A.S., Holt, K.G., 1980. Exploring a vibratory systems analysis of human movement production. Journal of Neurophysiology 43, 1183–1196.

Landau, W.M., 1974. Spasticity: the fable of a neurological demon and the emperor's new therapy. Archives of Neurology 31, 217–219.

Latash, M.L., 1993. Control of Human Movement. Human Kinetics, Urbana, IL.

Latash, M.L., Gutman, S.R., Gottlieb, G.L., 1991. Relativistic effects in single-joint voluntary movements. Biological Cybernetics 65, 401–406.

Latash, M.L., Penn, R.D., Corcos, D.M., Gottlieb, G.L., 1990. Effects of intrathecal baclofen on voluntary motor control in spastic paresis. Journal of Neurosurgery 72, 388–392.

Latash, M.L., Penn, R.D., 1996. Changes in voluntary motor control induced by intrathecal baclofen. Physiotherapy Research International 1, 229–246.

Levin, M.F., Feldman, A.G., 1994. The role of stretch reflex threshold regulation in normal and impaired motor control. Brain Research 657, 23–30.

Liddell, E.G.T., Sherrington, C.S., 1924. Reflexes in response to stretch (myotatic reflexes). Proceedings of the Royal Society of London, Series B 96, 212–242.

Matthews, P.B.C., 1959. The dependence of tension upon extension in the stretch reflex of the soleus of the decerebrate cat. Journal of Physiology 47, 521—546.

Morris, A.F., Vaughan, S.E., Vaccaro, P., 1982. Measurements of neuromuscular tone and strength in Down's syndrome children. Journal of Mental Deficiency Research 26, 41—46.

Musampa, N.K., Mathieu, P.A., Levin, M.F., 2007. Relationship between stretch reflex thresholds and voluntary arm muscle activation in patients with spasticity. Experimental Brain Research 181, 579—593.

Polit, A., Bizzi, E., 1979. Characteristics of motor programs underlying arm movements in monkeys. Journal of Neurophysiology 42, 183—194.

Powers, R.K., Marder-Meyer, J., Rymer, W.Z., 1988. Quantitative relations between hypertonia and stretch reflex threshold in spastic hemiparesis. Annals of Neurology 23, 115—124.

Rarick, G.L., Dobbins, D.A., Broadhead, G.G., 1976. The Motor Domain and its Correlates in Educated Handicapped Children. Prentice Hall, Englewood Cliffs, NJ.

Sainburg, R.L., Ghilardi, M.F., Poizner, H., Ghez, C., 1995. Control of limb dynamics in normal subjects and patients without proprioception. Journal of Neurophysiology 73, 820—835.

Spanò, M., Mercuri, E., Randò, T., Pantò, T., Gagliano, A., Henderson, S., Guzzetta, F., 1999. Motor and perceptual-motor competence in children with Down syndrome. European Journal of Paediatric Neurology 3, 7—14.

Stuart, D.G., Pierce, P.A., Callister, R.J., Brichta, A.M., McDonagh, J.C., 2001. Sir Charles S. Sherrington: humanist, mentor, and movement neuroscientist. In: Latash, M.L., Zatsiorsky, V.M. (Eds.), Classics in Movement Science. Human Kinetics, Urbana, IL, pp. 317—374.

Ulrich, B.D., Ulrich, D.A., 1993. Dynamic systems approach to understanding motor delay in infants with Down syndrome. In: Savelsbergh, G.J.P. (Ed.), The Development of Coordination in Infancy. Elsevier Science Publishers, Holland, pp. 445—457.

Watts, R.L., Köller, W.C., 2004. Movement Disorders: Neurological Principle and Practice, second ed. McGraw Hill, New York.

Zatsiorsky, V.M., Prilutsky, B.I., 2012. Biomechanics of Skeletal Muscles. Human Kinetics, Urbana, IL.

Reflexes

6

The notion of "reflex" is arguably one of the oldest, most broadly used, and least precisely defined notions in the field of movement studies. PubMed (the site of the US National Library of Medicine) produced at the end of 2013 over 100,000 entries for "reflex," and this number has been growing consistently by over 2500 per year. This notion is central for theories of motor control, studies of motor development and aging, clinical studies of movement disorders and rehabilitation, and other fields such as psychology. Despite its obvious importance and widespread use, this notion is rarely defined or defined in imprecise, descriptive terms.

6.1 Elements of history

It is likely that even the earliest ancestors of contemporary humans were aware of the fact that some actions happen very quickly, "without thinking," for example, withdrawing a hand from a hot object or blinking when a small object gets into one's eye. Automatic actions to stimuli were mentioned by such classics of medicine and philosophy as Hippocrates and Galen.

According to the historical review by Francois Clarac (2005), the pioneering scientific studies of automatic actions to stimuli were performed by two great scientists of the Renaissance, the French philosopher René Descartes (1594–1660) and the English physician Thomas Willis (1628–1678). In his book *De Homine*, Descartes illustrated an automatic withdrawal of the hand from fire with the clearly identified three main components of any reflex: the sensory signals (coming from the eyes in this example), the processing within the central nervous system (CNS; which he placed into the pineal gland), and the muscle reactions. Willis described a qualitatively similar scheme for quick actions, but he thought that the central processing of sensory signals happened in the striatum.

Jean Astruc (1684–1766), a French philosopher, is credited with introducing the notion of reflex as a noun. He used the metaphor of a mirror reflecting the light to describe how a signal into the brain can lead to a quick reaction. An important step toward defining reflex was made by Robert Whytt (1714–1766), a professor at the University of Edinburgh. In his monograph "On the Vital and Other Involuntary Motions of Animals," Whytt developed the principle, according to which sensory information is sometimes not perceived, since some actions are automatic and do not depend on the will. This view can be seen as the first step toward the idea of direct perception introduced about 200 years later by the great American psychologist James Gibson (1904–1979).

The English physician Marshall Hall (1790–1857) expressed a view that reflexes were produced by the spinal cord while psychic processes occurred in the higher

Biomechanics and Motor Control. http://dx.doi.org/10.1016/B978-0-12-800384-8.00006-5

structures of the CNS. This view received support in the studies of reflexes in decap-
itated frogs, in particular by the German physiologist Karl Pflüger (1829–1910). At
about the same time, one of the greatest physiologists of the nineteenth century,
Ivan Sechenov (1829–1905), wrote a book, *Reflexes of the Brain*, with the title suggesting that, indeed, there are many reflexes mediated by supraspinal structures.
Sechenov expanded the idea of reflexes to all motor reactions to external stimuli, independently of their time delay.

Over the past 150 years, many reflexes have been described and studied in depth.
Reflex testing has become common in clinical practice. Arguments about the importance of reflexes have become central to several discussions about the mechanisms
of behavior, especially voluntary motor behavior. Those include, in particular, the discussion on the role of conditioned reflexes in behavior and on the role of spinal reflexes
in everyday movements (both discussed in more detail later).

6.2 Current definitions of reflex

The current definitions of reflex, found in textbooks and on Internet sites, emphasize its
link to a stimulus, its involuntary nature, and its quickness. In particular, Wikipedia
defines reflex as an involuntary and nearly instantaneous movement in response to a
stimulus. Other definitions use related constructs such as "response not reaching the
level of consciousness," "action consisting of simple segments of behavior," "unique
correlation of response with the stimulus," etc.

While many of such definitions are intuitively appealing, they use other undefined
or loosely defined notions to define reflex. These definitions, however, point at
specific characteristics that may or may not be useful to separate reflexes from non-
reflexes. A recent paper coauthored by several prominent scientists (Prochazka et al.,
2000) asked a question whether *reflex* was a useful notion or if it was too ambiguous
to be used in research on biological movement. The authors did not come to a
consensus, but the discussion emphasized two main views: (1) reflex is a misnomer,
it implies an action that may look simple but otherwise not different from any other
action; and (2) reflex is a useful notion, it differs in an important way from voluntary
movement.

Figure 6.1 contrasts reflexes versus voluntary movements with respect to several
characteristics that have been most frequently mentioned for the definitions of reflex.
Note that the figure implies a continuous spectrum along each of the axes; it is not
black and white but potentially has gray segments.

The differences mentioned in Figure 6.1 are not indisputable. We discuss them
below one by one.

The first row of Figure 6.1: Reflexes are, by definition, linked to a stimulus. However, sometimes, stimuli are not obvious while some voluntary movements can be performed in response to a stimulus. For example, the notion of tonic stretch reflex is one
of the best-established ones (see Section 6.5.1). Its origin is primarily the activity of
length-sensitive receptors in the muscles, sensitive endings in muscle spindles. Since
muscle spindles are active all the time (or at least most of the time), there is no clear

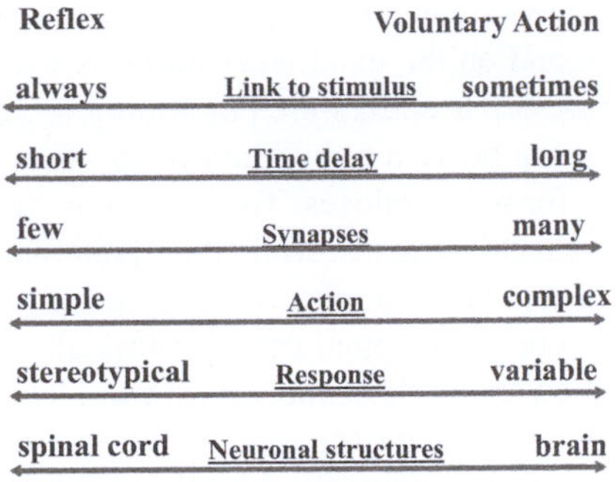

Reflex		Voluntary Action
always	Link to stimulus	sometimes
short	Time delay	long
few	Synapses	many
simple	Action	complex
stereotypical	Response	variable
spinal cord	Neuronal structures	brain

Figure 6.1 An illustration of the differences between reflexes and voluntary actions according to some of the common criteria.

identifiable stimulus that leads to tonic stretch reflex unless one views each and every action potential generated by spindles as a separate stimulus. On the other hand, some voluntary movements, such as pressing the brake pedal while driving, are commonly linked to a stimulus such as seeing the red streetlight.

The second row of Figure 6.1: Reflexes are commonly viewed as coming after a shorter time delay following a stimulus when compared to voluntary movements. While this is probably true for spinal reflexes in humans, there are reflex-like actions (see Section 6.5.4 and Chapter 7) that come at time delays comparable to those of the quickest voluntary movements, such as starting in a sprint race. Also, this criterion is applicable only to reflexes seen in response to a brief, identifiable stimulus. It is not easily applicable to ongoing steady-state reflexes such as the tonic stretch reflex. Muscle activation in response to high-frequency, low-amplitude vibration applied to the muscle or to its tendon (the tonic vibration reflex) typically takes a relatively long time, commonly over 1 s, to show discernible muscle activation. This time delay may be much longer than typical times of initiating a quick voluntary action to a stimulus.

The third row of Figure 6.1: Reflexes are viewed as involving a relatively short sequence of neural transmissions (synapses). This property of reflexes may be viewed as well established for monosynaptic and oligosynaptic (involving only few synapses) reflexes, while more complex, polysynaptic reflexes involve an unknown neural loop and an unknown number of synaptic connections. It is unknown whether the length of the neural loop is longer or shorter for those reflexes as compared to everyday voluntary actions.

The fourth row of Figure 6.1: Reflexes are commonly viewed as leading to a stereotypical motor response when compared to variable voluntary actions. This distinction is, however, not very strict. There are numerous studies showing effects of reflex modulation and even reversal, that is, reflex activation of a muscle group or of its antagonist muscles depending on conditions (see Section 6.6.4). Some reflex movements seen in animals with the spinal cord surgically separated from the brain show highly variable, adaptive patterns (e.g., the wiping reflex in frogs, Fukson et al., 1980; Berkinblit et al., 1986; see the next paragraph). Also, reflex-like reactions, such as M_{2-3} (preprogrammed reactions, see Chapter 7) show strong dependence on instruction to the person and on the external conditions when the stimulus occurs.

The fifth row of Figure 6.1: Reflexes are commonly seen as leading to relatively simple motor actions that can be used as building blocks for more complex, voluntary actions. This may be true for some reflexes. There are, however, very complex actions that most researchers would call reflexes, while their patterns may be highly complex, variable, and adaptive. One of the examples is the aforementioned wiping reflex in the spinal frog. In this preparation, the spinal cord is surgically separated from the brain and a stimulus is placed on the back of the sitting frog. The frog shows a sequence of beautifully coordinated wiping movements of the ipsilateral hindlimb that wipe the irritated spot in various directions (Berkinblit et al., 1986). These movements remain successful even if the hindlimb is loaded or one of its joints is blocked (Latash, 1993).

The sixth row of Figure 6.1 is related to involvement of the brain in the motor response. While spinal reflexes, by definition, do not involve brain structures, other established reflexes do. Examples include the pupillary reflex, the vestibular-ocular reflex, and many other reflexes seen in muscles controlled from the cranial nerve centers. Invoking the notions of consciousness or intentionality does not help much. For example, decerebrated (the neural axis is cut separating the cortex of the large hemispheres from the brain stem and the spinal cord) and spinal animals can show locomotor patterns that are not of a reflex origin (Shik et al., 1966; Orlovsky et al., 1966; reviewed in Orlovsky et al., 1999). To use the most striking example, a beheaded chicken can run and flap wings for some time. Spontaneous coordinated movements can also be seen in sleeping persons and those who are temporarily unconscious due to medications or an injury.

So none of the axes illustrated in Figure 6.1 allows introducing an unambiguous criterion for reflex. The wiping reflex example suggests that there is a hierarchy of characteristics. If a discrete response to a stimulus is produced by the spinal cord only, this response would be called a reflex by most, if not all, researchers, even if it takes a long time, shows a complex and variable pattern, can lead to nonlocal muscle activations, etc. The situation becomes less unambiguous when cyclic actions generated by the spinal cord are considered. For example, if a spinal cat is suspended above a treadmill such that its paws touch the treadmill, motion of the treadmill produces a cyclic, locomotor-like motion of the legs. According to the accepted view, this action is produced by a spinal central pattern generator (CPG; see Chapter 9), which can be driven by inputs from both supraspinal structures and peripheral receptors (as well as by certain drugs). Most researchers would not call this action a reflex even though it is mediated by the spinal cord and induced by a peripheral stimulus.

6.3 Preferred definition of reflex

Some years ago, one of the authors of this book suggested the following operational definition for reflex (Latash, 2008): A response to a stimulus mediated by at least one synapse within the CNS, which cannot be changed by "pure thinking," that is, a mental action not accompanied by identifiable motor events. Immediately after introducing this definition, a disclaimer was offered that many examples of actions called "reflexes" by researchers would fail this criterion. Consider the classical salivation reflex that was used by Ivan Pavlov in his famous experiments with conditioned reflexes (see Section 6.4.5 for more detail). When a hungry animal is shown food, it starts to salivate. On the other hand, if the reader starts thinking about a lemon, very likely salivation will start after a short time delay. Is this a reflex? Note that salivation in this example was produced by "pure thinking," without an identifiable external stimulus. If we accept that involuntary salivation is indeed a reflex, we have to accept that reflexes can be produced not only by external but also by internal stimuli. Other examples of such reflexes are those involved in changes in the heart rate produced by an emotional thought, changes in breathing, digestion, as well as urination and defecation triggered by the corresponding intrinsic stimuli.

Reflexes that are more commonly considered in movement science also sometimes fail to meet the criteria of the suggested definition. For example, consider the tonic vibration reflex mentioned earlier. This reflex can be suppressed voluntarily, which makes it fail the "no changes by mental action" test.

Our brief review here suggests that there is no unambiguous, universally acceptable definition for reflex. Nevertheless, this notion has been used in physiology, psychology, and clinical practice for a very long time. It is unlikely to be dropped or redefined in a major way. So, for now, we will stick to the operational definition suggested in the first paragraph of this section with a clear understanding that this definition will fail for some of the actions commonly classified as reflexes.

6.4 Classifications of reflexes

There are many classifications of reflexes. The following is an incomplete list of classifications based on some of the most commonly studied characteristics of reflexes (for review, see Rothwell, 1994; Latash, 2008).

6.4.1 Classification based on the number of synapses

One of the most commonly used classifications is based on the number of synapses within the CNS involved in the reflex loop. Within this classification, the neuromuscular synapses, which are always involved in any reflex motor action, are not counted. Since every synaptic transmission takes a certain time (on the order of 1 ms), fewer synapses generally mean shorter reflex delays. This is not an absolute rule because the reflex delay also involves transmission time along the afferent (sensory) and efferent (motor) pathways, which can add from a few milliseconds to a few dozen of milliseconds each to the overall reflex latency.

The shortest reflex loop involves only one synaptic transmission within the CNS (Figure 6.2). These *monosynaptic* reflexes are characterized by relatively short latencies. The best known and most commonly studied monosynaptic reflex originates from the primary sensory endings in muscle spindles. These afferents make direct

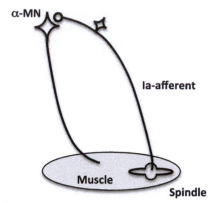

Figure 6.2 A schematic illustration of the monosynaptic reflex induced by stimulation of the primary spindle endings innervated by I-a afferents. The stimulation can be electrical (H-reflex) or mechanical (tendon tap or T-reflex).

projections onto alpha-motoneurons, which then produce muscle contractions. Monosynaptic reflexes are typically *autogenic*, which means that they are seen in the muscle, which houses the source of the afferent signal, and/or in its direct agonist. During early development, however, monosynaptic reflexes can also be seen in the antagonist muscles (Myklebust and Gottlieb, 1993). Such atypical monosynaptic reflexes emerge sometimes in adults in certain pathologies such as spasticity. These observations suggest that monosynaptic projections from primary spindle endings are not limited to the agonist muscles, but in healthy adults heterogenic projections are suppressed, possibly via the mechanism of presynaptic inhibition, and do not lead to observable monosynaptic reflexes.

Two methods are commonly used to induce and study monosynaptic reflexes in humans. The first involves applying a brief electrical pulse to the muscle nerve (e.g., *n. tibialis*). The second involves applying a brief, very quick stretch to the muscle. While both methods lead to monosynaptic reflex responses via the same loop, they have different names. The former is called the H-reflex, after Paul Hoffmann who was the first to describe this reflex in 1910. The latter has several names depending on the method of muscle stretch. It is called tendon jerk when tendon tap is used as the stimulus. Sometimes it is addressed as the tendon reflex or T-reflex. Another, less common, name is the M_1 response, in contrast to the later M_2 and M_3 responses produced by a quick joint perturbation (see Section 6.5.4).

One has to separate monosynaptic reflexes from monosynaptic projections. The former can be easily induced in some muscles but not in others. The latter may be common and widespread, but they may not be strong enough to produce reflex muscle contractions by themselves. This does not mean, however, that such monosynaptic projections are not important functionally during natural actions.

There are only a handful of reflexes for which the involved loop is known. All these involve only a few synapses within the CNS. Reflexes that involve a few (typically two or three) synapses are addressed as *oligosynaptic*. One of the best-known reflexes from this group is the reciprocal inhibition reflex (Figure 6.3). It originates

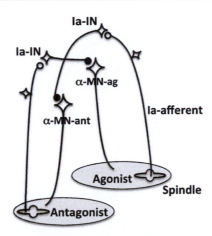

Figure 6.3 A scheme of reciprocal inhibition. Primary afferents (Ia-afferents) from spindles exert disynaptic inhibitory effects on alpha-motoneurons innervating the antagonist muscles. Excitatory projections are shown with open circles; inhibitory projections are shown with filled circles.

from the muscle spindles and involves two synapses within the CNS. First, the signals carried by primary afferents project on small interneurons (Ia-interneurons). Then, the Ia-interneurons make inhibitory projections on alpha-motoneurons innervating the antagonist muscle. Reciprocal inhibition was studied in depth by Sir Charles Sherrington who emphasized its functional importance—it prevents excessive activation of antagonist muscles during voluntary movements. Note that activation of a muscle typically leads to its shortening and a stretching of its antagonists. This stretch is expected to produce antagonist muscle activation via the stretch reflex loop (see Section 6.5.1) potentially counteracting the movement. Reciprocal inhibition prevents excessive antagonist activation during natural movements.

A group of oligosynaptic reflexes originates from Golgi tendon organs, sensory endings located at the junction between the muscle fibers and the tendon and innervated by Ib-afferents (Figure 6.4). The location of these endings makes them accurate sensors of tendon force. Ib-afferents project on Ib-interneurons in the spinal cord. These interneurons make inhibitory projections on the alpha-motoneurons that innervate the muscle housing the Golgi tendon organs. In addition, Ib-interneurons inhibit another group of interneurons, also inhibitory, that project onto the antagonist alpha-motoneuronal pool. As a result, activation of Golgi tendon organs located in a muscle leads to inhibition of that muscle's motoneurons and disinhibition (which is equivalent to excitation) of motoneurons innervating the antagonist muscle.

Most known reflexes involve an undefined number of synapses within the CNS and are called *polysynaptic*. These reflexes are typically characterized by longer time delays, more complex patterns of induced muscle activation, and sometimes by relatively slow changes in the muscle activation. Some of the typical examples of these reflexes such as the tonic stretch reflex, the flexor reflex, the crossed extensor reflex, and the tonic vibration reflex will be described in Section 6.5.

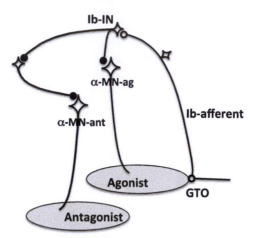

Figure 6.4 Reflex effects of Golgi tendon organs (GTO). Ib-afferents exert disynaptic inhibitory effects on alpha-motoneurons innervating the original muscle and its agonist and trisynaptic disinhibitory effects on alpha-motoneurons innervating the antagonist muscle. Excitatory projections are shown with open circles; inhibitory projections are shown with filled circles.

6.4.2   Classification based on response location

Reflex responses produced by a stimulus can be seen in a muscle or muscle group in close proximity to the stimulus site. Such reflexes are typically addressed as *autogenic*. Reflexes seen in remote muscles are addressed as *heterogenic*. Since the expression "close proximity" is not well defined, sometimes a reflex can be classified differently depending on the level of analysis. For example, reciprocal inhibition can be called a heterogenic reflex because it involves the antagonist ("remote") muscle. On the other hand, if one is interested in the distribution of reflex responses within a group of muscles serving a joint versus across joints, or within a limb versus across limbs, reciprocal inhibition can be classified as autogenic (because it causes changes in muscle activation affecting the same joint and the same limb).

6.4.3   Classification based on the time pattern

Reflexes that produce a quick, transient response to a stimulus are commonly addressed as *phasic*. In contrast, reflex contractions that show slow, steady-state contractions are addressed as *tonic*. Typical examples of phasic reflexes are all the aforementioned monosynaptic and oligosynaptic reflexes. Polysynaptic reflexes can be phasic or tonic.

Note that while monosynaptic projections of Ia-afferents can induce phasic monosynaptic reflexes, they can also contribute to tonic reflex contractions. In fact, one of the arguably most common and functionally significant reflexes, the tonic stretch reflex, (see Section 6.6.3) reflects afferent activity along all pathways including those making monosynaptic and oligosynaptic projections on motoneurons.

6.4.4 Classification based on involved neurophysiological structures

The neural loops of some of the mentioned reflexes are definitely limited to structures within the spinal cord; such reflexes are addressed as *spinal*. This is obvious for monosynaptic and oligosynaptic reflexes in the limbs. However, some of the more complex, polysynaptic reflexes also can be observed in spinal animals. In fact, some of these reflexes become stronger after transection of the spinal cord at a higher level. These include the tonic stretch reflex, flexor reflex, and crossed extensor reflex. In contrast, the tonic vibration reflex is usually not seen after complete spinal cord transection, which suggests that its loop involves supraspinal structures.

There are many *brain stem reflexes*. Some of them are in principle similar to the described reflexes for limb muscles. They are mediated by neurons in cranial nuclei, which play similar roles with respect to the control of head and neck muscles to those of the spinal cord neurons with respect to the control of limb and trunk muscles. Head and neck muscles can show mono-, oligo-, and polysynaptic reflexes similar to those described above. In addition, brain stem structures mediate a variety of reflexes that have no analogs at the spinal level. These include, for example, reflexes originating from sensors in the vestibular apparatus, such as the vestibular-ocular reflex that helps maintain the gaze direction during head movements. Another example is the pupillary reflex, which produced changes in the size of the pupil with changes in the overall level of luminance.

The issue of *transcortical reflexes* is less clear. While this expression is used sometimes to describe long-latency responses (M_3) to joint perturbation, these responses may not qualify as reflexes. We will consider them in more detail in Chapter 7. There are transcortical reflexes in some pathological states. For example, in myoclonus, a sensory stimulus applied to a limb can produce a giant evoked potential in the corresponding sensory cortical area and an involuntary phasic motor response. In a healthy person, a similar stimulus produces no response at all.

6.4.5 Classification based on animal's experience

The expression "conditioned reflex" is one of the most commonly used in the literature. PubMed produced over 20,000 hits for this entry. This notion formed the core of one of the most influential theories in the field of animal behavior introduced by Pavlov (1849–1936) and further explored by his numerous students and followers. According to Pavlov's views, the whole variety of everyday behaviors was built on two types of reflexes, *inborn* and *conditioned*. Pavlov performed ingenious experiments and showed that autonomic reflexes, such as salivation when the hungry animal saw food, could be modified in such a way that they emerged in response to a new, conditional, stimulus. Pavlov used the sound of a bell presented to the dogs prior to feeding and, after a series of repetitions, the animals started to salivate when they heard the bell even if food was not served. These observations formed the foundation for the main idea of Pavlov and his followers that conditioned reflexes, in combination with inborn reflexes, were the basic physiological elements for animal behavior, including motor behavior.

While Pavlov's views dominated the physiological science in the Soviet Union, they met strong opposition from Nikolai Bernstein (1896—1966), who viewed animals as active, not reactive, systems and claimed that no reflex combination could explain natural, everyday behavior (for a review see Meijer, 2002).

6.5 Examples of commonly studied reflexes in humans

6.5.1 Stretch reflex

Stretching a muscle leads to an increase in its electrical activity commonly addressed as the stretch reflex. It is generally assumed that the main source of afferent activity leading to the stretch reflex is the action potentials generated by the sensory endings in muscle spindles and transmitted to the spinal cord via the fast-conducting, myelinated afferent fibers. Muscle stretch, however, leads to changes in the activity level of many other sensory endings. For example, stretching a muscle by an external force leads to a nearly instantaneous increase in muscle force due to the peripheral mechanical properties of muscle fibers (see Chapter 2). This force increase is sensed by Golgi tendon organs. There is also unavoidable deformation of the skin and subcutaneous tissues leading to changes in the activity of several types of receptors located in those tissues. So stretch reflex can receive contributions from many types of receptors activated by muscle stretch.

Changes in muscle activation following muscle stretch depend on several factors. The first is the speed of stretch. A quick stretch can lead to a monosynaptic response of the muscle similar to the effects of tendon tap. This component is addressed as the *phasic stretch reflex* (Phasic SR in Figure 6.5). Slow muscle stretch does not lead to phasic changes in the muscle activity but induces a steady-state increase in the muscle

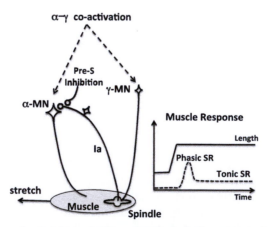

Figure 6.5 Quick muscle stretch can lead to a transient phasic response (phasic stretch reflex, Phasic-SR), followed by a steady-state increase in muscle activation (tonic stretch reflex, Tonic-SR). Muscle response depends on the activation level of gamma-motoneurons (γ-MN), which commonly show simultaneous changes in activation with alpha-motoneurons (α-MN): α-γ coactivation. Muscle response also depends on the level of presynaptic (Pre-S) inhibition.

activation level addressed as the *tonic stretch reflex* (Tonic SR in Figure 6.5). The tonic stretch reflex is a major component in the equilibrium-point (EP) hypothesis of motor control (see Section 6.6.3 and Chapter 12). There are also task- and instruction-dependent phasic changes in muscle activation that are sometimes addressed as long-latency reflexes or M_{2-3} (discussed in more detail in Chapter 7). Both tonic and phasic components of the stretch reflex can be seen in spinal animal preparations, proving that the basic reflex loops are within the spinal cord.

Another important factor that defines muscle response to stretch is the level of activity of the gamma-motoneurons, small neurons in the spinal cord innervating muscle fibers inside the muscle spindles (Figure 6.5). Changes in the gamma activity induce changes in the sensitivity of spindle endings to both velocity and amplitude of muscle stretch. Voluntary muscle activation is accompanied by an increase in the activity of both alpha-motoneurons (that send signals to power-generating muscle fibers) and gamma-motoneurons. This phenomenon, addressed as alpha-gamma coactivation, contributes to an increase in muscle responses to stretch with background muscle activation level.

While reflex testing is sometimes used as a means of estimating excitability of alpha-motoneuronal pools, results of such tests may be misleading due to the phenomenon of presynaptic inhibition. Presynaptic inhibition can affect selectively specific inputs into motoneuronal pools including those that contribute significantly to the stretch reflex, in particular, the Ia-afferents that carry information from the primary spindle endings sensitive to both muscle length and velocity (as illustrated in Figure 6.5). Changes in the level of presynaptic inhibition can lead to major modulation of reflex responses, in particular of the phasic stretch reflex, without any changes in the level of activation or excitability of the target motoneuronal pool.

The curves presented in Figure 6.5 look qualitatively similar to the behavior of a muscle under constant electrical stimulation. These phenomena are addressed as "dynamic force enhancement" and "residual force enhancement" (reviewed in Zatsiorsky and Prilutsky, 2012). Quantitatively, however, the muscle force increase induced by the stretch is very different in a muscle with intact reflexes and without reflexes (Nichols and Houk, 1976; see Figure 6.8). In particular, a muscle without reflexes shows a smaller response to stretch.

6.5.2 Flexor and crossed extensor reflexes

Several groups of sensory endings contribute to a complex pattern of reflex muscle activations collectively addressed as the *flexor reflex* and the *crossed extensor reflex*. These afferents are united under the label flexor reflex afferents or FRA. The FRA group includes, in particular, some of the cutaneous and subcutaneous receptors, secondary spindle endings, and free sensory endings scattered throughout muscles. In experiments, flexor reflex can be induced by a brief electrical stimulation of a mixed nerve such as the sural nerve or by mechanical stimulation, for example, scratching the sole of the foot. This stimulation produces, after a short time delay, activation of flexor muscles throughout the extremity leading to withdrawal of the extremity from the actual or virtual (in the case of electrical stimulation) stimulus. The contralateral extremity shows an opposite pattern consisting of activation of all the major extensor muscles—the crossed extensor reflex. The alternating, out-of-phase reflex actions in the two extremities resemble movements observed during locomotion.

This similarity led Sir Charles Sherrington to a theory that locomotion represented a sequence of alternating reflex muscle activations between the major flexor and extensor groups. When a paw touched the ground, the flexor reflex in the ipsilateral limb led to flexion in major leg joints while the extensor reflex in the contralateral limb accepted the weight of the animal. The induced reflex movements led to the contralateral paw touching the ground and the muscle activation pattern reversed. Despite the elegance of this hypothesis, it was later proven to be wrong (see Chapter 9).

6.5.3   Tonic vibration reflex

High-frequency (over 50 Hz), low-amplitude (up to 1 mm) muscle vibration leads to a host of sensory and motor events caused primarily by the unusually high level of activity of primary spindle endings that are highly sensitive to quick changes in muscle fiber length. One of the responses is the tonic muscle contraction that resembles voluntary muscle activation in several aspects. First, the electrical muscle activity does not show synchronized bursts typical of many reflexes. Second, humans can allow the reflex activity to develop or suppress it. This feature of the tonic vibration reflex violates the suggested definition of reflexes. Tonic vibration reflex can be observed in decerebrated animals but not in spinal animals. Either its reflex loop involves supraspinal structures or it relies crucially on disinhibition from those structures.

Muscle activation caused by the tonic vibration reflex is typically accompanied by profound suppression of monosynaptic reflexes in the muscle. This feature contrasts the facilitation of monosynaptic reflexes during voluntary muscle activation. Animal studies have shown that the suppression of monosynaptic reflexes is of a presynaptic origin (Gilles et al., 1969; Desmedt and Godeaux, 1978). This is also supported by observations of diminished suppression of monosynaptic reflexes by muscle vibration in patients with impaired presynaptic inhibition, in particular in spastic disorders (Ashby and Verrier, 1976).

Sensory effects induced by muscle vibration typically involve distorted perception of muscle length (kinesthetic illusions) corresponding to more stretched muscles, sometimes even leading to perception of anatomically impossible joint positions (Goodwin et al., 1972; Craske, 1977). When vibration is applied to a muscle participating in a functional task, the induced illusions can lead to complex motor effects such as postural deviations of standing persons (vibration-induced fallings) seen when vibration is applied to one of the leg muscles (Eklund and Hagbarth, 1967; Eklund, 1969).

6.5.4   Reflex-like actions: preprogrammed reactions

When a person is asked to keep a joint position against a constant external load, a quick change in the load leads to a sequence of responses in the muscles crossing the joint. If the perturbation is strong, a monosynaptic response can be seen (addressed as M_1)

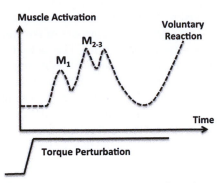

Figure 6.6 A quick joint perturbation leads to a sequence of reactions. The earliest (M_1) starts at a short delay, similar to that of the phasic stretch reflex. It is followed by two responses that come at the latencies of 50−90 ms.

followed by two more responses, sometimes not clearly distinguishable, that come at the latencies of 50−90 ms (Figure 6.6). These responses, addressed as M_2 and M_3, are followed by a voluntary correction of the joint position, which comes at a delay of about 150 ms. The M_2 and M_3 responses show modulation with instruction to the subject—they are strong if the subject is asked "to return to the initial position as quickly as possible" and much weaker if the instruction is "let the joint move" (Phillips, 1969; Tatton et al., 1978; Chan and Kearney, 1982). The modulation with instruction makes these responses fail one of the requirements for reflexes formulated earlier.

Responses at similar latencies are observed across a variety of actions. In particular, they are seen in postural muscles in response to unexpected motion of the supporting surface (e.g., when the bus starts moving), in leg and trunk muscles when the person steps on a sharp object or stumbles (the corrective stumbling reaction), and in hand muscles when a handheld object experiences an unexpected perturbation that endangers safety of the grip. M_2 and M_3 can be seen in remote muscles that are not directly affected by the perturbation. For example, they can be seen in arm muscles when a person stands in the bus and grabs a pole, and the bus suddenly starts to move.

The task specificity of the M_2 and M_3 responses and their modulation by instruction suggest that they represent brief voluntary actions that are prepared in advance (preprogrammed) and triggered by a sensory signal induced by the perturbation. So, they are sometimes addressed as preprogrammed or triggered reactions. The two responses involve different loops that include supraspinal structures. In particular, M_3 is believed to be transcortical. The difference in the loops is supported by the finding of an increase in M_3, but not M_2, under muscle fatigue (Balestra et al., 1992).

6.5.5   F-response: a nonreflex

Electrical stimulation of the nerves to some muscles leads to a quick muscle response after a delay similar to that of the H-reflex. However, no synaptic transmission within the CNS is involved in the loop of this response and so it does not qualify as a reflex.

This *F-response* is produced by backfiring of alpha-motoneurons in response to action potentials carried antidromically, from the site of the stimulation to the body of the neuron. Typically, neurons do not backfire. The lack of backfiring is due to the inactivation of sodium channels on the neuronal membrane, which leads to the phenomenon of refractoriness (lack of excitability). Some neurons, however, get out of the absolute refractory period while the local currents from the incoming action potential are still strong enough and can bring the membrane to the firing threshold. This is more likely to happen in neurons that are already excited by other inputs. So, the F-response magnitude can serve as an index of excitability of the target motoneuronal pool.

6.6   The role of reflexes in movements

The discussion on the role of reflexes in everyday movements has been ongoing for more than one hundred years. Two polar views may be represented as:

1. All movements are reflexes. The difference is in the complexity of the neural loops involved in the generation of a motor response to a stimulus. A typical example of this view is the mentioned Pavlov *theory of conditioned reflexes*, according to which all movements represent combinations of inborn and conditioned reflexes.
2. Reflexes are not important for typical everyday movements. They may be important for movements of lower animals (e.g., invertebrates), but not for humans. Possible reflex contributions to muscle activations have to be predicted by neural computational structures and taken into account during movement planning and generation. A typical example of this view would be the idea of *internal models*, that is, computational neural structures predicting interactions within the body and between the external world and the body.

6.6.1   Reflexes versus central pattern generators

At a more physiological level, the argument has been between the ideas that voluntary movements represent results of modulation of reflexes versus that voluntary movements are produced by neuronal structures within the CNS and are relatively independent of reflexes. In particular, cyclic movements, such as gait, have been associated with activity of CPG within the CNS (see Chapter 9). Both views have received substantial experimental support over the last hundred years.

The idea that movements were combinations of reflexes with properly modulated thresholds and gains received strong support in the first half of the twentieth century from the works by Sir Charles Sherrington and the aforementioned works by the Pavlov school. In particular, Sherrington advocated the idea that locomotion represented a cyclic process built on alternating flexion and crossed extension reflexes in contralateral limbs (see Section 6.5.2). While this particular view was later proven to be wrong, the overall idea that reflexes are an inherent part of the system for voluntary movement production survived and led to the introduction of the EP hypothesis by Anatol Feldman in the mid-1960s (Feldman, 1966; see Chapter 12).

The opposing view that movement patterns were produced by neural CPGs was developed in the early twentieth century by one of Sherrington's students, Thomas Graham Brown (1882–1965), who was also a great Scottish mountaineer. In the mid-1960s, the ideas of Graham Brown received strong support in classical studies by the Moscow school on the induced locomotion in decerebrate cats (Shik et al., 1966; Orlovsky et al., 1966). In particular, these experiments showed that electrical stimulation of a specific zone in the cat midbrain (the locomotor zone) could lead to locomotor movements of the legs placed on the treadmill. Moreover, when the cat was paralyzed with curare and no cyclic modulation of reflexes was possible, the stimulation produced cyclic bursts of activity in the alpha-motoneuronal pools innervating the leg muscles. This so-called *fictive locomotion* was a major argument in favor of neural structures within the CNS capable of producing patterns of signals to muscles even when no reflex contribution was possible. Relatively recently, studies in humans with a clinically complete spinal cord injury have supported the view that locomotor-like movements can be produced by structures within the human spinal cord (Shapkova, 2004).

Currently, there is little argument on the existence of CPGs and the fact that voluntary movements, whether cyclic or not, typically begin not with an external stimulus but with an internally generated intent. The argument on the role of reflexes in implementing the intent is still ongoing. The spectrum of opinions is very broad—from (1) "reflexes are an atavism, a nuisance; their effects have to be predicted by the CNS and compensated for," to (2) "reflexes are weak, they add maybe 5–10% to muscle activation," to (3) "reflexes are an important element in motor behavior that defines muscle mechanical properties," and to (4) "reflexes are crucial for voluntary movements, all movements in a healthy animal are produced via modulation of reflexes." The first opinion goes against the generally accepted view, originating from a classical paper by Bernstein (1935), that the CNS is in principle unable to predict with any degree of certainty the exact mechanical movement patterns due to the variable external conditions and states on neurons involved in movement production. If this is so, the contribution of reflex pathways to muscle activation is also impossible to predict.

It is very easy to present a counterexample to the second opinion. Imagine a person pressing "as strongly as possible" against a stop with a muscle group. The stop is unexpectedly removed, leading to a very fast movement that causes shortening of the muscle group. During that movement, the muscles show an abrupt cessation of their activation level at a short time delay (about 50 ms; Figure 6.7). This phenomenon is called the *unloading reflex*. Note that this reflex is able to cancel out 100% of the maximal muscle contraction; so, reflexes are not weak. Hence, we focus on the two latter views, (3) and (4).

6.6.2 Reflex-induced changes in muscle, joint, and limb mechanical properties

Reflex effects on mechanical characteristics of muscles and limbs have been well documented. In particular, in classical studies performed in the 1970s, T. Richard

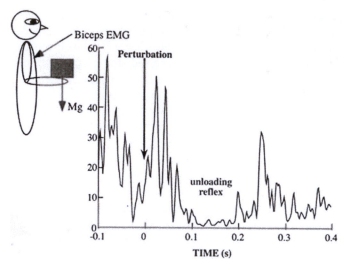

Figure 6.7 An illustration of the unloading reflex. The subject holds a load on the hand. When the load is suddenly removed, a quick elbow flexion follows accompanied by a nearly total silence of the biceps.
Reproduced by permission from Latash (2008).

Nichols and James Houk reported the force—length relations of a hindlimb muscle in a cat preparation (Nichols and Houk, 1976; Figure 6.8). The muscle with intact spinal reflexes showed a smooth, close-to-linear relation between muscle length and force during a slow, quasi-static stretch (thick line in Figure 6.8). After the corresponding dorsal roots were cut and afferent signals from the muscle could not reach the spinal cord, the force—length relation showed a major distortion (thin line in Figure 6.8). Since the initial activation level of the muscle was controlled under both conditions, the difference between the two curves in Figure 6.8 could be attributed to the action of reflexes in the intact muscle. One can conclude from these studies that reflexes contribute significantly to the muscle force—length relation, and they make it more linear.

One of the major sources of reflex effects on muscle activation is activity of the primary endings of muscle spindles. These receptors are highly sensitive to muscle velocity. As a result, their reflex effects lead to a dependence of muscle force on velocity, which adds to the force—velocity dependence inherent to skeletal muscles without reflexes (see Chapter 3). It is unknown how much velocity-sensitive reflexes contribute to the overall muscle behavior. The available estimates of nonreflex (peripheral) muscle damping suggest that muscles under typical external loading conditions are underdamped oscillators. However, very fast human movements typically do not end up in a visible oscillation, suggesting that human muscles (joints and limbs) are either overdamped or a special control pattern is used to produce fast movements (Latash and Gottlieb, 1991). The former view (Feldman et al., 1998; Gribble et al., 1998) implies very strong velocity-dependent reflexes that add significantly to joint damping, a view that still waits for direct experimental confirmation.

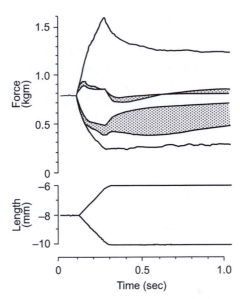

Figure 6.8 In the top graph, the outer lines show the mechanical responses of a muscle with reflexes to stretch (lower graph). The inner lines show matched responses of a muscle without reflexes (the shades show estimates of the range). Note the more symmetrical and linear response of the muscle with reflexes.
Reproduced by permission from Nichols and Houk (1976), © American Physiological Society.

While most studies on the role of reflexes in defining muscle mechanical properties have focused on the role of reflexes originating from muscle spindles, a series of recent studies have focused on the role of force-related reflex effects on muscle activation and mechanics. Some of the oligosynaptic reflexes from Golgi tendon organs have been described earlier (see Section 6.4.1). There are, however, polysynaptic reflexes induced by changes in muscle force that show more complex patterns with potentially significant effects on whole-limb mechanics. Such reflexes have been mapped in studies of the group of T. Richard Nichols (1989). Their patterns suggest a potentially important role in modulating mechanical resistance of the whole limb to external forces, in particular those emerging during locomotion.

6.6.3   Tonic stretch reflex and the control of movements

One of the most influential hypotheses in the field of motor control, the EP hypothesis (see Chapter 12 and Feldman, 1966, 1986), views central modulation of reflex parameters as the main mechanism of producing voluntary changes in muscle activation. According to the EP hypothesis, central control of a muscle can be adequately described as modulation of the threshold (λ) of the tonic stretch reflex (see Section 6.5.1). The notion of tonic stretch reflex relates to all the reflex effects on the alpha-motoneuronal pool innervating a muscle which are produced by muscle stretch. In particular, muscle spindle endings are viewed as the main contributor to the tonic stretch reflex, while effects from other receptor groups have also been documented.

Figure 6.9 illustrates a dependence of muscle force on muscle length, assuming that all the measurements are performed in steady states. Note that muscle force depends on length even when the muscle shows no signs of electrical activity (see Chapter 2). This dependence is shown in Figure 6.9 as the thin, dashed line with a relatively modest slope. At a certain muscle length (λ), nonzero muscle activation can be observed, and the force-length slope becomes steeper. With further stretch, both the activation level of the muscle and its force grow. Curves similar to the one in Figure 6.9 have been reported in decerebrated animal preparations when the animal could not intentionally change its descending command to the spinal cord. Electrical stimulation of different subcortical areas in the cat (Feldman and Orlovsky, 1972) has been shown to lead to a nearly parallel translation of the force—length dependence, supporting the idea that descending commands encode the threshold of the force—length dependences (threshold of the tonic stretch reflex).

Note that the primary endings of muscle spindles are highly sensitive to muscle velocity. This makes the tonic stretch reflex not exactly "tonic." In particular, if the tonic stretch reflex threshold of a muscle during its slow stretch is λ_0, during a quick stretch the muscle can show reflex activation at lower length values, $L < \lambda_0$. Similarly, when the muscle is shortening, it may show no signs of activation at length values $L > \lambda_0$. These effects have been modeled using the notion of effective threshold of the tonic stretch reflex, $\lambda^* = \lambda - \mu V$, where V is muscle velocity and μ is a constant.

The idea that sensitivity of alpha-motoneuronal pools to different reflex inputs can be modified in a task-specific way has led to the notion of *reflex gating*. One of the mechanisms of reflex gating is the presynaptic inhibition (illustrated earlier in Figure 6.5), which provides for selective changes in sensitivity of neurons to different inputs. Gating of reflex-like reactions, such as preprogrammed reactions (see above), has been well established.

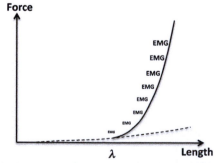

Figure 6.9 The thin, dashed line shows the dependence of muscle force on length in a quiescent muscle without reflexes. In a muscle with reflexes, at a certain length value (λ), the force—length dependence becomes much steeper (the thick curve). Muscle activation level (EMG) is zero to the left of λ and increases with length to the right of λ.

6.6.4   Reflex reversals

One of the more controversial features of reflexes is the phenomenon of *reflex reversal*. This term describes reflex activation to a standard stimulus that can be seen in different muscle groups, commonly in muscles opposing each other (antagonists), depending on the initial body configuration. Reflex reversals have been known for more than a hundred years. Early studies, in particular those of reflex deviations of an arm of the starfish to electrical stimulation of a sensory neuron, suggested a simple rule called the Uexküll rule: the longer muscle from an agonist–antagonist pair is more likely to show reflex activation. Later studies, however, showed exceptions to that rule. In particular, the tonic reflex contractions in response to vibration (tonic vibration reflex, Section 6.5.3) applied to a leg muscle in humans can be seen in different muscle groups depending on the body configuration and the presence or absence of pressure on the sole of the foot (Latash and Gurfinkel, 1976; Gurfinkel and Latash, 1978). The pattern of the reflex activation resembled muscle activation patterns observed when the person moves through those body configurations in the process of locomotion. These observations suggest that the neural loops of some of the more complex, polysynaptic reflexes may pass through structures involved in the production of voluntary movements, in particular, parts of the CPG for locomotion.

6.7   Pathological reflexes

The importance of descending pathways to the spinal cord for the modulation of reflexes can be seen under conditions when these pathways are physically damaged or malfunctioned due to another reason. In particular, severe spinal cord injury frequently leads to a condition termed *spasticity*, which is characterized by poorly controlled muscle responses to signals from peripheral receptors. Spasticity is a complex of signs and symptoms that include, in particular, an increase of reflexes to muscle stretch, both phasic and tonic. One of the most common definitions of spasticity (Lance, 1980) emphasizes the velocity-dependence of those atypically increased reflex responses.

Reflexes in spasticity can produce very strong lasting contractions leading to anatomical limits of joint rotation. In particular, a touch to the sole of the foot (and, in some cases, to other areas of the skin) can produce a flexion withdrawal reflex in all the major leg joints leading sometimes to a flexor spasm. Interactions of atypically strong reflexes lead to the clasp-knife phenomenon: If another person tries to move a joint of a person with spasticity into flexion, first there is strong resistance of the stretched extensor muscles produced by length and velocity-sensitive signals from spindle endings. When the extensor muscles produce very large forces, reflexes from Golgi tendon organs (see Figure 6.4) become so strong that they overcome the effects of the reflexes from spindle endings resulting in cessation of the extensor activity, and the joint collapses like a pocket knife.

Stimulation of the sole of the foot with a blunt instrument produces in healthy persons a flexion response in the toes. In newborn babies, this stimulation produces toe extension known as Babinski reflex. A response similar to Babinski reflex can also be seen in patients with spasticity. This is not the only example when an injury to descending pathways from the brain to the spinal cord leads to atypical reflex responses resembling those seen in newborns. Another example is the monosynaptic excitation of antagonist muscles produced by electrical stimulation of Ia-afferents. Note that healthy adults show only monosynaptic reflex contractions in agonist muscles, the H-reflex (see above).

Studies of reflexes are common in clinical evaluations. Their interpretation in terms of underlying neurophysiological mechanisms is sometimes ambiguous. For example, an increase/decrease in the amplitude of the H-reflex does not necessarily mean an increase/decrease in excitability of corresponding motoneuronal pools because of possible changes in presynaptic effects. Reflex interaction can give useful information regarding some of the mechanisms, for example, suppression of monosynaptic reflexes by muscle vibration has been used as an index of presynaptic inhibition (see Section 6.5.3).

Recently, the idea that the neural control of voluntary movements can be adequately described as modulation of thresholds for the tonic stretch reflex for the participating muscles (EP hypothesis, see Chapter 12) has been developed to address disordered movements, in particular, those seen in patients who suffer from spasticity (Levin and Feldman, 1994; Jobin and Levin, 2000; Musampa et al., 2007). This idea considers both impaired voluntary movements (paresis) and uncontrolled muscle contractions (spasms and increased muscle tone, see Chapter 5) as consequences of an impaired ability to change the tonic stretch reflex threshold for the involved muscles over its entire normal range. We will return to this promising theory in Chapter 12.

6.8 The bottom line

The notion of reflex has been and remains one of the central notions in both basic and clinical neurophysiology. This notion is not precisely defined, but the same can be said about the notion of voluntary movement, which is commonly used as opposing or complementary to that of reflex. The prevailing view that human actions can be classified into reflex and voluntary does not seem productive because of two reasons. First, there is no clear border between the two, and only actions at the very ends of the spectrum can be called reflex or voluntary with confidence. Second, voluntary movements always involve neural loops that are traditionally viewed as reflex.

Nevertheless, there are motor responses to stimuli that can be seen in reduced preparations (such as spinal and decerebrate animals) that can be viewed as reflexes with no or little controversy, even though such responses can show complex patterns and involve CPG. The notion of reflex is used broadly in clinical studies, and it is unlikely to disappear in the near future.

References

Ashby, P., Verrier, M., 1976. Neurophysiological changes in hemiplegia, possible explanation for initial disparity between muscle tone and tendon reflexes. Neurology 26, 1145–1151.

Balestra, C., Duchateau, J., Hainaut, K., 1992. Effects of fatigue on the stretch reflex in a human muscle. Electroencephalography and Clinical Neurophysiology 85, 46–52.

Berkinblit, M.B., Feldman, A.G., Fukson, O.I., 1986. Adaptability of innate motor patterns and motor control mechanisms. Behavioral and Brain Sciences 9, 585–638.

Bernstein, N.A., 1935. The problem of interrelation between coordination and localization. Archives of Biological Science 38, 1–35 (in Russian).

Chan, C.W.Y., Kearney, R.E., 1982. Is the functional stretch reflex servo controlled or preprogrammed? Electroencephalography and Clinical Neurophysiology 53, 310–324.

Clarac, F., 2005. The history of reflexes part 1: from Descartes to Pavlov. IBRO History of Neuroscience. http://www.ibro.info/Pub?Pub_Main_Display.asp?LC_Docs_ID=3155.

Craske, B., 1977. Perception of impossible limb positions induced by tendon vibration. Science 196, 71–73.

Desmedt, J.E., Godaux, E., 1978. Mechanism of the vibration paradox: excitatory and inhibitory effects of tendon vibration on single soleus muscle motor units in man. Journal of Physiology 285, 197–207.

Eklund, G., 1969. Influence of muscle vibration on balance in man. A preliminary report. Acta Societatis Medicorum Upsaliensis 74, 113–117.

Eklund, G., Hagbarth, K.E., 1967. Vibratory induced motor effects in normal man and in patients with spastic paralysis. Electroencephalography and Clinical Neurophysiology 23, 393.

Feldman, A.G., 1966. Functional tuning of the nervous system with control of movement or maintenance of a steady posture. II. Controllable parameters of the muscle. Biophysics 11, 565–578.

Feldman, A.G., 1986. Once more on the equilibrium-point hypothesis (λ-model) for motor control. Journal of Motor Behavior 18, 17–54.

Feldman, A.G., Orlovsky, G.N., 1972. The influence of different descending systems on the tonic stretch reflex in the cat. Experimental Neurology 37, 481–494.

Feldman, A.G., Ostry, D.J., Levin, M.F., Gribble, P.L., Mitnitski, A.B., 1998. Recent tests of the equilibrium-point hypothesis (lambda model). Motor Control 2, 189–205.

Fukson, O.I., Berkinblit, M.B., Feldman, A.G., 1980. The spinal frog takes into account the scheme of its body during the wiping reflex. Science 209, 1261–1263.

Gillies, J.D., Lance, J.W., Neilson, P.D., Tassinari, C.A., 1969. Presynaptic inhibition of the monosynaptic reflex by vibration. Journal of Physiology 205, 329–339.

Goodwin, G.M., McCloskey, D.I., Matthews, P.B., 1972. The contribution of muscle afferents to kinaesthesia shown by vibration induced illusions of movement and by the effects of paralysing joint afferents. Brain 95, 705–748.

Gribble, P.L., Ostry, D.J., Sanguineti, V., Laboissiere, R., 1998. Are complex control signals required for human arm movements? Journal of Neurophysiology 79, 1409–1424.

Gurfinkel, V.S., Latash, M.L., 1978. Reflex reversals in calf muscles. Physiologiya Cheloveka (Human Physiology) 4, 30–35.

Jobin, A., Levin, M.F., 2000. Regulation of stretch reflex threshold in elbow flexors in children with cerebral palsy: a new measure of spasticity. Developmental Medicine and Child Neurology 42, 531–540.

Lance, J.W., 1980. The control of muscle tone, reflexes, and movement: Robert Wartenberg lecture. Neurology 30, 1303−1313.

Latash, M.L., 1993. Control of Human Movement. Human Kinetics, Urbana, IL.

Latash, M.L., 2008. Neurophysiological Basis of Movement, second ed. Human Kinetics, Urbana, IL.

Latash, M.L., Gottlieb, G.L., 1991. Reconstruction of elbow joint compliant characteristics during fast and slow voluntary movements. Neuroscience 43, 697−712.

Latash, M.L., Gurfinkel, V.S., 1976. Tonic vibration reflex and position of the body. Physiologiya Cheloveka (Human Physiology) 2, 593−598.

Levin, M.F., Feldman, A.G., 1994. The role of stretch reflex threshold regulation in normal and impaired motor control. Brain Research 657, 23−30.

Meijer, O., 2002. Bernstein versus Pavlovianism: an interpretation. In: Latash, M.L. (Ed.), Progress in Motor Control, Structure-Function Relations in Voluntary Movement, vol. 2. Human Kinetics, Urbana, IL, pp. 229−250.

Musampa, N.K., Mathieu, P.A., Levin, M.F., 2007. Relationship between stretch reflex thresholds and voluntary arm muscle activation in patients with spasticity. Experimental Brain Research 181, 579−593.

Myklebust, B.M., Gottlieb, G.L., 1993. Development of the stretch reflex in the newborn: reciprocal excitation and reflex irradiation. Child Development 64, 1036−1045.

Nichols, T.R., 1989. The organization of heterogenic reflexes among muscles crossing the ankle joint in the decerebrate cat. Journal of Physiology 410, 463−477.

Nichols, T.R., Houk, J.C., 1976. Improvement in linearity and regulation of stiffness that results from actions of stretch reflex. Journal of Neurophysiology 39, 119−142.

Orlovsky, G.N., Deliagina, T.G., Grillner, S., 1999. Neuronal Control of Locomotion. From Mollusk to Man. Oxford University Press, Oxford.

Orlovsky, G.N., Severin, F.V., Shik, M.L., 1966. Locomotion evoked by stimulation of the midbrain. Proceedings of the Academy of Sciences of the USSR 169, 1223−1226.

Phillips, C.G., 1969. Motor apparatus of the baboon's hand. Proceedings of the Royal Society of London Series B 173, 141−174.

Prochazka, A., Clarac, F., Loeb, G.E., Rothwell, J.C., Wolpaw, J.R., 2000. What do reflex and voluntary mean? Modern views on an ancient debate. Experimental Brain Research 130, 417−432.

Rothwell, J.C., 1994. Control of Human Voluntary Movement, second ed. Chapman & Hall, London.

Shapkova, E.Yu., 2004. Spinal locomotor capability revealed by electrical stimulation of the lumbar enlargement in paraplegic patients. In: Latash, M.L., Levin, M.F. (Eds.), Progress in Motor Control-3. Human Kinetics, Champaign, IL, pp. 253−290.

Shik, M.L., Orlovskii, G.N., Severin, F.V., 1966. Organization of locomotor synergism. Biofizika 11, 879−886 (in Russian).

Tatton, W.G., Bawa, P., Bruce, I.C., Lee, R.G., 1978. Long loop reflexes in monkeys: an interpretive base for human reflexes. Progress in Clinical Neurophysiology 4, 229−245.

Zatsiorsky, V.M., Prilutsky, B.I., 2012. Biomechanics of Skeletal Muscles. Human Kinetics, Urbana, IL.

Preprogrammed Reactions

7

Reactions to unexpected stimuli at time delays (latencies) longer than those of typical reflexes (>40 ms) and shorter than those of the quickest simple reaction time (<100 ms) have been addressed using many names. These include long-loop reflexes, M_{2-3}, functional stretch reflexes, transcortical reflexes, triggered reactions, and preprogrammed reactions. The variety of the names reflects the different understanding of these phenomena. In particular, they have been viewed as examples of complex reflexes as well as examples of very quick voluntary actions. To avoid confusion, within this chapter we are going to address these phenomena as preprogrammed reactions. Later, arguments will be presented supporting this term.

7.1 Elements of history

Preprogrammed reactions were first discovered in humans and only later in animals. The first study by Hammond (1954) described reactions to sudden limb perturbations in humans that came at latencies longer than those of monosynaptic reflexes (these are under 40 ms even in persons with very long limbs) and shorter than those of the simple reaction time (these are longer than 90 ms, even in well-trained athletes). While the relatively short latency suggested a reflex origin of these reactions, they depended strongly on the instruction to the subject (see in detail later), which is a feature typical of voluntary actions.

Fifteen years later, reactions to limb perturbations at intermediate latencies were recorded in monkeys (Phillips, 1969). Since that time, similar reactions have been reported across a variety of experimental conditions, tasks, muscle groups, and species (reviewed in Forssberg, 1979; Prochazka et al., 2002). In particular, they have been studied in humans during such tasks as maintaining a joint (limb) posture, performing a quick joint movement, walking, standing, gripping an object, and speaking. In controlled laboratory conditions, two components of preprogrammed reactions to joint perturbation during a postural task were identified in muscle activation signals addressed as M_2 and M_3, or even M_{2-3}. The two components illustrated in Figure 7.1 are not always distinguishable, possibly because M_2 continues at the time when M_3 is initiated.

Two important features of M_2 and M_3 have been emphasized. First, these responses are large when the subject is instructed to compensate for the effects of the perturbation. In contrast, when the subject is instructed to "allow the motor to move your arm," these responses become much smaller in magnitude (compare the top and middle traces in Figure 7.1). Note that no such modulation is seen in the earliest response to the perturbation (M_1), which is a spinal reflex to muscle stretch with a strong monosynaptic component. Second, under certain manipulations, M_2 and M_3 show different

Biomechanics and Motor Control. http://dx.doi.org/10.1016/B978-0-12-800384-8.00007-7

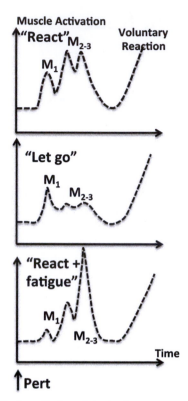

Figure 7.1 A perturbation (Pert) applied to a joint leads to a sequence of muscle responses. The first one (M_1) is a spinal reflex. The second and third are sometimes merged (M_{2-3}); they occur at latencies between 50 and 90 ms. The M_{2-3} responses (also addressed as preprogrammed reactions) are highly sensitive to instruction: They are large under the instruction to resist the perturbation (the top panel) and much smaller when the subject is instructed to let the motor move the joint (the middle panel). The different nature of the three responses is clearly seen in their changes under fatigue (the bottom panel): M_1 drops, M_2 shows small changes, and M_3 is increased.

behaviors, suggesting a difference in the involved neural pathways. For example, after a fatiguing exercise, M_1 is strongly suppressed, M_2 shows relatively small changes, and M_3 is increased substantially (compare the top and bottom traces in Figure 7.1; Balestra et al., 1992).

7.2   Current terminology

To discuss the terminology, we use Figure 7.2, which is similar to Figure 6.1 in Chapter 6. This figure shows schematically the location of preprogrammed reactions along the six axes. Preprogrammed reactions can always be linked to a stimulus, a perturbation.

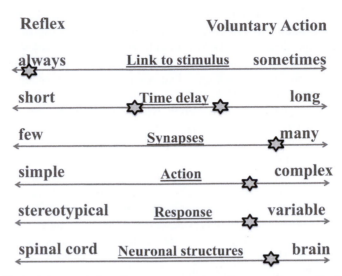

Figure 7.2 Typical characteristics of preprogrammed reactions are shown with stars along the axes that distinguish between reflexes and voluntary actions. Note that the preprogrammed reactions show characteristics typical of both reflexes and voluntary movements.

In that sense, they are similar to reflexes. Their time delay is intermediate between those for reflexes and voluntary actions; the latency of M_2 is closer to that of reflexes while the latency of M_3 is closer to that of quick voluntary movements. They involve an unknown number of synaptic transmissions, likely more than typical reflexes. Their patterns are rather complex and may involve many muscle groups broadly distributed over the body, for example, when a person stands in a bus holding onto a rail and the bus starts moving unexpectedly, preprogrammed reaction can be seen in leg, trunk, and arm muscles.

These responses are relatively stereotypical in controlled laboratory conditions, but their patterns show significant differences across populations (for example, between young and older adults, between toddlers and adolescents, between persons with and without Down syndrome, etc.). They involve supraspinal structures; in particular, M_3 is likely to involve cortical structures.

Among the frequently used terms for preprogrammed reactions are three expressions that use the word *reflex*. The most misleading one is arguably the *functional stretch reflex*. This term implies that the response is seen in the muscle stretched by the perturbation. However, this has been shown to be wrong many times. In the mentioned example of a person standing in the bus and holding onto the rail, a postural perturbation induced by bus acceleration leads to responses at about the same intermediate latency in muscles of the arm while the perturbation has no direct effects on the length of those muscles (more examples are presented later). Preprogrammed reactions have also been documented in muscles shortened by the perturbation (Marsden et al., 1979; Nashner and Cordo, 1981). All these observations suggest that these are not responses to muscle stretch, whether functional or not.

The term *transcortical reflex* may be applicable to M_3 but not M_2. The term *long-loop reflex* (sometimes, *long-latency reflex*) addresses adequately two of the features of preprogrammed reactions, their link to a stimulus and their long delay when compared to spinal reflexes. However, as can be seen in Figure 7.2, other features of preprogrammed reactions make them look more like voluntary actions than reflexes. In addition to the features illustrated in Figure 7.2, preprogrammed reactions are relatively immune to high-frequency muscle vibration, which suppresses monosynaptic reflexes and can also influence other reflexes. Even if one accepts that the term *reflex* is useful for studies of movements (see Chapter 6), applying it to a class of actions that show strong features of voluntary actions seems unjustified.

Addressing these responses such as M_2 (second muscle response) and M_3 (third muscle response) sounds more appropriate: These terms do not imply a specific neurophysiological mechanism of the responses, only their order of emergence. There are two caveats, however. First, using an abbreviation makes the term understandable only to those who already know quite a bit about the phenomenon. Second, and more importantly, both M_2 and M_3 can be seen in certain conditions in the absence of M_1. For example, in the above-mentioned example of a person standing in the bus, M_1 is not seen in most muscles that show M_2 and M_3. Typical postural perturbations in a person standing on a platform induced by a quick translation or rotation of the platform do not induce responses in the ankle muscles at latencies typical of M_1 (Diener et al., 1988). So M_2 and M_3 become the first and second muscle responses, which is confusing at best.

Two terms, *triggered reactions* and *preprogrammed reactions*, seem most appropriate because they reflect important features of these responses. First, the amplitude of these responses is known not to correlate with the magnitude of the perturbation that triggered the responses, while it may show correlation with the expected magnitude of the perturbation (based on previous few trials) (Houk, 1976). The fact of independence of these responses of the stimulus magnitude suggests that the stimulus represents a nongraded signal to the response generation and the response magnitude is defined prior to the stimulus based on some other factors. So the afferent signals produced by perturbation play the role of a trigger that informs the central nervous system only on the general location, nature, and direction of the perturbation. Given this feature, the term *triggered reaction* seems appropriate.

One of the most important features of these responses that distinguishes them from reflexes is the documented dependence of the magnitude of the responses on instruction to the subject (illustrated in Figure 7.1). This feature shows that a person can prepare for a perturbation in different ways depending on the task. The term *preprogrammed reaction* reflects this feature of the responses and does not seem to be ambiguous or misleading.

7.3   Definition and origins of preprogrammed reactions

The fact that preprogrammed reactions are widespread and can be seen across species, activities, and muscle groups suggest that they are functionally important. The predominant view is that these reactions represent an important line of defense of actions against perturbations that happen frequently during everyday activities. We suggest defining them as quick, frequently complex, actions that produce context-specific, urgent, crude corrections for the effects of perturbations.

Consider an example of a person slipping on ice. Immediately after the slip, there is a sequence of very quick actions of the whole body that look unique to the situation. These actions happen before the person realizes what is going on. Sometimes, they may lead to undesirable consequences. One of the authors of this book broke both elbows (on two different occasions) when he "automatically" extended an arm to support the body during the fall following a slip on ice. The fall happened anyway, but the broken elbow took a few months to heal. Existing estimates of the timing of such actions (Pavol et al., 2001) make them look like preprogrammed reactions triggered by the slip.

On the one hand, preprogrammed reactions frequently show complex patterns of muscle activation and joint motion, patterns that look unique and specific to the combination of the task and the perturbation. On the other hand, these reactions show poor modulation with the perturbation magnitude—the amount of compensation for the effects of the perturbation produced by preprogrammed reactions can vary from 0% to over 100% (overcompensation). Taken together, these features suggest that preprogrammed reactions emerge based on an algorithm (a set of rules) that generates situation-specific responses. The algorithm defines the pattern and direction of the response, which may look unique in each specific situation due to the different initial conditions and effects of the perturbation. The magnitude of the response is, however, not defined by the algorithm and may be viewed as preprogrammed.

7.3.1   Sensory source of preprogrammed reactions

A number of studies explored the issue of afferent signals that are likely to be the origin of signals triggering preprogrammed reactions. In particular, experiments with successive deafferentation of spinal segments and with selective blocking of the transmission along certain afferent systems have not provided conclusive evidence on the role of those afferent pathways in preprogrammed reactions. In most of those studies, the preprogrammed reactions persisted with only minor changes in their pattern and magnitude as long as the minimal sensory input was preserved informing on the occurrence of the perturbation.

Indeed, if one views these reactions as triggered responses to a signal provided by the perturbation, the source of this signal is not by itself significant. It is only necessary to provide minimal reliable information about occurrence of the perturbation and its effects on important task-related variables. So signals for the preprogrammed reactions can be provided by virtually any group of peripheral receptors sensitive to changes in load, position, pressure on the skin, and, in certain experimental conditions, also by visual, auditory, and vestibular receptors.

Elimination of one (or several, but not all) of the afferent sources should not be expected to lead to the elimination of the responses, although it can influence their certain features including, for example, latency, since the velocity of signal transmission and, possibly, time delays due to the central processing may be different for different sensory systems. Monkey experiments (Wylie and Tyner, 1981) with lifting an unknown load have demonstrated the presence of short-latency (as compared with the voluntary reactions) responses leading to "smoothening" of limb movements in cases of unexpected load changes. In these experiments, the preprogrammed responses disappeared after total deafferentation of the limb, and the residual compensation was based on visual information only.

7.3.2   Complex nature of preprogrammed reactions

The exact loop of the preprogrammed reactions is unknown. In one of the very first studies, Phillips (1969) offered a hypothesis that preprogrammed reactions represented a transcortical reflex, and this idea gained support in the years that followed (Allum, 1975; Cheney and Fetz, 1984; Day et al., 1991). However, there were reports on these reactions in decerebrated and even spinalized animals (Ghez and Shinoda, 1978; Tracey et al., 1980), although the observation of preprogrammed responses in spinal preparations has not been reproduced since the first reports. Discrepancies in reports were possibly due to two factors. First, as mentioned earlier, preprogrammed reactions are comprised of two responses with different latencies, M_2 and M_3. Second, the latency of the earlier response, M_2, is not much longer than the latency of spinal polysynaptic reflexes, in particular the tonic stretch reflex (see Chapter 6 for more details). Under experimental conditions when the preprogrammed response is observed in a muscle stretched by the perturbation, it may be difficult to distinguish it from the tonic stretch reflex. One method to distinguish the two responses is to observe them under a change in the instruction to the subject—preprogrammed reactions are highly sensitive to instruction while tonic stretch reflexes are not. However, for obvious reasons, this method is not applicable in spinal animal preparations.

One of the studies reported that the difference between the latencies of monosynaptic reflexes and preprogrammed reactions in arm and leg muscles was nearly the same (Darton et al., 1985). This result was used to argue against an involvement of a transcortical loop in preprogrammed reactions—such a loop is expected to lead to larger delays for leg muscles because of the longer transmission pathways. This argument can also be used against involvement of any supraspinal structure in preprogrammed reactions. There are three possible explanations for this result. First, it may be due to a purely spinal mechanism leading to preprogrammed reactions. Second, it may be due to the response mediated by the tonic stretch reflex and attributed to the preprogrammed reactions. Ultimately, the result may be due to a faster speed of transmission of action potentials along pathways involved in the leg response that cancels out the expected differences in the delay due to the different lengths of the pathways involved in the arm and leg responses. Given the available experimental material, we view the second explanation as most likely.

The current view on the mechanisms involved in preprogrammed reactions separates the M_2 and M_3. M_2 is viewed as a response mediated by supraspinal but subcortical structures, while M_3 is viewed as a transcortical response. These conclusions have been supported by studies of the two responses in neurological patients (Diener et al., 1985), their changes under certain drugs (Kofler, 2006), and neurophysiological studies in healthy human subjects (Nardone and Schieppati, 2008).

7.4   Examples of commonly studied preprogrammed reactions

7.4.1   Classical (lab conditions)

One of the most commonly used methods to study preprogrammed reactions is to ask a subject to hold a position in a joint against a bias external load and then to change the load quickly. In muscle biomechanics, a similar method is known as a controlled-release method (e.g., Hof, 1998). In a static force production task, the resistance is suddenly decreased or removed completely, the muscle is allowed to shorten by a certain amount, and after a stop the force is exerted statically again. The method is used to determine the elastic properties of the serial elastic components of the muscles.

Figure 7.3 illustrates responses to a perturbation in two muscles with opposing actions, an agonist–antagonist pair crossing the joint. Assume that the perturbation stretched the agonist muscle and shortens the antagonist one. The first response, M_1, is observed in the stretched muscle only. In contrast, the preprogrammed reactions can be seen in both muscles. The stretched muscle typically shows the "classical"

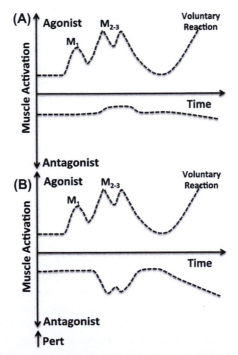

Figure 7.3 Joint perturbation that stretches one of the muscles ("agonist") of an agonist–antagonist pair leads to preprogrammed responses (M_{2-3}) in the stretched muscle and no major response or a drop in the baseline activation in its antagonist (panel A). In some populations with impaired motor coordination, an alternative, co-contraction pattern can be seen with similar preprogrammed responses in the opposing muscles (panel B).

pattern illustrated earlier in Figure 7.1. This pattern involves the M_2 and M_3 components of the preprogrammed reaction followed by a voluntary correction. Responses of the antagonist muscle shortened by the perturbation are more variable. This muscle can show a drop in its baseline level of activation at latencies typical of preprogrammed reactions (Figure 7.3(A)). Alternatively, it can show an increase in its baseline activity similar to the pattern seen in the agonist muscle (Figure 7.3(B)). The latter pattern, addressed as *co-contraction*, is more commonly seen in persons who show impaired motor abilities, for example, due to atypical development (as in Down syndrome, Latash et al., 1993).

Similar responses to quick perturbations can be seen when the main task involves performing a quick joint movement to a target rather than holding a fixed position. If the external load changes in the course of the movement, the involved muscles show deviations from their patterns seen in unperturbed trials, and the patterns and latencies of these deviations are similar to those seen in a position-holding task. In particular, an increase in the resisting load leads to an increase in the activation level of the agonist muscle (the one that is shortened during the movement) and a drop in the activation level of the antagonist. Similar to joint positional tasks, a co-contraction pattern can be seen in whole-body tasks (Horak et al., 1992; Misiaszek et al., 2000).

The flexibility of the preprogrammed reactions has been demonstrated in experiments with perturbations applied during quick thumb movements (Marsden et al., 1979). In those studies, the subjects performed thumb flexion against a constant load provided by a torque motor. The load could be increased or decreased unexpectedly in the movement course. An increase of the load led to an increase in the flexor activity at a latency characteristic of the M_{2-3} responses; a decrease of the load led to a drop in the flexor activity. In some trials, the thumb was unloaded by a different procedure—the hand was lifted by a force applied to the wrist, so that the thumb was moved away from the lever. In this case, the preprogrammed reaction flexor activity increased at the same characteristic latency. This inversion of the preprogrammed reaction showed that it was not driven by local changes in muscle mechanics but by the overall context of the task and perturbation.

7.4.2   Grip reaction

One of the commonly seen preprogrammed reactions is the response of the human digits to an external force applied to the handheld object. For example, imagine a person holding a vertically oriented object with a prismatic grasp, the four fingers opposing the thumb. The grip force is typically higher than the minimal force required to prevent slippage by a certain amount addressed as the *safety margin* (Johansson and Westling, 1984). If now an external force is applied unexpectedly to the object along its vertical axis, muscle responses leading to an increase in the grip force will be seen at a delay similar to that of typical preprogrammed reactions. Note that in this particular situation, the external force does not lead to any major change in any muscle length, which obviously does not allow viewing the responses as some kind of a stretch reflex. A series of studies by the group of Johansson (Johansson et al., 1992; Macefield et al., 1996; reviewed in Johansson, 1996) have

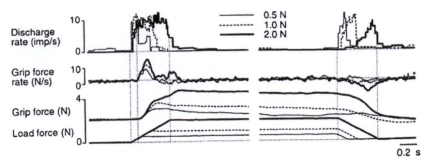

Figure 7.4 A person was holding an object using the precision grip. An unexpected load change (bottom lines) led to a burst of activity in fast-adapting skin receptors (top curves) followed by a change in the grip force (middle lines) at a delay similar to that of preprogrammed reactions. Similar but weaker effects were seen in response to load release (right curves).
Reproduced by permission from Macefield et al. (1996), © Springer.

shown that these quickest responses are likely to originate primarily from pressure-sensitive receptors in the skin of the fingertips. These results are illustrated in Figure 7.4, which shows responses of fast-adapting skin receptors and grip force changes induced by an unexpected load change at a fixed rate but over various magnitudes.

Similar gripping responses can originate from different afferent sources as well, as demonstrated in experiments by Traub et al. (1980). In those studies, the subject was asked to position the thumb and index finger just near the glass "as if going to grab it." Then, an unexpected perturbation was applied to proximal segments of the arm lifting it such that the glass was left below the hand. A preprogrammed reaction at a charac-teristic for the M_{2-3} response latency was seen leading to motion of the digits toward each other. Note that in this experiment no perturbation to the digits took place thus demonstrating that preprogrammed reactions can be triggered by sensory signals from remote body parts as long as the signal provides relevant information within the general context of the task.

7.4.3 Postural reactions

Arguably, one of the most common preprogrammed reactions during everyday activ-ities is related to perturbations of the vertical posture. As described in more detail in Chapter 14, vertical posture is inherently unstable. Unexpected perturbations of verti-cal posture are very common. For example, they happen when a person steps on an uneven surface, lifts an object with an unknown weight, contacts an external object or another person, stands in a vehicle (a train or a bus) that suddenly starts to move, and in many other everyday situations. Vertical posture is protected by several mech-anisms, both feed-forward and feedback. The latter act at different time delays, ranging to nearly instantaneous (mechanical responses of muscles and other tissues) to rela-tively long-latency ones such as voluntary corrections. Preprogrammed reactions play a very important role in postural control.

Postural adjustments triggered by external perturbations in standing persons have been mostly studied in response to unexpected translations and rotations of the platform on which the person stood. In young healthy subjects, a not-very-fast forward translation of the platform induces a body sway backward due to the inertial forces followed by an increase in the background activity of the ventral muscles (such as tibialis anterior, rectus femoris, and rectus abdominis) at a delay of about 70–80 ms, within the typical range for preprogrammed reactions. A backward translation of the platform results in a body sway forward and an increase in the background activity of the dorsal muscles (e.g., triceps surae and hamstrings) at about the same latency (see Figure 12.5). These muscle activation patterns show a characteristic distal-to-proximal sequence with the earliest responses in muscles crossing the ankle joint. This pattern of postural response and the accompanying kinematics has been called the "ankle strategy" (Horak and Nashner, 1986).

When the postural task become more challenging, for example, when a young healthy person balances on a narrow support surface or when the platform is translated very quickly, the order of muscle recruitment is reversed to proximal to distal, addressed as the "hip strategy" (Horak and Nashner, 1986; Woollacott and Shumway-Cook, 1986). Elderly persons are more likely to show the hip strategy in conditions when young persons typically show the ankle strategy (Figure 7.5, right panel).

The ankle-hip strategy dichotomy is practically useful but somewhat simplistic. More complex patterns of muscle and joint involvement depending on the task and perturbation have been reported (Allum et al., 1989). A different set of kinematic joint patterns (addressed as *eigenmovements*) has been introduced as the basis for the control of posture (Alexandrov et al., 2001). Each eigenmovement is a motion along a new variable defined as a linear combination of rotations in the three major joints (ankle, knee, and hip); the combinations are selected to ensure that equations of motion about these three new variables are decoupled.

Postural perturbations can lead to preprogrammed reactions in muscles that are not considered major contributors to postural stability. In particular, preprogrammed responses in arm muscles can be seen in response to a perturbation of the vertical posture when the person stands and grasps an object for additional support (Marsden et al., 1981). The pattern of these responses differs rather dramatically depending on the inertia of the object. When the object offers reliable support, the responses are seen in muscles that produce force on the object counteracting the postural effects of the perturbation. When the object has low inertia, the responses could invert and emerge in the antagonistic muscle groups. Then the movement pattern resembles that observed in a person holding a cup of coffee while standing in a bus in response to a postural perturbation produced by a sudden acceleration of the bus.

7.4.4 Corrective stumbling reaction

Another common group of preprogrammed reactions are those seen during locomotion in response to a mechanical stimulus applied to the foot or in response to an electrical stimulus applied to a nerve innervating the foot area (Lisin et al., 1973; Forssberg et al.,

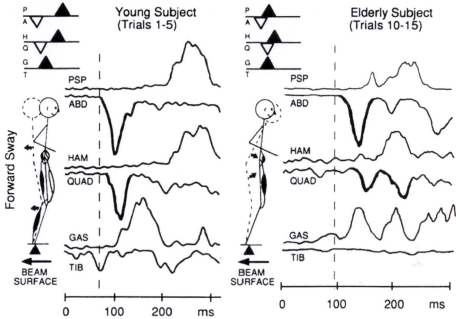

Figure 7.5 Responses of postural muscles and overall body kinematics in response to a quick backward displacement of the platform. Note the patterns typical of the "ankle strategy" in the young person (left panel) and patterns typical of the "hip strategy" in the older person (right panel). TIB, tibialis anterior; GAS, gastrocnemius; QUAD, quadriceps; HAM, hamstrings; ABD, rectus abdominis; PSP, paraspinal muscles.
Reproduced by permission from Horak et al. (1992), © Elsevier.

1975; Duysens and Pearson, 1976; Prochazka et al., 2002). Such responses have been documented in both human and animal studies. In cats, the responses can be induced even by a puff of air directed at the paw. The latency of these responses is about 50–80 ms after the stimulus application, and their patterns depend on the phase of stepping. The latency and the variable pattern of the responses allow considering them as preprogrammed reactions.

In particular, when applied in the swing phase, the stimulus typically induces a flexion response in the major leg joints with the leg producing a movement that looks like withdrawing the leg from the stimulus or stepping over an invisible obstacle. When the same stimulus occurs in the support phase, the response is mostly seen in leg extensor muscles such that the support phase is accelerated. An illustration of the corrective stumbling reaction to electrical stimulation of the sural nerve in different phases of walking is shown in Figure 7.6. Note the predominance of flexor responses in the swing phase and extensor responses in the single-support phase. The former pattern helps lift the foot over the nonexistent "obstacle." The latter pattern may be viewed as useful when a person steps on a sharp object. In this case, lifting the foot is impossible since it supports the weight of the body, and a reasonable strategy is to accelerate the step to minimize the time the foot experienced the offending stimulus.

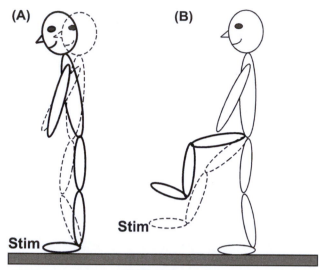

Figure 7.6 An illustration of the corrective stumbling reaction to electrical stimulation of the sural nerve. Note the predominance of extensor responses in the single-support phase (A) and of flexor responses in the swing phase (B). The dashed lines show the configuration of the segment (body) prior to the stimulus application. The bold lines show the configuration after the response. Reproduced by permission from Latash (2012).

7.5 The role of preprogrammed reactions in movements

Preprogrammed reactions, by definition, are seen in response to unexpected perturbations. Perturbations (changes in external forces) happen frequently during everyday movements, and, more frequently than not, they are unpredictable. Typical examples include manipulation of an object with the weight that happens to differ from what the person expects, stepping on an inclined surface or on an uneven surface unexpectedly, stumbling over an obstacle during locomotion, being pulled by the dog on the leash, bumping into another person in a crowded place, etc. In all those situations, it is imperative to ensure stability of task-specific important variables. This issue is considered in more detail in Chapter 11 where we discuss how elements of a redundant motor system can be organized in a task-specific way to ensure stability in certain directions in the space of variables produced by the elements (elemental variables).

The presence of preprogrammed reactions across many tasks suggests that they play an important role in ensuring stability of performance. One may even say that each and every movement from the everyday repertoire is associated with a mechanism that leads to the generation of preprogrammed reactions if a perturbation occurs. Since all natural actions are performed by redundant sets of muscles, there are potentially an infinite number of patterns of preprogrammed reactions that could be equally able to generate a required stabilizing action. On the one hand, in most situations, patterns of preprogrammed reactions are qualitatively similar across young, healthy persons. On the other hand, in some situations (for example, during slipping on ice, see Section 7.3), the observed patterns of quick corrections look unique for each specific person and situation. Taken together, these observations suggest a certain rule, an algorithm that leads to the generation of preprogrammed reactions. In situations when the initial conditions can differ dramatically, such as when slipping on ice, the resulting patterns may be highly variable. In more reproducible initial conditions, the patterns may be expected to show high consistency.

Preprogrammed reactions typically generate only approximate corrections for the effects of the perturbations. While the corrections are in an appropriate direction, their magnitude correlates poorly with the perturbation magnitude. There is, however, correlation of the magnitude of a preprogrammed correction with the average magnitude of perturbations experienced over a few previous trials (Houk, 1976). While in most situations preprogrammed reactions contribute to stability, under some conditions, preprogrammed reactions may themselves turn into perturbations. For example, when a sequence of strong postural perturbations is followed by a relatively small postural perturbation, the preprogrammed reaction is likely to lead to overcompensation of the effects of the last perturbation. If the subject is standing under conditions of reduced postural stability (for example, on a narrow surface or on a slippery surface), the preprogrammed reaction can lead to a loss of balance.

7.6 Atypical preprogrammed reactions

Atypical preprogrammed reactions may be seen in many populations characterized by impaired stability of movements including healthy elderly persons, persons with atypical development, and patients with various neurological disorders. Commonly, atypical preprogrammed reactions represent simultaneous activation responses in agonist–antagonist muscle pairs (Horak and Nashner, 1986; Figure 7.7). Such responses increase the apparent stiffness of the joint crossed by the muscles (on *apparent stiffness*, see Chapter 2) and reduce kinematic effects of the perturbation. Another advantage of co-contraction patterns is that they reduce the kinematic effects of a perturbation independently of its direction. On the other hand, a co-contraction pattern is ineffective in generating a change in the joint torque that would counteract the effects of the perturbation and may be viewed as energetically wasteful.

Some neurological disorders, such as Parkinson's disease and cerebellar disorders, are associated with exaggerated preprogrammed responses to perturbations while the ability to modulate these responses is impaired (Horak et al., 1992; Horak and Diener, 1994). Figure 7.8 presents an illustration of responses in ankle joint muscles to an unexpected platform translation in a healthy person (left traces) and a patient with cerebellar ataxia (right traces). Note the typical latency of the responses (50–100 ms) and the exaggerated responses in both ankle flexors and extensors in the patient. Such increased responses have been discussed as adaptive to the underlying loss of movement stability (Latash and Anson, 1996)—while they make movements look rigid and jerky, without these responses stability could be totally lost.

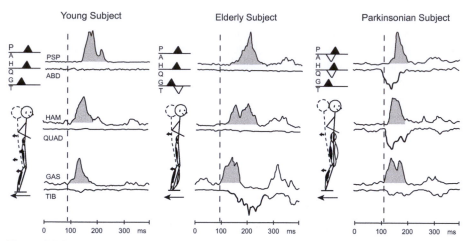

Figure 7.7 Responses of postural muscles and overall body kinematics in response to a quick backward displacement of the platform. Note the patterns typical of the "ankle strategy" in the young person (left panel), slightly changed patterns in a healthy older person (middle panel), and synchronous responses in muscles at all the major leg joints with visible antagonist co-contraction in the patient with Parkinsonism (right panel). TIB, tibialis anterior; GAS, gastrocnemius; QUAD, quadriceps; HAM, hamstrings; ABD, rectus abdominis; PSP, paraspinal muscles.
Reproduced by permission from Horak et al. (1992), © Elsevier.

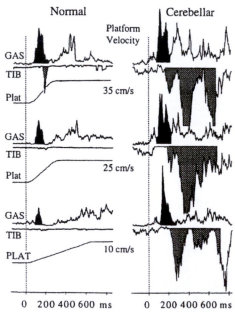

Figure 7.8 Responses of postural muscles to platform displacement at different velocities over the same amplitude in a healthy person (left) and a patient with a cerebellar disorder (right). Note the higher responses in the healthy person for quicker platform translations. In the patient, note the much stronger, prolonged response in both muscles crossing the ankle joint. TIB, tibialis anterior; GAS, gastrocnemius; Plat, platform.
Reproduced by permission from Horak and Diener (1994), © American Physiological Society.

7.7   The bottom line

There is little argument that the responses to perturbations at delays in between those typical of spinal reflexes and voluntary reactions exist and play an important role in ensuring stability of everyday movements. These responses can involve muscles throughout the body that can help counteract effects of the perturbation on important mechanical variables. Patterns of these responses may be highly variable, even to a standard perturbation, depending on stability conditions and the state of the person's body. In particular, they show major changes in many populations characterized by impaired stability of movements. It does not matter what term is used to address these responses. We believe, however, that any term that includes the word *reflex* is at least partly misleading. The term *preprogrammed reaction* seems most adequate to address these responses because it reflects an important feature of the responses, that is, their poor modulation with amplitude of the perturbation. Preprogrammed reactions include at least two components mediated by different neural pathways, likely with and without involvement of cortical structures. Their exact loops remain unknown.

References

Alexandrov, A.V., Frolov, A.A., Massion, J., 2001. Biomechanical analysis of movement strategies in human forward trunk bending. I. Modeling. Biological Cybernetics 84, 425–434.

Allum, J.H.J., 1975. Response to load disturbances in human shoulder muscles: the hypothesis that one component is a pulse test information signal. Experimental Brain Research 22, 307–326.

Allum, J.H.J., Honneger, F., Pfaltz, C.R., 1989. The role of stretch and vestibulospinal reflexes in the generation of human equilibrating reactions. Progress in Brain Research 80, 399–409.

Balestra, C., Duchateau, J., Hainaut, K., 1992. Effects of fatigue on the stretch reflex in a human muscle. Electroencephalography and Clinical Neurophysiology 85, 46–52.

Cheney, P.D., Fetz, E.E., 1984. Corticomotoneuronal cells contribute to long-latency stretch reflexes in the rhesus monkey. Journal of Physiology 349, 249–272.

Darton, K., Lippold, O.C.J., Shahani, M., Shahani, U., 1985. Long-latency spinal reflexes in humans. Journal of Neurophysiology 53, 1604–1618.

Day, B.L., Riescher, H., Struppler, A., Rothwell, J.C., Marsden, C.D., 1991. Changes in the response to magnetic and electrical stimulation of the motor cortex following muscle stretch in man. Journal of Physiology 433, 41–57.

Diener, H.C., Ackermann, H., Dichgans, J., Guschlbauer, B., 1985. Medium- and long-latency responses to displacements of the ankle joint in patients with spinal and central lesions. Electroencephalography and Clinical Neurophysiology 60, 407–416.

Diener, H.C., Horak, F.B., Nashner, L.M., 1988. Influence of stimulus parameters on human postural responses. Journal of Neurophysiology 59, 1888–1905.

Duysens, J., Pearson, K.G., 1976. The role of cutaneous afferents from the distal hindlimb in the regulation of the stepcycle in thalamic cats. Experimental Brain Research 24, 245–255.

Forssberg, H., 1979. Stumbling corrective reaction: a phase dependent compensatory reaction during locomotion. Journal of Neurophysiology 42, 936–953.

Forssberg, H., Grillner, S., Rossignol, S., 1975. Phase dependent reflex reversal during walking in chronic spinal cat. Brain Research 85, 103–107.

Ghez, C., Shinoda, Y., 1978. Spinal mechanisms of the functional stretch reflex. Experimental Brain Research 32, 55—68.

Hammond, P.H., 1954. Involuntary activity in biceps following the sudden application of velocity to the abducted forearm. Journal of Physiology 127, 23P—25P.

Hof, A.L., 1998. In vivo measurement of the series elasticity release curve of human triceps surae muscle. Journal of Biomechanics 31, 793—800.

Horak, F.B., Diener, H.C., 1994. Cerebellar control of postural scaling and central set in stance. Journal of Neurophysiology 72, 479—493.

Horak, F.B., Nashner, L.M., 1986. Central program of postural movements: adaptation to altered support-surface configurations. Journal of Neurophysiology 55, 1369—1381.

Horak, F.B., Nutt, J.G., Nashner, L.M., 1992. Postural inflexibility in parkinsonian subjects. Journal of Neurological Science 111, 46—58.

Houk, J.C., 1976. An assessment of stretch reflex function. Progress in Brain Research 44, 303—314.

Johansson, R.S., 1996. Sensory control of dextrous manipulation in humans. In: Wing, A., Haggard, P., Flanagan, R. (Eds.), Hand and Brain. Academic Press, San Diego, pp. 381—414.

Johansson, R.S., Riso, R., Häger, C., Bäckström, L., 1992. Somatosensory control of precision grip during unpredictable pulling loads. I. Changes in load force amplitude. Experimental Brain Research 89, 181—191.

Johansson, R.S., Westling, G., 1984. Roles of glabrous skin receptors and sensorimotor memory in automatic control of precision grip when lifting rougher or more slippery objects. Experimental Brain Research 56, 550—564.

Kofler, M., 2006. Levetiracetam suppresses long-loop reflexes at the cortical level. Muscle and Nerve 33, 785—791.

Latash, M.L., 2012. Fundamentals of Motor Control. Academic Press, NY.

Latash, M.L., Anson, J.G., 1996. What are normal movements in atypical populations? Behavioral and Brain Sciences 19, 55—106.

Latash, M.L., Almeida, G.L., Corcos, D.M., 1993. Preprogrammed reactions in individuals with Down syndrome: the effects of instruction and predictability of the perturbation. Archives of Physical Medicine and Rehabilitation 73, 391—399.

Lisin, V.V., Frankstein, S.I., Rechtman, M.B., 1973. The influence of locomotion on flexor reflex of the hind limb in cat and man. Experimental Neurology 38, 180—183.

Macefield, V.G., Rothwell, J.C., Day, B.L., 1996. The contribution of transcortical pathways to long-latency stretch and tactile reflexes in human hand muscles. Experimental Brain Research 108, 147—154.

Marsden, C.D., Rothwell, J.C., Traub, M., 1979. Long latency stretch reflex of the human thumb can be reversed if the task is changed. Journal of Physiology 293, 41P—42P.

Marsden, C.D., Merton, R.A., Morton, H.B., Rothwell, J.C., Traub, M.M., 1981. Reliability and efficacy of the long-latency stretch reflex in the human thumb. Journal of Physiology 316, 47—60.

Misiaszek, J.E., Stephens, M.J., Yang, J.F., Pearson, K.G., 2000. Early corrective reactions of the leg to perturbations at the torso during walking in humans. Experimental Brain Research 131, 511—523.

Nardone, A., Schieppati, M., 2008. Inhibitory effect of the Jendrassik maneuver on the stretch reflex. Neuroscience 156, 607—617.

Nashner, L.M., Cordo, P.J., 1981. Relation of automatic postural responses and reaction-time voluntary movements of human leg muscles. Experimental Brain Research 43, 395—405.

Phillips, C.G., 1969. Motor apparatus of the baboon's hand. Proceedings of the Royal Society of London, Series B 173, 141−174.

Pavol, M.J., Owings, T.M., Foley, K.T., Grabiner, M.D., 2001. Mechanisms leading to a fall from an induced trip in healthy older adults. Journal of Gerontology, Series A: Biological Sciences and Medical Sciences 56, M428−M437.

Prochazka, A., Gritsenko, V., Yakovenko, S., 2002. Sensory control of locomotion: reflexes versus higher-level control. Advances in Experimental Medicine and Biology 508, 357−367.

Tracey, D.J., Walmsey, B., Brinkman, J., 1980. "Long-loop" reflexes can be obtained in spinal monkeys. Neuroscience Letters 18, 59−65.

Traub, M.M., Rothwell, J.C., Marsden, C.D., 1980. A grab reflex in the human hand. Brain 103, 869−884.

Woollacott, M.H., Shumway-Cook, A., 1986. The development of postural and voluntary motor control systems in Down's syndrome children. In: Wade, M.G. (Ed.), Motor Skill Acquisition and the Mentally Handicapped: Issues in Research and Training. N-Holland, Amsterdam, pp. 45−71.

Wylie, R.M., Tyner, C.F., 1981. Weight-lifting by normal and deafferented monkeys: evidence for compensatory changes in ongoing movements. Brain Research 219, 172−177.

Efferent Copy

8

The problem of interaction between perception and action is one of the central ones in fields such as psychology, neurophysiology of perception, and motor control. This problem is bidirectional. Sensory signals play an important role in movements. The most obvious example is muscle reflexes (see Chapter 6) that originate from sensory endings and lead to changes in muscle activation levels. Other examples include adjustments of movement patterns to changes in external conditions reflected in activity levels of sensory signals of different modalities. Such adjustments can be intentional; then, these effects are commonly referred to as sensory-based corrections of movements. They can also be more direct, without involving a change in one's subjective representation of the environment (cf. Gibson's notion of *direct perception*; Gibson, 1979). There are also effects in the opposite direction: Efferent (motor-related) processes affect how one perceives signals from sensory endings. Arguably, one of the most famous examples is the inability of humans to tickle themselves (Blakemore et al., 2000). The notion of efferent copy has been one of the central notions in theories that attempt to interpret results of numerous observations, showing that what one does affects what one feels.

8.1 Elements of history

In a classical book *Reflexes of the Brain*, published originally in 1863, the great Russian physiologist Ivan Sechenov (1829–1905) wrote that humans do not see but look, do not hear but listen. He implied the active process of extracting information from sensory signals provided by different sensory systems. At about the same time, Hermann von Helmholtz (1821–1894), one of the greatest German physicists and physiologists, paid attention to an observation known to most inquisitive elementary school students. If a person closes one eye and presses on the other eyeball with a finger, the world seems to move. A similar eye rotation produced naturally does not lead to such an illusory perception; the person knows very well that the eye moves and the world remains stationary. Note that in both cases, light that enters the eyeball through the lens moves over the retina, but the changes in the activity of light-sensitive retinal receptors induced by this motion lead to qualitatively different percepts depending on the cause of eye motion. Von Helmholtz drew a conclusion that signals to muscles involved in voluntary eye movements somehow affected visual perception in combination with the signals from light-sensitive receptors. Another observation by von Helmholtz supporting the idea that a copy of motor commands to eye muscles is used during visual processing at some brain level was the fact that, during changes in the gaze direction produced by saccades, the external world is not perceived as moving.

Biomechanics and Motor Control. http://dx.doi.org/10.1016/B978-0-12-800384-8.00008-9

Erich von Holst (1908–1962), a German physiologist and psychologist, is credited with introducing the term *efferent copy* (sometimes, *efference copy*). The paper by von Holst and Mittelstaedt published in 1950 described results of an experiment on flies with the head rotated by 180° and introduced the notion of *Efferenzkopie*, a copy of the current motor command, which is used by the nervous system to estimate sensory feedback expected from the action and correct the action if there is a discrepancy between the expected and actual sensory signals. The notion of efferent copy was also central for the interpretation of the so-called *posture-movement paradox*—voluntary movements do not induce resistance expected from posture-stabilizing mechanisms (for more details, see Chapter 12). Later, an American physiologist, Roger Wolcott Sperry (1913–1994), introduced a different term, *corollary discharge*, basically synonymous with efferent copy, based on his studies of the optokinetic response.

The notion of efferent copy has been used over the past half century to interpret numerous observations including the control of eye movements, kinesthetic illusions induced by muscle vibration, and aspects of feed-forward control of actions. Only recently, has this notion been revisited critically (Feldman, 2009, 2011) within a hypothesis on the control of movements with changes in referent body configurations (see Chapter 12). Feldman's analysis has shown that the traditional views on efferent copy and its role in action and perception have to be modified.

8.2 Current terminology

In the mentioned paper, Von Holst and Mittelstaedt (1950) offered the following scheme (Figure 8.1). They considered two components of sensory signals, which they addressed as *reafference* and *exafference*. The former was assumed to reflect effects of voluntary movements on sensory signals while the latter reflected sensory effects produced by changes in external forces (and other factors). Their analysis started with an assumption that the neural control of a voluntary action could be adequately described as specifying a level of activation of involved alpha-motoneuronal pools ("motor command" in Figure 8.1). A copy of the command ("efferent copy") is used to estimate an expected change in sensory signals and is sent to a neural structure that also receives signals reflecting the change of the sensory feedback produced by the action (reafference). The reafference signals are compared to the efferent copy, and, if there is a mismatch, a correction signal (ΔRA in Figure 8.1) is sent to the motoneurons. This scheme seems to solve the posture-movement paradox because only unexpected changes in sensory signals are allowed to exert effects on muscle activation. This solution of the posture-movement problem has been addressed as the *principle of reafference*. Note that exafference, by definition, has no matching efferent copy. Hence, exafference always produces posture-stabilizing effects on the motoneurons.

Within this scheme, efferent copy has two functions. First, it is used to extract perceptual information from afferent signals: Exafference always contributes to perception while reafference does not, unless it deviates significantly from its expected level based on the efferent copy (ΔRA). Second, it plays an important role in the modification of commands to muscles in response to changes in sensory information.

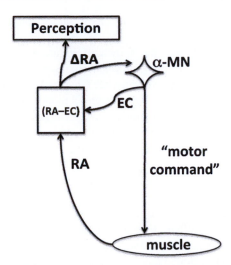

Figure 8.1 In the classical efferent copy scheme, neural processes result in the generation of a "motor command" to muscles. A copy of the command (efferent copy; EC) is used to estimate an expected change in sensory signals and sent to a neural structure that computes the difference between EC and the change of the sensory feedback produced by the action (reafference; RA). This difference (ΔRA) produces a correction signal sent to the motoneurons and participates in perception.

Schemes similar to the one in Figure 8.1 are used in more recent studies based on the ideas of internal models (Wolpert and Ghahramani, 2000; see also Chapter 13). Efferent copy is an important component of the assumed direct internal model, which estimates sensory feedback changes induced by the ongoing efferent process. The internal model is supposed to involve computational processes reflecting the interactions among elements of the neuromotor system within the body and also with the environment.

Recent developments of this approach have used the concept of Bayesian probability (reviewed in Wolpert, 2007). Bayesian probability reflects a particular state of knowledge (or belief) based on the previous experience. This apparatus is related to evaluating probability of a hypothesis based on some prior probability, which is then updated given new, relevant data. The Bayesian approach is more sophisticated than the one illustrated in Figure 8.1, but it shares certain basic features with the classical notion of efferent copy. In particular, within this approach, the motor command is used in state estimation, for example, with the help of a predictor, such as the Kalman filter (Goodwin and Sin, 1984), which is a Bayesian estimator for a time-varying system. Probability of a body state is defined not only by likelihood estimated based on current sensory signals but also by a prior reflection of earlier events, for example, experience stored in memory. So, the box with (RA−EC) in Figure 8.1 is replaced with a sophisticated computational mechanism. Its functioning is assumed to be linked to the concepts of optimal estimation of states of the world and our own bodies and then determining optimal actions; as such, this concept is tightly linked to the optimal

control theory and ideas of internal models as tools for estimates of the body state
(reviewed in Diedrichsen et al., 2010).

A few problems with the scheme shown in Figure 8.1 have been emphasized
(Feldman, 2009). First, some of the posture-stabilizing mechanisms are not mediated
by sensory feedback loops (for more details, see Chapter 14). Muscle forces depend on
muscle length and velocity, and this dependence is modulated by the level of muscle
activation (see Chapters 2 and 3). As a result, any perturbation is compensated (partly)
by immediate changes in muscle forces always directed against the perturbation. When
muscle activation level increases, forces generated by these so-called *preflexes* (Loeb,
1999) in response to a standard perturbation increase. Preflexes act independently of
the cause of a change in muscle length, voluntary or induced by a change in the
external force field.

More importantly, the scheme in Figure 8.1 is unable to handle one of the basic
experimental observations. When a person makes a voluntary movement from one
position to another (for simplicity, assume that the external force is always zero),
the muscles are quiescent prior to the movement, they show transient changes in the
activation level immediately prior to and during the movement, and they return to
quiescence a few seconds after the movement termination. So, activation signals to
the muscles are close to zero at both initial and final positions, which means that the
efferent copy signals in Figure 8.1 are also the same (about zero). At the final position,
however, muscle length values are different from the initial position and the position-
related change in the afferent signals (reafference) is expected to move the limb back to
the initial posture because this change is not balanced by a change in the efferent copy
(Figure 8.2). Actually, as described later, the activity of length-sensitive receptors

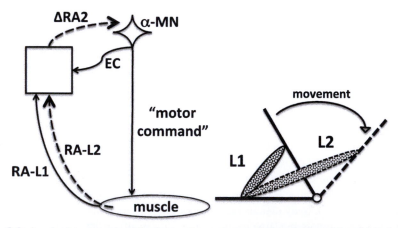

Figure 8.2 A voluntary movement leads to a change in muscle length (from L1 to L2 in the right
drawing). Since muscles are quiescent both prior to and after the movement, the EC is expected
to be the same while the reafference (RA) differs. This is expected to create a mismatch signal
(ΔRA2) leading to a nonzero change in the activation of motoneurons that moves the limb
toward its initial position. Experiments show that this is not true.

in human muscles remains nearly unchanged at different positions at the steady state (Hulliger et al., 1982). Since humans are able to relax muscles at different joint positions as shown, for example, in a classical study by Wachholder and Altenburger (1927), the scheme in Figure 8.1 is clearly inadequate.

A major problem with the scheme in Figure 8.1 seems to be the direct association of efferent copy with commands to or from alpha-motoneurons. As described in more detail in Chapter 12, the central nervous system is in principle unable to predefine the total activation signal to alpha-motoneurons because of the presence of reflex feedback loops. The general idea that the process of generation of a voluntary action plays an important role in kinesthesia remains valid. However, its implementation with the help of the classical notion of efferent copy is flawed. To resolve the problem, one has to revisit one of the central notions in motor control, that of *motor command*. This issue will be addressed in more detail later in this chapter.

8.3   Kinesthetic perception

Kinesthetic perception is formed based on signals from receptors (sensory endings) sensitive to a range of mechanical variables. While there seems to be enough sensor types sensitive to most relevant variables such as muscle length and velocity, muscle force, and joint angle, all of these signals carry ambiguous information as discussed below. Disambiguation of this information requires participation of neural signals generated within the central nervous system, in particular related to the production of voluntary movements. By the end of this section, we will have to decide whether the term *efferent copy*, defined according to Figure 8.1, is adequate to describe the processes of extracting kinesthetic information from the two types of signals, those associated with afferent and efferent processes.

8.3.1   Afferent contribution to proprioception

A number of sensory endings in peripheral tissues show activation levels reflecting mechanical variables relevant for kinesthetic perception. The sensory endings are parts of neural cells, proprioceptor neurons, with the body located in the spinal ganglia, close to the spinal cord (for the endings in the extremities and the trunk). The action potentials generated within a sensory ending travel along the peripheral branch of the T-shaped axon of the corresponding neuron and then are carried along the central branch into the central nervous system where they can have effects on activation of alpha-motoneurons (including reflexes, Chapter 6) and participate in perceptual processes (Figure 8.3).

Most sensory endings in natural conditions change their frequency of action potential generation with deformation. The location of an ending with respect to peripheral tissues makes it sensitive to specific variables such as displacements, velocities, and forces. Three types of sensory endings that have been most commonly associated with kinesthetic perception are those located in muscle spindles, Golgi tendon organs, and articular receptors.

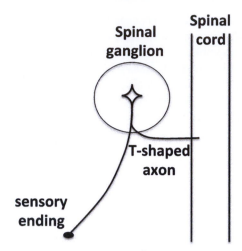

Figure 8.3 The sensory endings are parts of neural cells, proprioceptor neurons, with the body located in the spinal ganglia, close to the spinal cord. They generate action potentials in response to adequate stimuli. The action potentials travel along the peripheral branch of the T-shaped axon of the corresponding neuron and then are carried along the central branch into the central nervous system.

Muscle spindles are rather sophisticated structures scattered inside the skeletal muscles (Figure 8.4). They consist of a shell linked with connective tissue threads to power-producing muscle fibers (addressed as extrafusal fibers). Inside, the spindle contains muscle fibers of a different type and function (intrafusal fibers) that have no connection to tendons and cannot contribute to muscle force generation. The location of the spindles in parallel to the extrafusal fibers makes their sensory endings sensitive to movement kinematics, muscle fiber length, and velocity. Two types of sensory endings are located on the intrafusal fibers. Primary sensory endings are sensitive to both muscle length and velocity; they are innervated by the thickest, fastest-conducting afferent fibers (group Ia, conduction speed of up to 120 m/s). Secondary sensory endings are sensitive to muscle length only; they are innervated by afferent fibers from group II that are thinner and conduct at lower speeds, about 40—60 m/s.

Intrafusal fibers within muscle spindles are innervated by specialized neurons (gamma-motoneurons, Figure 8.4) that induce contractions of those fibers and thus change their mechanical properties, in particular, their deformation under the action of external forces provided by motion of extrafusal fibers (see Chapter 2). As a result, activity of gamma-motoneurons changes the characteristics of the sensory endings within the spindle, namely their sensitivity to relevant variables such as muscle length and velocity. There are two types of gamma-motoneurons, gamma-static and gamma-dynamic, that change the sensitivity of the endings to muscle length and velocity, respectively.

There are several factors that complicate using signals from muscle spindles for kinesthetic perception. First, most natural motions in the human body are joint rotations. While joint rotation is coupled to changes in the length of the "muscle-plus-tendon"

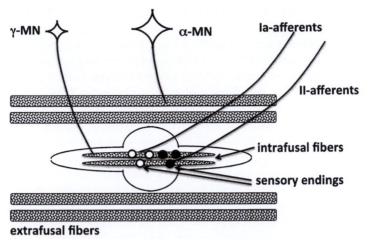

Figure 8.4 Muscle spindles are oriented in parallel to the power-producing extrafusal muscle fibers innervated by alpha-motoneurons (α-MN). They contain intrafusal muscle fibers inner-vated by gamma-motoneurons (γ-MN) and sensory endings innervated by primary (Ia) and secondary (II) afferent fibers. The endings are sensitive to muscle length (II) and to both length and velocity (Ia).

complex, muscle spindles generate information on changes in the length of muscle fibers only. For example, during a muscle contraction in isometric conditions, muscle fibers shorten and the tendons stretch (compare the two panels in Figure 8.5). In addition, any voluntary contraction is associated with an increase in the activation level of gamma-motoneurons (referred to as *alpha-gamma co-activation*). It has been shown that in isometric conditions, an increase in muscle force is accompanied by an increase in the frequency of action potentials generated by spindle endings (Vallbo, 1974). The phenomenon of alpha-gamma co-activation also leads to counterintuitive changes in the

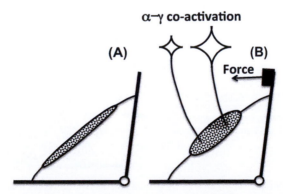

Figure 8.5 During isometric force production, two factors contribute to changes in the activity of spindle endings. First, there is co-activation of alpha- and gamma-motoneurons. Second, muscle fibers shorten while the tendon is stretched (compare the (A) and (B) drawings).

activity of spindle endings during voluntary movements. For example, the activity of both primary and secondary endings in humans remains nearly unchanged during a slow movement against a constant external load, while such a movement obviously leads to changes in the length of a muscle containing the spindles (Hulliger et al., 1982).

Golgi tendon organs are sensory endings located in series to the extrafusal muscle fibers close to the junction between the fibers and the tendon (Figure 8.6(A)). This location makes Golgi tendon organs sensitive to tendon force. Unlike muscle spindle endings, Golgi tendon organs have no direct neural control over their sensitivity to muscle force, no mechanism that would be similar to that of gamma-motoneurons. While signals from Golgi tendon organs can be used to extract information on tendon force, they do not directly supply information on the joint moment of force, which depends also on the lever arm. Note that moment of force is the adequate variable to describe effects of forces during rotational actions (see Chapter 1). Panel B of Figure 8.6 illustrates two joint angles that differ in the lever arm (L_1 and L_2). If the muscle force is the same, Golgi tendon organs show the same levels of activation while the two moments of force are different ($M_1 > M_2$).

Sensory endings located in the joints, articular receptors, are sensitive to joint rotation. Their location makes them perfect candidates to supply information about joint position. However, there are relatively few articular receptors sensitive to joint motion in its mid-range, while most of them generate action potentials when the joint approaches one of its limits of rotation. In addition, signals from articular receptors also change with such factors as joint capsule tension (for example, produced by co-contraction of muscles crossing the joint) and joint inflammation. This makes these signals unlikely candidates for the role of the primary source of information on joint position. Observations of the nearly unchanged joint position sense in persons after total joint replacement also suggest that these receptors do not provide crucial kinesthetic information.

This brief review suggests that afferent signals, by themselves, produce ambiguous information that has to be deciphered to become useful for kinesthetic perception.

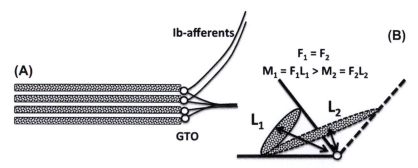

Figure 8.6 (A) Golgi tendon organ (GTO) is sensitive to tendon force. It is innervated by Ib-afferents. (B) A change in the joint position is associated with a change in the muscle force lever arm (L). So, the same force-related sensory signals correspond to different moments of force (M) produced by the muscle.

Later, in Section 8.4, we will discuss how processes associated with the generation of voluntary action can be used to disambiguate the afferent information without using the traditional notion of efferent copy (as illustrated in Figure 8.1).

8.3.2   Kinesthetic illusions induced by muscle vibration

The importance of signals from muscle spindles for kinesthetic perception has been shown, in particular, in experiments with high-frequency, low-amplitude muscle vibration (Eklund and Hagbarth, 1967; Goodwin et al., 1972; Feldman and Latash, 1982; Roll and Vedel, 1982). Primary endings are particularly sensitive to vibration because of their sensitivity to muscle velocity. In fact, vibration can drive virtually all the primary spindle endings within the muscle, which means that each ending generates at least one action potential at the frequency of the vibration. During vibration, many studies reported kinesthetic illusions leading to perception of body configurations corresponding to an increase in the length of the muscle subjected to vibration. Even anatomically impossible joint positions can be perceived (Craske, 1977). When subjects were asked to report the perceived joint rotational velocity, the illusions became even stronger: The magnitudes of velocity were so large that the joint could be expected to make several revolutions about its axis over the time of vibration application (Sittig et al., 1985).

Several studies have shown that effects of muscle vibration on kinesthetic perception can be modified and even reversed if the subject is simultaneously involved in another activity (Feldman and Latash, 1982). These observations have provided further support to the idea that signals from peripheral receptors are interpreted within a reference frame provided by the ongoing actions.

8.3.3   Vibration-induced effects on posture and locomotion

When a person performs a whole-body action, vibration applied to a muscle participating in the action can induce major changes in the action. One of the phenomena of this group is the so-called *vibration-induced fallings* (VIFs; Eklund and Hagbarth, 1967; Eklund, 1969). VIFs represent deviations of the body of a person standing with the eyes closed from a comfortable posture caused by vibration of leg muscles (VIFs have also been reported in response to vibration of other muscles such as muscles of the neck and arms, Roll et al., 1989; Latash, 1995). For example, vibration of the ankle extensors (applied to the Achilles tendon) produces a visible deviation of the body backward. This phenomenon has been interpreted as follows. The vibration produces an unusually high level of activation of spindle endings in the ankle extensors. These afferent signals are interpreted by the central nervous system as caused by an increase in the length of those muscles. In natural conditions, such a change in muscle length is associated with a body deviation forward. To keep balance, this illusory body deviation is corrected, causing the actually observed deviation backward.

Another phenomenon of this group is the effects of vibration on locomotion. These effects are much more complex and do not allow a simple interpretation based on overestimation by the central nervous system of the length of the muscle subjected to vibration. In particular, when a person marches in place, vibration applied to the hamstrings produced an unintentional forward progression (Ivanenko et al., 2000). During walking, vibration applied to different muscles could increase or decrease the speed of progression significantly, while application of vibration to muscles of one leg only can produce walking along an arc.

8.3.4 Kinesthetic illusions induced by muscle stimulation

If one assumes that the processes associated with the production of voluntary action participate in kinesthetic perception, another type of illusion can be expected when movements are produced not by voluntary muscle activation but by electrical stimulation of a muscle. In such a situation, a veridical sensory input provided by afferent signals may be interpreted wrongly given that the induced movements are not associated with neural signals that produce such movements naturally. For example, during electrical stimulation of a muscle its contraction is not associated with an increase in the level of activity of gamma-motoneurons that are activated during voluntary muscle contractions.

Such illusions have indeed been observed (Feldman and Latash, unpublished). Figure 8.7 shows schematically two relations between the actual joint position and its perception measured using a 0–10 scale reported by the subject. The scale was developed in advance using practice trials. During voluntary movements with eyes closed, there is a good correspondence between the actual and reported joint positions (solid line). During movements produced by the muscle nerve electrical simulation, the joint displacement was consistently overestimated (the dashed line) corresponding to perception of shorter length of the muscle subjected to the stimulation. This observation can be interpreted as a result of the mismatch between the level of activity of

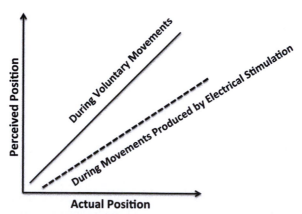

Figure 8.7 During voluntary movements, perceived position matches the actual one (the solid line). When movements are produced by electrical stimulation of the muscle nerve, there is a consistent perceptual bias corresponding to perception of shorter length of the muscle subjected to the stimulation. The axes show displacement of the effector (the index finger), actual (X-axis), and perceived (Y-axis) measured from a certain initial position.

gamma-motoneurons and the actual level of muscle activation produced by the stimulation—in natural conditions, the same muscle activation level would be accompanied by an increase in the activation of the gamma system.

8.4 Efferent copy within a general scheme of motor control

Based on the evidence reviewed earlier in this chapter, there is little argument that neural processes related to movement production participate in processes related to perception of the position (and force) of the moving effector. Some of the problems associated with the classical notion of efferent copy, as introduced by von Holst and Mittelstaedt, have been briefly reviewed in Section 8.2. These problems suggest that the role of efferent processes in perception has to be reconsidered. In the next two sections, we review on how this issue is handled within two main approaches to motor control, the one that can be addressed as "force control" and the other, which is commonly called "control with referent configurations" (for more details, see Chapter 12).

8.4.1 Within schemes of control with forces and muscle activations

These schemes of the neural control of movements are based on two pillars, classical mechanics and control theory. The former states that, in order to produce an accurate movement of a material object, appropriate force time patterns have to be applied to the object. The latter offers schemes based on computations performed by a controller based on the information about the current state of the body (delivered by sensory signals) and on the current efferent processes. Hence, most control schemes developed within this approach assume that the central nervous system computes requisite forces that have to be generated by muscles and updates these computations based on cascades of *internal models*—hypothetical neural processes that model (emulate) the numerous input−output relationships describing the interactions within the body and between the body and the environment (reviewed in Wolpert et al., 1998; Kawato, 1999; for more detail, see Chapter 13). Models of two main types are being used: direct and inverse. Direct models act as predictors of future body states while inverse models precompute requisite force profiles required to move the effector to the target and also estimate neural signals at different levels of the neural hierarchy that would be required to produce those forces.

While both classical mechanics and control theory are established fields with strong associated computational apparatus, several arguments have been presented against this approach (reviewed in Ostry and Feldman, 2003; Latash, 2012). First, as emphasized by Bernstein in the 1930s (Bernstein, 1935), the dependence of muscle force on its length and velocity due to both peripheral muscle properties and reflexes (Chapters 2, 3, and 6), makes it impossible for the central nervous system to predict or prescribe patterns of forces and displacements by all muscles taking part in natural movements with any degree of accuracy. Second, control of a time-varying multielement physical system can only be implemented by changing its parameters but not direct precursors of its output (e.g., mechanical) variables (e.g., Glansdorf and Prigogine, 1971). Third, if one accepts that the goal of motor control is to understand laws of nature that participate in the production of natural biological movements, assuming computations within the central nervous system is equivalent to giving up and trying to address the problem with tools developed for the control of inanimate, man-made objects.

While computational processes within these schemes take place in undefined neural structures, sometimes the cerebellum is implicated (Wolpert et al., 1998), and they are commonly assumed to create neural signals reflecting either muscle forces or their direct precursors such as levels of muscle activation or total presynaptic inputs into alpha-motoneuronal pools (reviewed in Shadmehr and Wise, 2005). These signals are assumed to participate in perceiving the body configuration (in combination with processed afferent signals) and predicting changes in the body configuration in a near future. While the involved computational processes may be rather sophisticated, the idea of comparing signals to muscles with sensory feedback about muscle state in these schemes is, in principle, the same as in the classical scheme in Figure 8.1 (e.g., Wolpert and Ghahramani, 2000). As a result, it suffers from the same problems.

One of the main problems is the experimental fact that, at different joint positions occupied voluntarily, muscle forces and levels of activation may be about the same (assuming that the external load is the same and changes in the moment arm are small). This means that the different position-related sensory signals can be accompanied by the same muscle activations without leading to any error signal that would cause the joint to move. The scheme in Figure 8.1, however, suggests that a change in the reafference signal not matched by a change in the efferent copy signal has to result in a nonzero ΔRA causing a change in muscle activations and movement. Moreover, if a change in the external load moves the joint away from its position, muscle activations and changes in muscle forces are observed resisting the deviation. This is true for any joint position, suggesting that voluntary movements are associated with changes in the set point about which the joint is stabilized. These effects, however, do not require any comparison of an efferent copy to the reafference but result from the functioning of physiological posture-stabilizing mechanisms including peripheral muscle properties and reflex loops as assumed in the equilibrium-point hypothesis and its recent development as the referent configuration hypothesis (Feldman, 2009; see in detail in Chapter 12).

8.4.2 Within the referent configuration hypothesis

The alternative approach to motor control originates from the equilibrium-point hypothesis (Feldman, 1966, 1986; for more details, see Chapter 12). Within this approach, the brain uses signals associated with depolarization of neuronal pools (typically, subthreshold depolarization) that, given the external force field, define a set of referent values for task-specific salient variables referred to as the *referent body configuration*. Further, a sequence of few-to-many transformations results in neural signals to the motoneuronal pools innervating all the muscles that participate in the movement (Latash, 2010). Such a scheme does not predefine muscle activation patterns, forces, and displacements but ensures stability with respect to the task-specific salient variables. No computations within the central nervous system are assumed within this approach. As shown in this section, kinesthetic perception can be naturally incorporated into the scheme of control with referent body configurations (Feldman and Latash, 1982; Feldman, 2009).

Consider first the control of a single muscle. Active muscle force and muscle activation level depend on muscle length (Figure 8.8). According to the equilibrium-point hypothesis (see Chapter 12), central control of a muscle can be adequately described as setting a value of the threshold for the dependence of active muscle force on muscle length, addressed as the threshold of the tonic stretch reflex (λ), which plays the role of the referent configuration for a single-muscle system. Setting a value of λ contributes to solving the problem of perceiving muscle length and force. Indeed, for a given λ, only points on the F(L) dependence in Figure 8.8 are possible as equilibrium states of the system "muscle + reflexes + load." A few such points are shown with black circles in Figure 8.8. Other states (e.g., those shown with open circles) are impossible. So, the efferent process reduces the problem of perception from finding a point on the two-dimensional force—length plane to a search for a point on the one-dimensional F(L) curve.

Consider now changes in different proprioceptive signals along the F(L) curve. All of them are expected to increase with an increase in the distance of points on this curve from λ. Indeed, as one moves along the F(L) curve from λ to the right, muscle length increases, which is expected to lead to an increase in the activity level of muscle spindle endings. At the same time, muscle force increases, which leads to an increase in

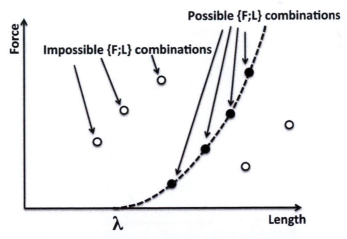

Figure 8.8 According to the equilibrium-point hypothesis, neural control of a muscle can be adequately described as setting its threshold (λ) of the tonic stretch reflex. This results in a dependence of active muscle force on muscle length. For a given λ, only points on the F(L) curve (black circles) are possible as equilibrium states of the system, while other states (open circles) are impossible. So, the problem of perception is reduced from identifying a point on the force—length plane to searching for a point on the one-dimensional F(L) curve.

the activity of Golgi tendon organs. Muscle activation level also increases along the F(L) curve produced by higher activation levels of the corresponding alpha-motoneurons. The higher activity of alpha-motoneurons is expected to lead to higher activation levels of gamma-motoneurons (alpha-gamma co-activation), also contributing to an increase in the activity of spindle endings. As a result, any of the aforementioned signals can be used to identify a point on the F(L) curve. Hence, the abundance of sensory signals does not create problems related to ambiguity in interpreting each signal individually, out of the context of the current value of λ. In contrast, it makes perceiving a point on that curve robust to possible problems with any one of the signals, for example, due to a pathology or surgery. Any of the mentioned sensory signals can be used to identify a point on the F(L) curve and produce simultaneous perception of the muscle force and length. One can say that there are two components to perceive muscle length (L_P), efferent and afferent, E_P and A_P. The former reflects the current value of λ, while the latter reflects deviations of the actual muscle length from λ: $L_P = E_P + A_P$ (see Figure 8.9).

This view can be generalized to perception of position of any effector. Setting a referent value (R) for this positional variable is associated with setting a dependence of this variable on external load. For a given value of load (L), position of the effector (P) is defined by both R and deviation of the effector from R produced by L, which is reflected in sensory signals (S): $P = R + S$ (Figure 8.10). In other words, setting R generates a reference frame for measuring position-related sensory signals. Note that R and λ are measured in positional units (e.g., meters), and sensory signals reflecting the deviation of the equilibrium point (EP in Figure 8.9) from the current value of R (λ) can also be expressed in positional units.

Perception of motion can be produced by changes in either component of position sense. For example, a change in the external load (panel A in Figure 8.10) produces a change in the sensory component of position sense: $R = \text{const.}$, $\Delta P = P_2 - P_1 = \Delta S = S_2 - S_1$. A voluntary action ($\Delta R = R_2 - R_1$) can lead to

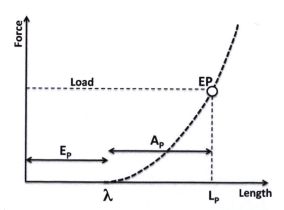

Figure 8.9 Two components of perceiving muscle length (L), efferent (E_P) and afferent (A_P). The former reflects the current value of λ, while the latter reflects deviations of the actual muscle length from λ: $L_P = E_P + A_P$. EP, equilibrium point.

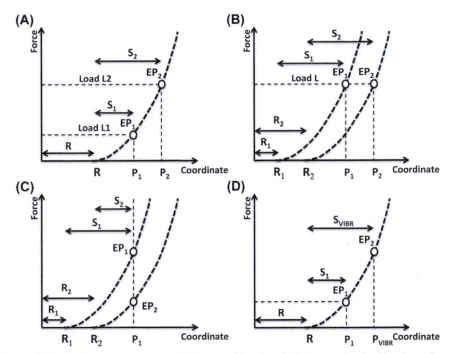

Figure 8.10 Setting a referent value (R) for a positional variable is associated with a dependence of its position on the external load. Position of the effector (P) is defined by both R and deviation of the effector from R produced by L and reflected in sensory signals (S): $P = R + S$. (A) A change in external load from L1 to L2 leads to motion resulting in a different value of S (S_2) and perception of a new position (P_2). (B) A voluntary movement in isotonic conditions leads to a change in R with no major changes in S. (C) In isometric conditions, there are counter-directional changes in R and S resulting in an unchanged perception of position. (D) Muscle vibration leads to larger S (S_{VIBR}) resulting in a kinesthetic illusion ($P_{VIBR} > P_1$).

different perceptual consequences depending on the external load. In isotonic conditions, no major change in S is expected, and $\Delta P = \Delta R$ (panel B in Figure 8.10). In isometric conditions, however, there will be counter-directional changes in S and R balancing each other: $\Delta P = 0$ ($\Delta R = -\Delta S$) (panel C in Figure 8.10).

Within this general scheme, a number of kinesthetic illusions can be interpreted. Illusions can be produced by a mismatch in the actual and perceived values of either of the two contributors to P. For example, a change in the afferent signals produced by muscle vibration is expected to create a signal corresponding to a larger deviation from the referent position: $S_{VIBR} > S$ (panel D in Figure 8.10). This is expected to lead to perception of longer muscle length ($P_{VIBR} > P$) corresponding to experimental observations (see Sections 8.3.2 and 8.3.3). Direct muscle stimulation produced changes in muscle force corresponding to a shift of the F(L) characteristic to the left. This is expected to be interpreted as a smaller value of R and results in perception of positions corresponding to the shorter length of the muscle ($P_{STIM} < P$), also corresponding to experimental observations (Section 8.3.4).

There are both common features and important differences between the schemes illustrated in Figures 8.1 and 8.8. Both schemes assume that position sense emerges based on neural signals reflecting both efferent and afferent processes. The scheme with efferent copy assumes that neural control can be adequately described as setting signals to muscles and invokes a corrective feedback loop based on a mismatch between the efferent copy signal and the reafference (both converted into commensurate units). No such assumption is made within the alternative view, which equates neural control of an effector not with setting muscle activations but with defining a threshold value of a positional variable, which allows muscle activations to vary with external load and participates in both motor and perceptual functions.

8.5   The bottom line

There is a general agreement that both afferent (sensory) and efferent (motor) processes participate in both perception and generation of movements. Shall we call efferent signals associated with the control of movement and also participating in perception of position (for example, those that define λ or R) *efferent copy*? The term has been already defined within a framework, assuming that the brain is able to prescribe patterns of activation of alpha-motoneurons. If one accepts this view (which we see as incompatible with experimental data), the term can be used. If one accepts the alternative view of control with referent body configurations, addressing a hypothetical copy of λ or R (in Figures 8.9 and 8.10, respectively) taking part in processes associated with perception as *efferent copy* is misleading. No error signal is computed within this approach (note that an error signal is central to the classical notion of efferent copy), and the efferent process may be associated with specifying a reference frame within which afferent signals are interpreted. What would be a good term for designating "signals associated with the control of movement and participating in perception of position"? Maybe "origin of the reference frame" can be used. This expression is less elegant that *efferent copy*, but maybe in this case elegance can be traded for exactness.

References

Bernstein, N.A., 1935. The problem of interrelation between coordination and localization. Archives of Biological Science 38, 1–35 (in Russian).

Blakemore, S.J., Wolpert, D., Frith, C., 2000. Why can't you tickle yourself? Neuroreport 11, R11–R16.

Craske, B., 1977. Perception of impossible limb positions induced by tendon vibration. Science 196, 71–73.

Diedrichsen, J., Shadmehr, R., Ivry, R.B., 2010. The coordination of movement: optimal feedback control and beyond. Trends in Cognitive Science 14, 31–39.

Eklund, G., 1969. Influence of muscle vibration on balance in man. A preliminary report. Acta Societatis Medicorum Upsaliensis 74, 113–117.

Eklund, G., Hagbarth, K.E., 1967. Vibratory induced motor effects in normal man and in patients with spastic paralysis. Electroencephalography and Clinical Neurophysiology 23, 393.

Feldman, A.G., 1966. Functional tuning of the nervous system with control of movement or maintenance of a steady posture. II. Controllable parameters of the muscle. Biophysics 11, 565−578.

Feldman, A.G., 1986. Once more on the equilibrium-point hypothesis (λ-model) for motor control. Journal of Motor Behavior 18, 17−54.

Feldman, A.G., 2009. Origin and advances of the equilibrium-point hypothesis. Advances in Experimental Medicine and Biology 629, 637−643.

Feldman, A.G., 2011. Space and time in the context of equilibrium-point theory. Wiley Interdisciplinary Reviews: Cognitive Sciences 2, 287−304.

Feldman, A.G., Latash, M.L., 1982. Inversions of vibration-induced senso-motor events caused by supraspinal influences in man. Neuroscience Letters 31, 147−151.

Gibson, J.J., 1979. The Ecological Approach to Visual Perception. Houghton Mifflin, Boston, MA.

Glansdorf, P., Prigogine, I., 1971. Thermodynamic Theory of Structures, Stability and Fluctuations. Wiley.

Goodwin, G.M., McCloskey, D.I., Matthews, P.B., 1972. The contribution of muscle afferents to kinaesthesia shown by vibration induced illusions of movement and by the effects of paralysing joint afferents. Brain 95, 705−748.

Goodwin, G.C., Sin, K.S., 1984. Adaptive Filtering Prediction and Control. Prentice Hall, Englewood Cliff, NJ.

Hulliger, M., Nordh, E., Vallbo, A.B., 1982. The absence of position response in spindle afferent units from human finger muscles during accurate position holding. Journal of Physiology 322, 167−179.

Ivanenko, Y.P., Grasso, R., Lacquaniti, F., 2000. Influence of leg muscle vibration on human walking. Journal of Neurophysiology 84, 1737−1747.

Kawato, M., 1999. Internal models for motor control and trajectory planning. Current Opinions in Neurobiology 9, 718−727.

Latash, M.L., 1995. Changes of human vertical posture due to vibration of shoulder muscles. Human Physiology (Fiziologiya Cheloveka) 21 (1), 125−128.

Latash, M.L., 2010. Motor synergies and the equilibrium-point hypothesis. Motor Control 14, 294−322.

Latash, M.L., 2012. Fundamentals of Motor Control. Academic Press, New York, NY.

Loeb, G.E., 1999. What might the brain know about muscles, limbs and spinal circuits? Progress in Brain Research 123, 405−409.

Ostry, D.J., Feldman, A.G., 2003. A critical evaluation of the force control hypothesis in motor control. Experimental Brain Research 153, 275−288.

Roll, J.P., Vedel, J.P., 1982. Kinaesthetic role of muscle afferents in man, studied by tendon vibration and microneurography. Experimental Brain Research 47, 177−190.

Roll, J.P., Vedel, J.P., Roll, R., 1989. Eye, head and skeletal muscle spindle feedback in the elaboration of body references. Progress in Brain Research 80, 113−123.

Shadmehr, R., Wise, S.P., 2005. The Computational Neurobiology of Reaching and Pointing. MIT Press, Cambridge, MA.

Sittig, A.C., Denier van der Gon, J.J., Gielen, C.C., 1985. Separate control of arm position and velocity demonstrated by vibration of muscle tendon in man. Experimental Brain Research 60, 445−453.

Vallbo, A., 1974. Human muscle spindle discharge during isometric voluntary contractions. Amplitude relations between spindle frequency and torque. Acta Physiological Scandinavica 90, 310−336.

Von Holst, E., Mittelstaedt, H., 1950/1973. Daz reafferezprincip. Wechselwirkungen zwischen Zentralnerven-system und Peripherie, Naturwiss 37, 467−476. The reafference principle. In: The Behavioral Physiology of Animals and Man. The Collected Papers of Erich von Holst. Martin, R., (translator), University of Miami Press, Coral Gables, FL, 1, pp. 139−173.

Wachholder, K., Altenburger, H., 1927. Do our limbs have only one rest length? Simultaneously a contribution to the measurement of elastic forces in active and passive movements. Pflügers Archiv für die gesammte Physiologie des Menschen und der Thiere 215, 627−640. Cited after: Sternad, D., 2002. Foundational experiments for current hypotheses on equilibrium point control in voluntary movements. Motor Control 6, 299−318.

Wolpert, D.M., Ghahramani, Z., 2000. Computational principles of movement neuroscience. Nature Neuroscience 3 (Suppl.), 1212−1217.

Wolpert, D.M., 2007. Probabilistic models in human sensorimotor control. Human Movement Science 26, 511−524.

Wolpert, D.M., Miall, R.C., Kawato, M., 1998. Internal models in the cerebellum. Trends in Cognitive Science 2, 338−347.

Central Pattern Generator

It has been known for centuries that movements with a rhythmic structure of kinetic and/or kinematic variables can be produced in the absence of a similarly structured input from the brain. For example, a chicken with the head chopped off runs for some time and flaps wings. These rhythmic movements are obviously produced by the spinal cord, possibly as a result of interactions with feedback signals from peripheral receptors. Experimental studies in the twentieth century (see below) have also shown that a variety of rhythmic patterned outputs can be produced without sensory feedback, including actions such as locomotion, respiration, and whisking. Taken together, these observations suggest the existence of autonomous neural structures within the central nervous system able to produce patterned (in particular, rhythmic) activity. Such structures have been addressed as central pattern generators (CPGs).

The experimental demonstration of the existence of CPGs in various species has provided important support for the general view that at least some of the natural biological movements are not produced by reflexes (e.g., as suggested by the theory of I. P. Pavlov and his school, reviewed in Windholz, 1987; Meijer, 2002) but originate within the nervous system. This alternative view was summarized by Bernstein (1967) as the principle of activity. According to this principle, functional voluntary movements are initiated by processes within the central nervous system.

The physiological nature of CPGs varies across species from interactions among currents in individual neurons (endogenous oscillator neurons) to interactions among sets of neurons (network-based CPGs). Within this brief review, we focus on the latter type of CPGs typical of mammals and other animals with a relatively developed central nervous system. In those animals, CPGs are commonly assumed to be located within the spinal cord or within a set of neural ganglia and to involve relatively small and autonomous neural networks. There are questions that remain open with respect to the notion of CPGs. In particular (1) Is any movement not triggered by a sensory input a product of a CPG? (2) Are all rhythmic movements produced by corresponding CPGs? (3) Can apparently nonrhythmic, discrete actions be produced by a CPG? (4) Is there a hierarchy of CPGs for whole-body actions, individual extremities, and individual joints? (5) What is the adequate description of the output of a CPG? (6) How do CPGs interact with sensory-based feedback loops?

9.1 Elements of history

While the mentioned example of a beheaded chicken has probably been known since the times chicken were domesticated, the first scientific evidence for the existence of CPGs came from classical experiments by Graham Brown (1911, 1914). Those studies

Biomechanics and Motor Control. http://dx.doi.org/10.1016/B978-0-12-800384-8.00009-0

showed, in particular, that the basic pattern of stepping could be produced by the spinal cord without the need of descending commands from the cortex and without the need of input from peripheral sensory receptors into the spinal cord. The latter conclusion conflicted with the theory developed by Sir Charles Sherrington that locomotion was based on alternating sequences of flexion and crossed-extension reflexes (see Chapter 6).

The first modern evidence of the CPG came from experiments on the locust (Wilson, 1961). Those studies showed, in particular, that the isolated, deafferented (deprived of afferent, sensory, input) ganglia of the locust could produce a rhythmic output resembling that observed during the locust flight. Rhythmic activity of the locust ganglia could also be induced by neuroactive chemicals.

In the 1960s, a group of Moscow physiologists (Shik et al., 1966; Orlovsky et al., 1966) studied the locomotion of cat preparations, spinalized and decerebrated, with the paws placed on the treadmill (Figure 9.1). Their main findings included the following. Electrical stimulation of an area in the midbrain (*mesencephalic locomotor region, MLR*) of decerebrate cats could produce rhythmic movements of the legs resulting in walking, trotting, or galloping on the treadmill. The movement rhythm was unrelated to the frequency of the stimulation. Varying the strength of the stimulation could produce different gaits, characterized by different relative phases of movement of individual legs. Later studies showed that similar motor effects could be produced by electrical stimulation within an area called the *locomotor strip* on the border between the medulla and the spinal cord and including the cervical segments of the spinal cord (Shik and Iagodnitsyn, 1978). Locomotor-like movements could also be produced by neuroactive chemicals, such as DOPA, injected into the spinal cord. Such movements could also be seen in the absence of any direct neural stimulation if the treadmill was moved by the motor. Note that in all those studies, the animals were unable to stand on their own, and their body weight had to be supported by a system of belts. When the neuromuscular transmission in an animal was blocked by curare, the electrical

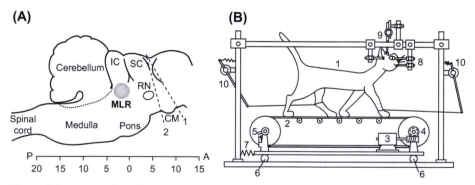

Figure 9.1 (A) Location of the mesencephalic locomotor region (MLR) in the brain stem of the cat. IC − inferior colliculi, SC − superior colliculi. The broken lines show the surgery for the premammillary (1) and postmammillary (2) preparations. (B) The schematics of the experiment. Reproduced by permission from Shik et al. (1966).

stimulation of the MLR could not produce rhythmic movements. Nevertheless, the outputs of the alpha-motoneuronal pools innervating leg muscles showed rhythmic bursts of action potentials, proving that the spinal cord could produce rhythmic neural activity in the absence of rhythmic afferent (sensory) inflow from the legs. This phenomenon has been termed *fictive locomotion*. Later it was shown that CPG also worked in a deafferented cat (Grillner and Zangger, 1984).

Taken together, those observations provided strong support for the existence of spinal CPGs for locomotion in cats. They also showed that the CPG could be driven by a nonpatterned input from supraspinal structures as well as by an input from peripheral receptors. On the other hand, the CPG did not require either of those inputs as shown in experiments on induced locomotion of spinal animals and on fictive locomotion in curarized and deafferented preparations.

Further, CPGs have been studied in a variety of species with both relatively simple locomotor organs powered by only two muscles with opposing actions (agonist—antagonist pair), as in locust and *Clione* (a floating sea slug) and very complex multisegment locomotor organs characterized by strong mechanical interactions. An isolated pair of ganglia and even a single ganglion of *Clione* can generate rhythmic activity similar to that observed during swimming (1—3 Hz; Arshavsky et al., 1985). Such activity is associated with groups of interneurons firing out of phase.

In crustaceans, CPGs consist of a chain of ganglia (Chrachri and Clarac, 1989). A single ganglion can generate rhythmic activity with alternating activity in neurons controlling opposing movements, while there is a consistent phase shift between adjacent ganglia. In crayfish and lobster, an isolated chain of ganglia can generate a rhythmic pattern with characteristics similar to that during swimming (Mulloney et al., 1987) with a phase delay between adjacent ganglia. Qualitatively similar observations were made in toad tadpole—a chain of ganglia could generate rhythmic activity with a phase shift along the body (Tunstall and Roberts, 1994) leading to a fictive swimming pattern, which was similar to that during normal swimming.

One of the favorite objects of study of CPGs has been the lamprey. When the spinal cord is removed from the lamprey, the cord can survive for days in vitro. It also has relatively few neurons and can be easily stimulated to produce a fictive swimming motion reflecting the activity of a CPG. In the early 1980s, Ayers et al. (1983) proposed that there was a CPG responsible for most undulating movements in the lamprey including swimming forward and backward, burrowing in the mud, and crawling on a solid surface.

As mentioned earlier, spinalized cats can show locomotion-like movements, for example, in response to a DOPA injection into the spinal cord. There are other CPGs in cats, for example, for scratching, which produce rhythmic patterns in response to an electrical or chemical stimulation. The pattern of neural activity produced by these CPGs differs from that observed during stepping. Paw shake is another example of a rhythmic activity that is likely produced by a spinal CPG in cats.

Attempts to produce locomotion in spinalized monkeys resulted in inconsistent reports. Some studies reported that DOPA could not activate CPG in spinal macaque. However, it could in other monkeys (*Callithrix jacchus*; Hultborn et al., 1993).

The first reports on a spinal CPG for locomotion in humans included rhythmic leg movements induced by electrical stimulation of the lumbar portion of the spinal cord in patients with clinically complete spinal cord injury who showed no signs of voluntary movements and no sensation below the level of injury (Shapkov et al., 1995, 1996). Calancie et al. (1994) claimed to observe the first clear example of a central rhythm generator for stepping in the adult human. Later studies have provided more evidence for rhythmic air stepping induced by electrical stimulation of the spinal cord in patients (Dimitrijevic et al., 1998; Shapkova, 2004; Gerasimenko et al., 2008). The ability of humans to walk with absent or limited sensory signals from peripheral receptors has also been confirmed in patients with a rare disease, large-fiber peripheral neuropathy, sometimes referred to as *deafferented persons* (Lajoie et al., 1996). Some of these patients have no proprioceptive sensation and no touch sensation below the neck level. Nevertheless, they are able to walk under continuous visual guidance.

Related studies used muscle vibration in healthy persons who were instructed not to produce voluntary movements but not to resist if such movements started on their own (Gurfinkel et al., 1998; Selionov et al., 2009). Vibration-induced air stepping was observed in about half of the subjects, while the other half showed no such responses to leg muscle vibration. Taken together, the mentioned recent studies provide strong support for the existence of spinal CPGs in humans.

9.2   Current terminology

CPG is commonly defined as an autonomous neural system, a single cell or a network, able to produce a rhythmic output pattern without sensory feedback. The term *autonomous* in this definition means that the CPG does not require a patterned input from other sources within the body, in particular from other neural structures, to continue producing the rhythmic output. According to this definition, a single cell (e.g., a neuron) can be viewed as a CPG if it can generate action potentials rhythmically. For example, pacemaker cells in the frog's heart (and in the hearts of other species) can produce rhythmic heart contractions even after the heart is removed from the body.

Neurons within the central nervous system of mammals, including alpha-motoneurons, can also be turned into a state of continued rhythmic generation of action potentials (reviewed in Heckman et al., 2003). This can happen if persistent inward currents (PICs) in dendrites are strong enough to create a new equilibrium potential on the membrane, which is above the threshold for action potential generation. Figure 9.2 illustrates this phenomenon with the voltage—current characteristic of the membrane in the absence of PICs (dashed curve; one stable equilibrium potential) and when PICs are strong (solid curve; two stable equilibrium potentials, one of which is above the threshold, Th, shown by the dashed vertical line).

Addressing pacemaker cells and neurons in a state of continued action potential generation as CPGs is counterintuitive. Traditionally, this term has been used for *neurons or autonomous neura*l systems that produce functional rhythmic motor

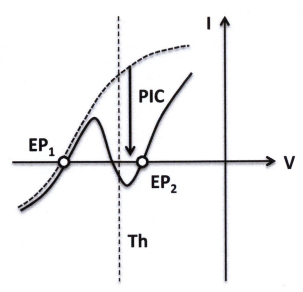

Figure 9.2 The current–voltage characteristic of a neuronal membrane has one equilibrium potential (EP$_1$) below the threshold for action potential generation (Th), the dashed curve. Persistent inward currents (PIC) can lead to emergence of another stable equilibrium potential (EP$_2$) above the threshold (the solid curve). The neuron can turn into a generator of action potentials.

actions that can be initiated and stopped by the animal *and/or by sensory signals*. Heart pacemaker cells normally act all the time, while neurons in the state above the threshold produce action potentials at a high frequency that does not correspond to any functional rhythmic action.

Accepting this definition leads to a few more consequences. First, not any open-loop action can be viewed as based on a CPG. For example, such discrete movements as reaching for an object, uttering a word, turning the body while standing, and many others are not characterized by rhythmic activity, even when the movement is very fast and shows no corrections to sensory signals. Using the term CPG for such actions is counterintuitive. Second, even if one produces a rhythmic action, for example, tapping a rhythm with the hand, involvement of a CPG is not obligatory because many such actions involve neural processes distributed throughout the whole central nervous system and cannot be attributed to a neuron or a set of neurons acting autonomously. Third, many functional rhythmic actions require ongoing motor activity that does not necessarily show rhythmicity and may be observed independently of the rhythmic actions. An example would be postural activity necessary for functional walking but also observed during a variety of nonrhythmic actions. Using the notion of CPG for such supporting actions would mean expanding the suggested definition to nearly all actions. Note that the mentioned studies on induced locomotion in spinalized and decerebrated cats required that the weight of the cat's body was supported by a system of belts.

9.3   Various CPGs

Anatomical details of CPGs are specifically known in only a few cases. CPGs have been shown to exist in the spinal cords of various vertebrates and to generate rhythmic patterns based on relatively small and autonomous neural networks.

9.3.1   The half-center model

One of the first models was introduced by Graham Brown (1911) and developed in many experimental and modeling studies (reviewed in Orlovsky et al., 1999). This model assumes the existence of two groups of neurons with mutual inhibitory projections. A typical scheme is presented in Figure 9.3, which illustrates the *two-half-centers* model. Two groups of neurons produce outputs to locomotor organs. They also inhibit each other via interneurons. When this system receives excitation from hierarchically higher neural structures, one of the main neuronal groups (for example N1) wins and suppresses the activity of the other group (N2). If one assumes that each of the main neuronal groups can fatigue relatively quickly, the active N1 group will stop generating action potentials after some time and disinhibit the N2 group. The N2 group will

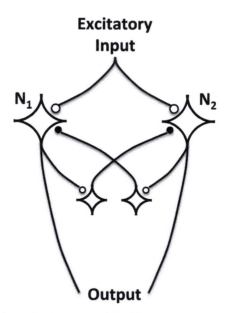

Figure 9.3 A simple schematic of the two-half-center model. Two groups of neurons, N_1 and N_2, produce outputs to locomotor organs. They also inhibit each other via interneurons. When they receive excitation, one group wins and generates action potentials until it fatigues. Then the other group is disinhibited and produces an episode of activation. Excitatory and inhibitory projections are shown with open and filled circles, respectively.

become active for some time, while the N1 group is inhibited and recovers from the assumed fatigue. Then, a switch in the opposite direction happens produced by fatigue of N2 neurons. Frequency of the rhythmic firing of the scheme in Figure 9.3 would depend on a number of factors including the strength of the mutual inhibitory projections and the rate of fatigue.

It is not necessary to assume fatigue of neurons to make the two-half-center model produce rhythmic activity. This can also happen if the scheme in Figure 9.3 is supplemented with delayed excitation between the N1 and N2 neuronal groups. Experimental studies of the lobster somatogastric ganglion have shown that a pair of neurons with mutual inhibitory projections can indeed produce alternating bursts of action potentials (Miller and Selverston, 1982). Experiments on tadpoles of *Xenopus* (Arshavsky et al., 1993) have suggested a somewhat more complicated scheme (Figure 9.4) with an interaction of excitatory and inhibitory projections in both of the two symmetrical half-centers.

It is also possible that each of the half-centers is able to generate rhythmic activity. Then, interactions between the half-centers define the relative phase of their activation bursts but not the frequency of firing. Studies of the locomotor system of *Clione* have supported the existence of swimming rhythmicity produced both within each half-center and by interactions between the two half-centers (Panchin et al., 1995).

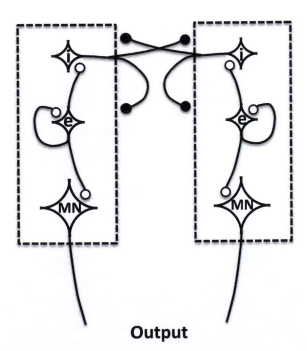

Output

Figure 9.4 A more complex two-half-center model involving symmetrical schemes with both inhibitory (i) and excitatory (e) interneurons. The output to effectors (muscles) is produced by the two groups of motoneurons (MN).

9.3.2 More complex CPGs

Several designs of CPGs have been described in animals with relatively few neurons. In addition to the mentioned schemes with inhibitory interactions between antagonistic groups of interneurons (in *Clione*, tadpole, and lamprey), more complex triphasic schemes with both excitatory and inhibitory interactions among three groups of interneurons have been described in *Tritonia*. Even more complex schemes with interactions among multiple interneurons have been described in the leech.

While the neuronal mechanisms of spinal CPGs in mammals (e.g., cats) are unknown, the two-half-center model has been used successfully to describe some of the salient features of locomotor patterns (Lundberg, 1981). A more sophisticated model has been suggested which is able to describe both stepping and scratching rhythmic movements produced by the same small set of neurons in response to different inputs (Gelfand et al., 1988). This scheme is illustrated in Figure 9.5 as an interaction of flexor and extensor neuronal units (FL and EX), with the flexor unit consisting of two parts, FL_1 and FL_2. An excitatory input into the FL unit produces a pattern resembling that during scratching, while an input into the EX unit produces a different pattern resembling that during stepping. This model suggests that apparently different rhythmic behaviors may be based on a single neuronal network.

A number of studies on cats with the spinal cord transected at a thoracic level have shown that the hindlimbs of those animals can produce locomotor movements when

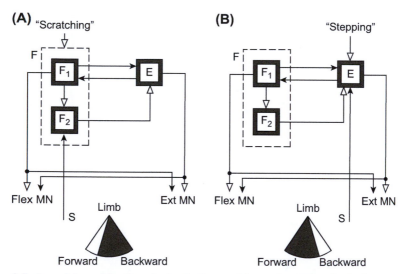

Figure 9.5 A model capable of producing both scratching and stepping rhythmic movements. The two regimes are induced by different tonic descending excitations shown by open arrow inputs. In the model, the flexor unit consists of two parts, F_1 and F_2, while the extensor unit consists of only one part, E. Inhibitory projections are shown with black arrows and excitatory projections with open arrows. The inhibitory signal S reflects the limb position. It comes from different ranges of motion (shown by black sectors) in the two regimes.
Reproduced by permission from Gelfand et al. (1988), © Wiley.

placed on a treadmill, particularly if the hindquarters of the cat's body are suspended providing partial weight support (reviewed in Rossignol, 1996). The performance of such animals could be improved considerably by activation of the noradrenergic system of the spinal cord and by regular training. The frequency of the stepping adapted to the speed of treadmill motion with appropriate changes in the relative phase of limb movement leading to gait changes. These observations suggested the existence of a separate CPG (or a subgroup of CPGs) for the hindlimbs of the cat independent of the CPG for the forelimbs.

9.3.3 Descending control of CPGs

The pioneering experiments on induced locomotion in decerebrated cats performed by the Moscow group of Shik, Orlovsky, and Severin (Shik et al., 1966, 1967) led to the discovery of the *mesencephalic locomotor region* (MLR), a small area in the midbrain (Figure 9.1). When the weight of the cat's body was supported by belts and its paws were placed on a treadmill, electrical or chemical stimulation of the MLR led to locomotion in such preparations. The speed and gait of the induced locomotion could be changed by varying the strength of electrical stimulation with transitions from walking to trotting and to galloping. In later studies, the possibility to induce locomotion by the electrical stimulation of the MLR in intact cats was also demonstrated (Sirota and Shik, 1973; Mori, 1987). The MLR receives inputs from the output nuclei of the basal ganglia, which is the likely pathway involved in the initiation and termination of locomotion.

It was also shown that locomotion could be produced in animals with the destroyed MLR (Orlovsky, 1969), suggesting that the MLR is only one of the areas involved in the control of locomotion in cats. Another area that can induce locomotion is the *subthalamic locomotor region*. Both the MLR and the subthalamic locomotor region project on neurons in the reticular formation (bilaterally), and the axons of the neurons that form the reticulospinal tract produce activation of the spinal CPGs.

The cerebellum is an enigmatic structure of the brain. On the one hand, its role in the coordination of voluntary movements is well established (reviewed in Houk, 2005). In particular, the cerebellum has been implicated in the coordination of supraspinal neural structures involved in motor behavior and spinal CPGs. On the other hand, animals with half of the cerebellum, or even the whole cerebellum, removed from the body recover surprisingly well and are able to show much of their motor repertoire. In addition, recent studies have emphasized the importance of the cerebellum in nonmotor functions such as memory, language, emotional state, spatial cognition, etc.

In humans and other primates, a major injury to the motor cortex has profound effects on locomotion, and even simple locomotor tasks become impossible. In contrast, in cats and dogs, the spinal cord is able to produce the basic locomotor patterns without an input from the motor cortex (reviewed in Kalaska and Drew, 1993). These animals with the motor cortex destroyed or functionally inactivated are able to walk and run on a flat surface. If the environmental constraints become more challenging, however, these animals fail to adapt their locomotor patterns in a functional way. In particular, the motor cortex plays a major role in visually guided gait modifications and in compensations for large perturbations of the locomotor pattern. On the other hand, the rhythmic activity of the spinal networks influences the motor cortex, which shows rhythmic modulation of its activity. This modulation is likely to play a role in tasks that require action in appropriate phases of the locomotor cycle, for example, as if stepping over an obstacle or during accurate stepping on targets.

9.3.4 Interactions among multiple CPGs

It is currently assumed that each limb of a quadrupedal animal has dedicated neural structures (a limb CPG) that define the rhythm of stepping as well as the relation between the stance and swing phases. During locomotion, the CPGs for individual limbs are coordinated with each other resulting in a common frequency of limb motion and appropriate phase relations that adjust to the speed of locomotion.

Recent studies of the rhythmic leg motion induced by the electrical stimulation of the spinal cord in patients with clinically complete spinal injury have suggested a hierarchy of spinal networks (CPGs) from a single joint to a single limb and to a multilimb level (Shapkova and Latash, 2005). Changing the parameters of the stimulation could produce in such patients a bilateral out-of-phase motion of both legs resembling that observed during walking or running as well as less typical patterns. The latter involved single-leg rhythmic motion with no visible rhythmic activity in the contralateral leg, and even single-joint rhythmic activity within a leg with only minimal displacements in other joints of the leg, possibly produced by mechanical coupling and the action of biarticular muscles. These observations have led to a scheme (Figure 9.6) with single-joint CPGs united into a leg CPG, and individual limb CPGs united into a functional structure able to produce locomotion.

Although humans are bipedal animals, they show accompanying arm movements during locomotion that suggest the presence of four CPGs for individual limbs. Traditionally, the arm movements have been described as counteracting the rotational action of the out-of-phase leg movement on the trunk (Umberger, 2008; Zehr et al., 2008), but a recent study questioned this assumption (Shapkova et al., 2011) and suggested that the arm movements were produced by an autonomous set of CPGs, not coupled in an obligatory way to the CPGs responsible for the leg movements.

9.3.5 Sensory effects on CPGs

Functionally important rhythmic actions are based on neural structures involving both open-loop and closed-loop, sensory-based circuits. Some of the mentioned CPGs can produce rhythmic activity with minimal or no feedback (e.g., in *Clione*, leech, and tadpole). There are, however, CPGs that cannot function under the open-loop conditions, for example, the CPG in the stick insect. Possibly, this difference reflects the fact that motion with impacts on the ground depends more on sensory feedback when compared to swimming in a medium without such discrete events. Most CPGs, in particular those in the cat, can generate rhythmic locomotor-like patterns

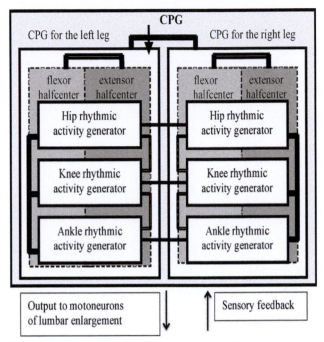

Figure 9.6 A hierarchy of CPGs based on observations of various rhythmic patterns produced in patients with clinically complete spinal cord injury.
Reproduced by permission from Shapkova and Latash (2005), © Marin Drinov Academic Publishing House.

without rhythmic sensory feedback. This has been shown, in particular, in experiments with *fictive locomotion* in animals paralyzed with curare. In such animals rhythmic bursts of activation from alpha-motoneurons innervating leg muscles were observed in response to electrical stimulation of the MLR while no movements happened and, hence, no rhythmic modulation of afferent sensory signals was possible. However, functional locomotion requires sensory feedback.

Sensory inputs can have rather dramatic effects on CPGs leading to qualitatively different movement patterns. For example, the neural network including the CPG responsible for swimming in the *Tritonia diomedea* can produce a reflex withdrawal response to a weak sensory input. A stronger sensory input produces an escape swimming response and crawling after the escape swimming has ceased.

In mammals, the effect of sensory inputs can vary depending on the phase of the rhythmic pattern in which it occurs. One of the well-known examples is the corrective stumbling reaction (Figure 9.7). This reaction is seen in response to a mechanical stimulus applied to the paw or electrical stimulation of a nerve innervating the paw area. If the stimulus is applied during the early stance phase, it leads to a short-latency response in limb extensors producing an accelerated step. A similar stimulus applied during the early swing phase produces a flexion response with a movement pattern as if stepping over an obstacle. Sensory effects on the gait can also be produced by signals from the vestibular apparatus and from neck proprioceptors (Gottschall and Nichols, 2007).

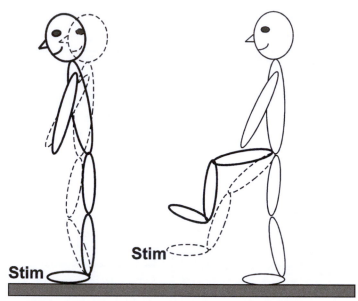

Figure 9.7 An illustration of the corrective stumbling reaction. If a stimulus to the foot is applied during the stance phase, a short-latency response in major leg extensors is seen ("accelerating the step"). The same stimulus applied during the swing phase leads to a flexion response ("stepping over an obstacle").
Reproduced by permission from Latash, 2012.

These effects may be responsible for changes in gait when the animal walks uphill or downhill when compared to level walking.

Sensory input from the limbs may truncate or extend individual phase durations. For example, walking with a pebble in one of the shoes alters the entire gait pattern, even though the stimulus is strong only during the stance phase of the corresponding leg. During that leg's swing phase, the stimulus is minimal or absent, but that phase is extended resulting in the typical limping pattern. Effects of such annoying stimuli on gait have been recently suggested as a tool to improve gait in patients with asymmetrical gait patterns, for example, following cortical stroke. The idea of this *discomfort-induced therapy* (Aruin and Kanekar, 2013) is to add sensory discomfort to the relatively intact foot leading to gait adjustments toward a more symmetrical pattern.

9.4   Cyclic versus discrete actions

As seen from the presented brief review, CPGs are assumed to produce a basic rhythmic pattern that can be adjusted by sensory signals. It is also evident from everyday experience that voluntary actions can be superimposed on rhythmic patterns produced by CPGs, commonly involving the same muscle groups and extremities. A typical example would be stepping over a puddle or a stone during walking or kicking the football while continuing to run. Other examples include speaking or swallowing while breathing, and hunting or escaping movements performed by various animals during running, flying, or swimming.

There is an ongoing argument on the nature of discrete and rhythmic movements (reviewed in Hogan and Sternad, 2007). Views have been expressed that rhythmic movements represent concatenations of discrete movements, that discrete movements represent truncated rhythmic movements, and that rhythmic and discrete movements represent two primitives based on independent, qualitatively different processes. The latter view seems to be compatible with the general notion of CPGs as autonomous neural structures specialized for the production of rhythmic movements. No CPGs have been postulated or observed specialized for discrete actions, although some of the involved neural networks can generate discrete actions in response to appropriate sensory stimuli. For example, the mentioned observations on *T. diomedea* suggest that one and the same network of neurons in these animals can produce a discrete action in response to a weak sensory stimulus and a rhythmic swimming motion in response to a stronger stimulus.

Should the notion of CPG be expanded to include autonomous neural structures for any actions, rhythmic or discrete? Alternatively, one may assume existence of a hierarchically higher neuronal network that receives sensory information and is able to trigger different types of outputs, directed either at a CPG or at a different network generating a discrete action (Figure 9.8). The CPG and the network producing the

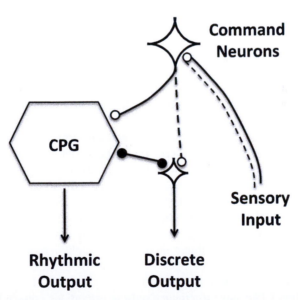

Figure 9.8 Two networks, a CPG and a network producing a discrete action, receive excitatory projections from a hierarchically higher command network. The CPG and the discrete-action network are inhibiting each other. A weak sensory stimulus (dashed lines) produces an excitatory output directed primarily on the discrete-action network, which suppresses the CPG. A stronger sensory stimulus (solid lines) excites primarily the CPG, and the discrete-action network is suppressed. Open and closed circles show excitatory and inhibitory projections, respectively.

discrete action are inhibiting each other such that only one response is possible. The question whether there are CPGs for discrete actions remains open. Note that expanding the notion of CPGs to discrete actions would make the neurophysiological mechanisms of all reflexes (e.g., of the flexor reflex and crossed extensor reflex, see Chapter 6) qualify as CPGs. This sounds counterintuitive and would effectively eliminate the specificity of the notion of CPG. This would also contradict the earlier definition that requires CPGs to be able to produce patterned neural activity in the absence of sensory feedback.

9.5   The role of CPGs in movements

The notion of CPG is central for the ongoing argument on the role of reflexes in voluntary movements (Chapter 6). According to Bernstein's principle of activity (Bernstein, 1966), movements represent an interaction of processes that originate within the central nervous system with those induced by afferent information generated by sensory endings. Patterns of neural activity produced by CPGs clearly belong to the first group of processes. Effects of the afferent information form at least two groups. First, afferent signals can affect the functioning of a CPG. In particular, they can trigger its activation as shown in experiments with walking of spinalized cats induced by treadmill motion (Shik et al., 1966). Second, activation patterns of alpha-motoneurons (and, hence, muscle activations) are defined by both the input generated within the central nervous system (e.g., from a CPG) and reflex effects from peripheral sensory endings. The basic temporal structure of alpha-motoneuronal activity can be seen in animals paralyzed with curare pointing at a CPG as the main source of the rhythmic activity. On the other hand, the observations of gross changes in the locomotion patterns of deafferented animals and humans with large-fiber peripheral neuropathy suggest that functional rhythmic motor patterns rely on intact sensory feedback.

Traditionally, the output of a CPG has been studied in terms of patterns of activation of alpha-motoneurons and muscles. Alternatively, the output of a CPG may be viewed as defining a rhythmic trajectory of a variable that serves as a referent point for the involved effector, a referent configuration for that effector (see Feldman, 2009 and Chapter 12). While this approach is highly attractive, its experimental confirmation has been limited to a handful of studies. In particular, the observation of global minima in muscle activation patterns throughout the body during rhythmic actions (Feldman et al., 1998) suggests that a rhythmic input into an alpha-motoneuronal pool may indeed correspond to rhythmic changes in the referent muscle length (associated with the threshold for its tonic stretch reflex).

In everyday life, CPGs have to produce muscle activations compatible with other ongoing motor tasks. For example, the effects of a CPG for locomotion on muscle activations have to be compatible with postural stability. The mechanisms that ensure at least minimally acceptable interactions between the locomotion CPG and the postural control system can be seen in some animals soon after birth, for example, in foals. In humans and some other animals, babies at birth are helpless and can perform only a few reflex-like actions such as grasping and sucking. While signs of CPG functioning can be seen relatively early in newborns leading to rhythmic alternating leg movements, it takes months for the development of adequate postural control that would allow functional walking to start. It takes even longer time to ensure that attempts to walk do not lead to falling down because of the poor interaction between the CPG-induced activity and postural control.

9.6   The bottom line

The existence of CPGs for a variety of rhythmic movements may be viewed as proven for a wide range of animals, from sea slugs and insects to humans. The neuronal mechanisms of CPGs have been deciphered only for animals with relatively simple nervous systems, up to the lamprey. In mammals, CPG remains a black-box notion. Not all rhythmic movements are products of CPGs, while some discrete movements may be produced by neural networks involved in CPGs in response to an appropriate sensory stimulus. The border between CPGs as autonomous structures producing patterned (rhythmic) outputs and distributed neural networks that participate in the production of voluntary movements is not clearly defined. CPGs for individual effectors (joints, body segments, and extremities) are likely to be organized in a hierarchical way that ensures proper relative timing of their involvement. CPGs need intact sensory feedback to produce functional behaviors. Adequate description of CPG functioning has to consider the reflex feedback effects on muscle activations; it may be viewed as defining a rhythmic trajectory of a referent value for a spatial variable that describes the involved effector. Muscle activity produced by a CPG may interfere with concurrent motor tasks; ability to avoid such interference may be partly inborn but, more commonly, it emerges in the process of motor development and requires practice.

References

Arshavsky, Y.I., Beloozerova, I.N., Orlovsky, G.N., Panchin, Y.V., Pavlova, G.A., 1985. Control of locomotion in marine mollusk Clione limacina. II. Rhythmic neurons of pedal ganglia. Experimental Brain Research 58, 263–272.

Arshavsky, Y.I., Orlovsky, G.N., Panchin, Y.V., Roberts, A., Soffe, S.R., 1993. Neuronal control of swimming locomotion: analysis of pteropod mollusk Clione and embryos of the amphibian Xenopus. Trends in Neuroscience 16, 227–233.

Aruin, A.S., Kanekar, N., 2013. Effect of a textured insole on balance and gait symmetry. Experimental Brain Research 231, 201–208.

Ayers, J., Carpenter, G.A., Currie, S., Kinch, J., 1983. Which behavior does the lamprey central motor program mediate? Science 221, 1312–1314.

Bernstein, N.A., 1966. Essays on the Physiology of Movements and Physiology of Activity. Meditsina, Moscow.

Bernstein, N.A., 1967. The Co-ordination and Regulation of Movements. Pergamon Press, Oxford.

Calancie, B., Needham-Shropshire, B., Jacobs, P., Willer, K., Zych, G., Green, B.A., 1994. Involuntary stepping after chronic spinal cord injury. Evidence for a central rhythm generator for locomotion in man. Brain 117, 1143−1159.

Chrachri, A., Clarac, F., 1989. Synaptic connections between motor neurons and interneurons in the fourth thoracic ganglion of the crayfish Procambarus clarkii. Journal of Neurophysiology 62, 1237−1250.

Dimitrijevic, M.R., Gerasimenko, Y., Pinter, M.M., 1998. Evidence for a spinal central pattern generator in humans. Annals of the New York Academy of Sciences 860, c360−376.

Feldman, A.G., 2009. Origin and advances of the equilibrium-point hypothesis. Advances in Experimental Medicine and Biology 629, 637−643.

Feldman, A.G., Levin, M.F., Mitnitski, A.M., Archambault, P., 1998. 1998 ISEK Congress keynote lecture: multi-muscle control in human movements. International Society of Electrophysiology and Kinesiology. Journal of Electromyography and Kinesiology 8, 383−390.

Gelfand, I.M., Orlovsky, G.N., Shik, M.L., 1988. Locomotion and scratching in tetrapods. In: Cohen, A.H., Rossignlo, S., Grillner, S. (Eds.), Neural Control of Rhythmic Movements in Vertebrates. Wiley, New York, NY, pp. 167−199.

Gerasimenko, Y., Roy, R.R., Edgerton, V.R., 2008. Epidural stimulation: comparison of the spinal circuits that generate and control locomotion in rats, cats and humans. Experimental Neurology 209, 417−425.

Gottschall, J.S., Nichols, T.R., 2007. Head pitch affects muscle activity in the decerebrate cat hindlimb during walking. Experimental Brain Research 182, 131−135.

Graham Brown, T., 1911. The intrinsic factors in the act of progression in the mammal. Proceedings of the Royal Society of London Series B 84, 309−319.

Graham Brown, T., 1914. On the nature of the fundamental activity of the nervous centres; together with an analysis of the conditioning of the rhythmic activity in progression and a theory of the evolution of function in the nervous system. Journal of Physiology 48, 18−46.

Grillner, S., Zangger, P., 1984. The effect of dorsal root transection on the efferent motor pattern in the cat's hindlimb during locomotion. Acta Physiologica Scandinavica 120, 393−405.

Gurfinkel, V.S., Levik, Y.S., Kazennikov, O.V., Selionov, V.A., 1998. Locomotor-like movements evoked by leg muscle vibration in humans. European Journal of Neuroscience 10, 1608−1612.

Heckman, C.J., Lee, R.H., Brownstone, R.M., 2003. Hyperexcitable dendrites in motoneurons and their neuromodulatory control during motor behavior. Trends in Neuroscience 26, 688−695.

Hogan, N., Sternad, D., 2007. On rhythmic and discrete movements: reflections, definitions and implications for motor control. Experimental Brain Research 181, 13−30.

Houk, J.C., 2005. Agents of the mind. Biological Cybernetics 92, 427−437.

Hultborn, H., Peterson, N., Brownstone, R., Nielsen, J., 1993. Evidence of fictive spinal locomotion in the marmoset (Callithrix jacchus). Society for Neuroscience Abstracts 19, 539.

Kalaska, J., Drew, T., 1993. Visuomotor coordination. Exercise and Sport Science Reviews 21, 397−436.

Latash, M.L., 2012. Fundamentals of Motor Control. Academic Press, NY.

Lajoie, Y., Teasdale, N., Cole, J.D., Burnett, M., Bard, C., Fleury, M., Forget, R., Paillard, J., Lamarre, Y., 1996. Gait of a deafferented subject without large myelinated sensory fibers below the neck. Neurology 47, 109−115.

Lundberg, A., 1981. Half-centres revisited. In: Szentagothai, J., Palkovits, M., Hamori, J. (Eds.), Regulatory Functions of the CNS. Principles of Motion and Organization, Advances in Physiological Sciences, 1. Pergamon Press, Budapest, pp. 155−167.

Meijer, O., 2002. Bernstein versus Pavlovianism: an interpretation. In: Latash, M.L. (Ed.), Progress in Motor Control. Vol. 2: Structure-Function Relations in Voluntary Movement. Human Kinetics, Urbana, IL, pp. 229−250.

Miller, J.P., Selverston, A.I., 1982. Mechanisms underlying pattern generation in lobster stomatogastric ganglion as determined by selective inactivation of identified neurons. IV. Network properties of pyloric system. Journal of Neurophysiology 48, 1416−1432.

Mori, S., 1987. Integration of posture and locomotion in acute decerebrate cats and in awake, freely moving cats. Progress in Neurobiology 28, 161−195.

Mulloney, B., Acevedo, L.D., Bradbury, A.G., 1987. Modulation of the crayfish swimmeret rhythm by octopamine and the neuropeptide proctolin. Journal of Neurophysiology 58, 584−597.

Orlovsky, G.N., 1969. Spontaneous and induced locomotion of the thalamic cat. Biophysics 15, 1154−1162.

Orlovsky, G.N., Severin, F.V., Shik, M.L., 1966. Locomotion evoked by stimulation of the midbrain. Proceedings of the Academy of Science of the USSR 169, 1223−1226.

Orlovsky, G.N., Deliagina, T.G., Grillner, S., 1999. Neuronal Control of Locomotion. From Mollusk to Man. Oxford University Press, Oxford.

Panchin, Y.V., Arshavsky, Y.I., Deliagina, T.G., Popova, L.B., Orlovsky, G.N., 1995. Control of locomotion in a marine mollusk Clione limacine. IX. Neuronal mechanisms of spatial orientation. Journal of Neurophysiology 73, 1924−1937.

Rossignol, S., 1996. Neural control of stereotypic limb movements. In: Rowell, L.B., Sheperd, J.T. (Eds.), Handbook of Physiology. Oxford University Press, New York, NY, pp. 173−216.

Selionov, V.A., Ivanenko, Y.P., Solopova, I.A., Gurfinkel, V.S., 2009. Tonic central and sensory stimuli facilitate involuntary air-stepping in humans. Journal of Neurophysiology 101, 2847−2858.

Shapkov, Y.T., Shapkova, E.Y., Mushkin, A.Y., 1995. Spinal generators of human locomotor movements. In: Fourth IBRO World Congress of Neuroscience, Kyoto, Japan.

Shapkov, Y.T., Shapkova, E.Y., Mushkin, A.Y., 1996. Evoked locomotion in paraplegic patients. In: Abstracts of the 9-th International Symposium on Motor Control, p. 88 (Borovets, Bulgaria).

Shapkova, E.Y., 2004. Spinal locomotor capability revealed by electrical stimulation of the lumbar enlargement in paraplegic patients. In: Latash, M.L., Levin, M.F. (Eds.), Progress in Motor Control-3. Human Kinetics, Champaign, IL, pp. 253−290.

Shapkova, E.Y., Latash, M.L., 2005. The organization of central spinal generators in humans. In: Gantchev, N. (Ed.), From Basic Motor Control to Functional Recovery − IV. Marin Drinov Academic Publishing House, Sofia, Bulgaria, pp. 141−149.

Shapkova, E.Y., Terekhov, A.V., Latash, M.L., 2011. Arm motion coupling during locomotion-like actions: an experimental study and a dynamic model. Motor Control 15, 206−220.

Shik, M.L., Iagodnitsyn, A.S., 1978. Reactions of cat hindbrain "locomotor strip" neurons to microstimulation. Neirofiziologiia (Neurophysiology) 10, 510−518.

Shik, M.L., Orlovskii, G.N., Severin, F.V., 1966. Organization of locomotor synergism. Biofizika 11, 879−886.

Shik, M.L., Severin, F.V., Orlovskii, G.N., 1967. Structures of the brain stem responsible for evoked locomotion. Sechenov Physiological Journal of the USSR 53, 1125−1132.

Sirota, M.G., Shik, M.L., 1973. The cat locomotion elicited through the electrode implanted in the mid-brain. Sechenov Physiological Journal of the USSR 59, 1314−1321.

Tunstall, M.G., Roberts, A., 1994. A longitudinal gradient of synaptic drive in the spinal cord of Xenopus embryos and its role in co-ordination of swimming. Journal of Physiology 474, 393–405.

Umberger, B.R., 2008. Effects of suppressing arm swing on kinematics, kinetics, and energetics of human walking. Journal of Biomechanics 41, 2575–2580.

Wilson, D.M., 1961. The central nervous control of flight in a locust. Journal of Experimental Biology 38, 471–490.

Windholz, G., 1987. Pavlov as a psychologist. A reappraisal. Pavlovian Journal of Biological Sciences 22, 103–112.

Zehr, E.P., Balter, E.J., Ferris, D.P., Hundza, S.R., Loadman, P.M., Stoloff, R.H., 2008. Neural regulation of rhythmic arm and leg movement is conserved across human locomotor tasks. Journal of Physiology 582, 209–227.

Part Three

Motor Control Concepts

Redundancy and Abundance

10

The words *redundancy* and *abundance* have many different meanings in various areas outside the fields of motor control and biomechanics. Before trying to define and analyze these terms in specific ways that would make them useful for analysis of biological movements, it makes sense to look at the meanings of these words in other areas and try to summarize the main similarities and differences in these two words.

Redundancy is defined as the use of words, components, or data that could be omitted without loss of meaning or function. For example, in engineering, redundancy is the duplication of critical components of a system with the intention of increasing its reliability in the form of a backup or fail-safe or the inclusion of extra components that are not strictly necessary to functioning, in case of failure in other components. In information theory, redundancy is the number of bits used to transmit a message minus the number of bits of actual information in the message. In linguistics, redundancy means information expressed more than once. Overall, this word means something extra, something that is not normally needed, but may become useful under special conditions.

Abundance is defined as a very large quantity of something, more than one needs, typically with a positive connotation of plentifulness and prosperity. There are more specialized meanings of abundance. For example, abundance as an ecological concept refers to the relative representation of a species in a particular ecosystem. Relative species abundance is calculated by dividing the number of species from one group by the total number of species from all groups. In number theory, abundance is a property of abundant numbers, that is, numbers for which the sum of its proper divisors is greater than the number itself. The integer 12 is the first abundant number. Its proper divisors are 1, 2, 3, 4, and 6 for a total of 16. The amount by which the sum exceeds the number is the abundance. For the number 12, abundance equals 4. Sūrat al-Kawthar ("Abundance") is the 108th sura of the Qur'an, which refers to a river where true believers will drink only once and will never experience hunger or thirst again. Overall, the word *abundance* is similar to *redundancy* in meaning something extra, but unlike redundancy, this something is being used joyfully.

The problem of motor redundancy is seen as a central problem of motor control. Nikolai Bernstein (1967) is usually credited for drawing attention to the fact that most everyday motor tasks may be successfully performed using a large number (even an infinite number) of solutions. He illustrated this notion using the example of pointing with the index finger at a target without moving the trunk. The human arm has at least seven major axes of joint rotation illustrated in Figure 10.1: three rotations in the shoulder joint, one in the elbow joint, two in the wrist, and one more shared between the wrist and the elbow (pronation-supination). In fact, this is an incomplete count because it ignores movement of the scapula and of the hand/finger joints. On the other hand, location of a point target can be described with three

Biomechanics and Motor Control. http://dx.doi.org/10.1016/B978-0-12-800384-8.00010-7

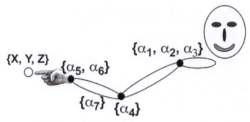

Figure 10.1 An illustration of the kinematic redundancy of the human arm. There are three axes of rotation in the shoulder (α_1, α_2, and α_3), one in the elbow (α_4), two in the wrist (α_5 and α_6), and one shared between the elbow and the wrist (α_7). Pointing tasks are associated with three constraints corresponding to target coordinates in the external space (X, Y, and Z). Modified by permission from Latash (2012).

coordinates. So, finding a joint configuration that brings the tip of the index finger into the target is equivalent to solving three equations with at least seven unknowns:

$$x = \sum_{i=1}^{7} k_{ix}\alpha_i$$

$$y = \sum_{i=1}^{7} k_{iy}\alpha_i \qquad (10.1)$$

$$z = \sum_{i=1}^{7} k_{iz}\alpha_i$$

where x, y, and z are target coordinates, α_i are joint angles, and k_i are expressions that link joint angles to endpoint coordinates given the anatomy of the limb. This problem has an infinite number of solutions. Nevertheless, each time a movement is produced, a single solution is being realized. How is this solution selected from the infinite set?

This problem is not limited to analysis of multijoint kinematics. Further we will review briefly similar problems that emerge at all levels of analysis of the human neuromuscular systems. They emerge when the number of degrees of freedom at the level of elements is greater than the number of constraints associated with typical actions. We will define degrees of freedom as variables at a selected level of analysis that can be changed independently, at least theoretically. This is not a trivial concept, and we will discuss to it later in this chapter.

10.1 Elements of history

Aristotle (384–322 BCE) was arguably the first to pay attention to a feature of biological movements called *coordination*. He equated coordination with harmony and attributed both to the design of creation. The notion of coordination is very close to the issues of redundancy and abundance. Indeed, a movement can be called coordinated only if there are other movement patterns that can solve the problem in a less coordinated or uncoordinated manner. So, the notion of coordination is applicable only to systems that can solve motor tasks in different ways, that is, to redundant systems.

For centuries, the issue of coordination of biological movements had been all but ignored despite the numerous examples of exquisite coordination (in athletes, dancers, musicians, and other highly trained professionals) and poor coordination (in patients with injuries to the neuromuscular system, persons with developmental abnormalities, healthy elderly, and normally developing clumsy toddlers). The revolution in the approach to coordination can be dated back less than 100 years to the famous experiment on blacksmiths performed by Bernstein in the mid-1920s (Bernstein, 1926). In that experiment, Bernstein equipped the blacksmiths with a set of electric bulbs attached to different body parts and filmed their professional movement of hitting the chisel with the hammer using an ingenious apparatus, a great-grandfather of the contemporary motion analysis systems (for details, see Bernstein, 1930). One of his main conclusions was that those perfectly trained subjects still used variable joint trajectories across repetitive hits to ensure a relatively invariant trajectory of the hammer in space. This experiment was followed by a series of studies of movements by experts in various fields, from concert pianists to elite runners and gymnasts.

Those studies led Bernstein to the formulation of the problem of motor redundancy, which is commonly referred to as the Bernstein problem (Turvey, 1990): How does the central nervous system (CNS) select specific patterns from the infinite sets equally able to solve the problem? Bernstein used the expression "elimination of redundant degrees of freedom" for the ability of the CNS to produce specific movement patterns in the presence of choice. He viewed the problem of elimination of redundant degrees of freedom as the central problem of motor control. Note that the formulation of the problem suggested a particular bias—it accepted axiomatically that the CNS selected unique solutions for the problems of motor redundancy. In other words, it adds missing equations to systems like Eqn (10.1) to make a unique solution possible or uses other approaches, for example, looking for an optimal solution based on some criterion.

Two approaches to the Bernstein problem in its classical formulation started to emerge in the middle of the twentieth century. The first was the idea that the CNS adds constraints to the neuromotor system and thus limits the area of possible solutions. Note that some of the constraints may be related to implicit task components such as preserving integrity of the body and not falling down. A classical example is the size principle of recruitment of motor units (known also as the Henneman principle; Henneman et al., 1965). The presence of many motor units within each muscle is one of the examples of motor redundancy if one considers overall level of muscle activation as the task variable and frequency of firing of individual motor units as degrees of freedom (Figure 10.2). For any desired muscle activation level between zero and maximal possible one, there exist an infinite number of combinations of motor units recruited at specific frequencies that can match the required activation level. The size principle states that motor units are recruited in a specific order, determined by the size of their alpha-motoneurons (three characteristics of a motor unit change in parallel, the size of the body of the alpha-motoneuron, the diameter of its axon, and the number of innervated muscle fibers), from the smallest to the largest (Figure 10.2).

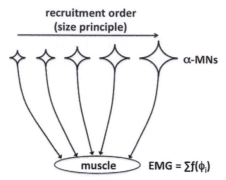

Figure 10.2 Alpha-motoneurons are recruited according to the size principle. The overall muscle activation levels (its electromyogram, EMG) is a function of the frequency of firing of the recruited motor units.

This principle does not specify, however, the frequencies of firing of the recruited motor units. As a result, it does not lead to a unique solution for the problem, but it does impose constraints on the possible patterns of recruitment and may be viewed as alleviating the problem of motor redundancy.

The other approach is optimization (for a review of early studies, see Nelson, 1983). We are going to discuss it in detail later in the chapter. The main idea of this approach is to accept a function (addressed as a *cost function* or an *objective function*) of variables produced by individual degrees of freedom and select a solution that minimizes (or maximizes) the value of this function. Assuming this approach implies that either (a) the brain really performs computational operations involved in optimization or (b) the brain does not optimize anything, it works according to some unknown physical principles but researchers can describe its behavior using optimization methods. This approach can lead to a unique solution, but its application in research hinges on a few assumptions that look rather arbitrary, such as assuming a particular cost function.

The next important step in addressing the Bernstein problem was made by the famous mathematician Israel Gelfand (1913−2006) and his younger colleague, the brilliant physicist Michael Tsetlin (1924−1966). Gelfand and Tsetlin (1961, 1966) introduced elements of formal description of redundant systems involved in the production of movements based on the principle of nonindividualized control—elements of the system were not assumed to be moved by the controller as limbs of a marionette but united into task-specific or intention-specific *structural units*. The presence of structural units was expected, in particular, to alleviate the problem of redundancy. Introducing this notion was the first step toward the *principle of abundance* formulated about 50 years later (Gelfand and Latash, 1998, 2002). Gelfand and Tsetlin also introduced the *principle of minimal interaction*, which stated that the interaction between elements at the lower level of the hierarchy was organized so as to minimize the input to each of the elements. For the sake of illustration, imagine a set of active neurons with only excitatory synapses among the cells that form a closed system, that is, they do not receive inputs from other parts of the brain and sensory organs. If one cell becomes inactive it ceases to send excitation to other cells. This may be sufficient to deactivate another cell. As a result, a cascade of deactivation occurs and the activity of the entire network decreases or even comes to a complete halt. Such a "physiological" minimization may be associated with a decrease in many biomechanical variables, such as, for instance, force, power and energy expenditure. As a result of evolution and learning, the "physiological" minimization could be associated with minimal values of some biomechanical variables to a larger degree than with minimization of other variables.

Overall, this line of thinking switched the emphasis from trying to find a unique (maybe optimal) solution for each problem of motor redundancy to defining rules leading to families of solutions equally able to solve the task. Realization of a particular solution from such a family may happen by chance, based on some probability distribution, under the action of time-varying factors that are not controlled in typical studies (e.g., levels of excitability of individual neurons).

10.2   Current terminology

10.2.1   Elemental variables and degrees of freedom

The notion of redundancy can only be applied when the number of degrees of freedom at a selected level of description of a system is larger than the number of degrees of freedom that specify its overall functioning. If we assume that performance of the neuromechanical system for the production of voluntary movements is defined by a task, it becomes necessary to postulate the existence of at least two hierarchical levels involved in movement production, a hierarchically higher level specifying the task and the hierarchically lower level implementing the task with more than necessary elements. To be more exact, for a system to be considered redundant, the number of variables produced by the elements at the lower level has to be larger than the number of variables that specify the task. These, so-called *elemental variables* are equivalent to degrees of freedom, both representing independent coordinates that specify a state of the system, at the lower level of the assumed hierarchy.

The notion of elemental variables is nontrivial because of a possible, sometimes hidden, coupling among these variables that is not task specific but can be seen in any action. If two or more elemental variables are strongly coupled in an obligatory fashion, this two-element system can be described as a single degree-of-freedom system. For example, in studies of movement kinematics, rotations about separate axes of joint rotation are commonly viewed as elemental variables (Scholz and Schöner, 1999; Scholz et al., 2000). Note, however, that joints are coupled by biarticular and multiarticular muscles. In addition, there are interjoint and even interlimb reflexes (reviewed in Nichols, 1994, 2002) that also make individual joint rotations nonindependent of each other. Nevertheless, it is commonly assumed that humans can move any single joint at a time if they wish to do so; while this assumption sounds intuitively correct, unfortunately, no experimental evidence for or against this assumption exists.

Coupling among variables produced by apparent elements can be obvious, for example, during finger motion or force production. Finger force and motion interdependence has been documented in multiple studies and addressed as enslaving or lack of individuation (Kilbreath and Gandevia, 1994; Zatsiorsky et al., 2000; Schieber and Santello, 2004; Kim et al., 2006). It has been assumed, however, that the phenomenon of enslaving by itself does not lead to an obligatory change in the number of elemental variables, that is, hypothetical variables manipulated at some level of the CNS during multifinger actions. These hypothetical variables have been addressed as *finger modes* and assumed to reflect desired involvement of a finger in a multifinger task. For example, one can try to press with one finger only or with two, three, or four fingers. In all these situations, a smaller or larger subset of finger modes will be used, while all four fingers will always produce force. So, depending on the level of analysis, a task may be seen as involving one, two, three, or four elemental variables at the level of finger modes, while the number of elemental variables at the level of finger forces remains unchanged, always four.

A similar notion has been introduced for analysis of multimuscle systems. Since the end of the nineteenth century, it has been assumed that brain structures do not control muscles one by one but unite them into groups (Hughlings Jackson, 1889). Such muscle groups have recently been addressed as *muscle synergies* (see Chapter 16 for more details; for a review, see Tresch and Jarc, 2009) or *muscle modes* (Krishnamoorthy et al., 2003). Within a muscle mode (we will use this term to avoid confusion with synergies described later), a change in a single variable by a hypothetical hierarchically higher level leads to proportional changes in activation of all the muscles involved in that mode.

As the presented examples suggest, identification of elemental variables is not always trivial. We may safely assume, however, that such variables exist and that their number allows multiple solutions for typical tasks reflected, in particular, in the variable involvement of elements during repetitive attempts at virtually any task.

10.2.2   More examples of the problem of motor redundancy

The examples of multijoint reaching and producing a desired level of muscle activation by a set of motor units have already been described briefly. Similar problems of redundancy emerge at other levels of analysis of the system for movement production. For example, consider the task of producing a value of static joint torque (see Chapter 1) in a simple joint with only one rotational degree of freedom (Figure 10.3). Any joint is

$$M_{TOTAL} = M_1 + M_2 + M_3 + M_4 + M_5 + M_6$$

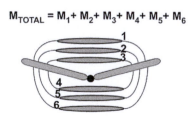

Figure 10.3 Redundancy at the single-joint level. Moment of force in a joint (M_{TOTAL}) is the algebraic sum of the moments of force produced by the individual muscles ($M_1 - M_6$).

spanned by at least two muscles with opposing actions, while typically joints are spanned by multiple muscles. For example, the elbow joint is spanned by at least three flexors and three extensors (as illustrated in Figure 10.3). Each muscle produces a moment of force that contributes to the joint torque; hence, potentially the problem of sharing a value of joint torque among a set of n muscles is equivalent to solving an equation with n unknowns:

$$T_J = \sum_{i=1}^{n} d_i F_i, \tag{10.2}$$

where T_J is the joint torque, F_i are forces of individual muscles, and d_i are corresponding lever arms.

A similar problem emerges at a single-neuron level. Indeed, whether the neuron will produce an action potential or not depends on whether its membrane potential reaches a particular threshold value or not. Changes in the membrane potential are produced by numerous, potentially hundreds or even thousands, of synaptic inputs, both inhibitory and excitatory. The dimensionality of this problem is truly enormous.

10.2.3   State redundancy and trajectory redundancy

So far, we have reviewed examples of the problem of motor redundancy that emerge at the level of steady states. All these problems are characterized by a mismatch between the dimensionality at the level of elements and the number of task-related constraints. Another group of problems of redundancy emerges when one considers trajectories leading from an initial state to a target state of a system. Consider a single joint with a single axis of rotation (Figure 10.4). If the task is to move the joint from an initial to a final position, there are an infinite number of joint angle time profiles (we will address these as *trajectories*) able to perform the task successfully. Three of them are illustrated in Figure 10.4. Only one trajectory is realized in each attempt to perform this task. In the aforementioned classical study of blacksmith by Bernstein, two major observations were made. First, the intertrial variability of the trajectories of

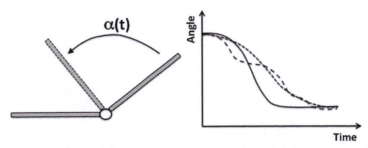

Figure 10.4 An illustration of the trajectory redundancy. A single-joint movement from a starting position to a target may be associated with variable joint angle time profiles (trajectories). Three possible trajectories are illustrated with different lines.

the markers placed over the main arm joints was claimed to be higher than the variability of the trajectories of the hammer. Second, even the hammer trajectory was not perfectly reproduced across trials. This happened despite the fact that the subjects in this study had been perfectly trained over many years to perform this task.

The two types of the problem of motor redundancy, state redundancy and trajectory redundancy (we use the term *trajectory* in the meaning of a time profile of a mechanical variable), have been discussed within the two main approaches. The first approach is based on the assumption that a single solution is selected in each trial from the infinite set based on some criteria. For the problem of state redundancy, this approach is equivalent to elimination of redundant degrees of freedom. The second approach is based on the principle of abundance. It assumes that no single solution is selected but individual solutions are realized by chance from a family of solutions equally able to solve the problem.

10.2.4   Marginal redundancy

There are tasks and associated levels of analysis when the problem of motor redundancy seems to disappear. Consider a task involving two elements and two constraints. For example, imagine that two fingers are involved in the task of accurate total force production. In addition, the fingers act on the opposite sides of a pivot and they have to produce equal in magnitude and opposite moments of force (Figure 10.5). Formally, this problem corresponds to solving two equations with two unknowns:

$$F_{TOT} = F_1 + F_2$$
$$M_{TOT} = F_1 d_1 + F_2 d_2 = 0$$

(10.3)

where M_{TOT} stands for the total moment of force, and d_1 and d_2 are the lever arms of the two forces. Each of the two equations has an infinite number of solutions

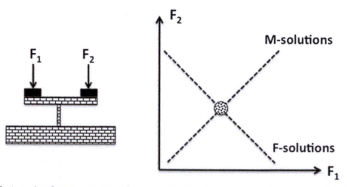

Figure 10.5 A task of accurate total force production by two effectors is associated with a solution space (F-solutions) in the space of elemental variables, F1 and F2. A task of producing a zero total moment of force is associated with another solution space (M-solutions). Since both constraints allow small errors, there is an infinite number of solutions (the small circle). This is a marginally redundant task.

corresponding to all the points on the two slanted lines shown in Figure 10.5. The point of intersection of the two lines is the only solution that satisfies both criteria.

In actual motor tasks, however, task constraints are not absolute and allow making errors. These errors, however small they may be, create subspaces in the space of elemental variables that contain an infinite number of points corresponding to an infinite number of solutions that perform the task within an acceptable error margin. For the task illustrated in Figure 10.5, total force has to be very close to F_{TOT}; however, small deviations, for example, those that are under the level of detection by the available sensors, are inevitable. The $M_{TOT} = 0$ constraint is also not absolute, for example, because the area of contact with the pivot is never zero. This means that all the points within an area illustrated in Figure 10.5 by the circle are acceptable solutions for the task. Since the number of solutions is infinite, the task belongs to the group of redundant ones. Tasks that are formally nonredundant but practically redundant have been addressed as *marginally redundant* (Latash et al., 2001).

10.2.5 Principle of abundance

Abundance is a relatively new term in the field of motor control. It follows rather directly the traditions set by Gelfand and Tsetlin (1966) who suggested two ways of solving problems. One of them was based on what they addressed as the *principle of economy*, that is, using a minimal number of elements to avoid the problem of redundancy. This approach is closely related to the idea of elimination of redundant degrees of freedom. Indeed, eliminating redundant degrees of freedom leaves only a minimal, necessary number of degrees of freedom available, the problem becomes nonredundant, and a unique solution emerges. The alternative is using many more elements than necessary for each task, never constraining the system to a single solution, and allowing elements to contribute to the task in varying patterns. This alternative was later reformulated as the *principle of abundance* (Gelfand and Latash, 1998; Latash, 2012)

The principle of abundance considers the apparently redundant degrees of freedom as useful and even vital for such aspects of motor behavior as its ability to perform several tasks simultaneously and stability (discussed in more detail in Chapter 11). According to this principle, hierarchically higher levels of the CNS use all the available degrees of freedom and facilitate families of solutions equally able to solve the task. This is accomplished by imposing task-specific constraints on the hierarchically lower levels and allowing solutions to emerge given the actual state of the system "body + environment," which never repeats exactly over repetitive trials leading to the natural patterns of variability. As discussed later (and also in Chapter 11), while the concept of abundance is based on the body morphology, its application to specific tasks is based on neural mechanisms that adjust the abundance afforded by the design of the body to the tasks and intentions.

The two different approaches can be illustrated using, as an example, the organization of a scientific laboratory composed of researchers (elements). Assume that the laboratory has to solve a specific problem, for example, to design a new stadium.

The laboratory may be organized based on the principle of economy. Then, a minimal number of researchers are hired, and each researcher is assigned a unique, specific function, so that redundancy is avoided. Such a laboratory would be able to perform the assigned task, but it will be unable to perform any other task and it would fail at the original task if one of the members of the laboratory suddenly falls ill or retires. Based on the principle of abundance, more than necessary researchers are hired and asked to solve the problem. Each researcher is expected to contribute to the team's effort. Such a group might take more time to solve the original task but it may be expected to show important advantages. First, it may be expected to reorganize successfully if the task is modified, for example, if it involves designing a new shopping mall. Second, the group may also be expected to continue successful functioning if one of the members quits.

10.3   Optimization

The basic idea of optimization is readily compatible with the aforementioned principle of economy applied to a redundant system. If multiple (an infinite number of) solutions are possible for a task, a single solution is selected based on an a priori criterion. In research such criteria are typically formulated as requirements to satisfy the task constraints and simultaneously minimize (or maximize) the value of a cost function (also addressed as an objective function) of the redundant set of elemental variables. While the range of cost functions used in the field of motor control is rather broad, they share a common underlying assumption—the CNS is typically considered lazy and frugal; it tries to spend as little as possible of a resource that it views as precious.

Optimization approaches have been used with respect to problems of both state redundancy and trajectory redundancy. A typical example of the former would be defining how to share a desired pressing force across the four fingers of the hand. A typical example of the latter would be defining a trajectory of the end effector, for example, of the hand, during a reaching movement from a certain initial position to a certain target within a specific movement time.

One of the common approaches is *minimization of the norm* of the solution, that is, minimization of the sum $\sum \Delta EV^n$, where ΔEV stands for a change in an elemental variable associated with an action and n is the power. This approach may be seen as minimizing the estimated price of moving all the involved elemental variables to a final state, assuming that moving an elemental variable is associated with a price increase proportional to its change in power n. For $n = 2$, this criterion is equivalent to minimizing the integral of the Euclidean distance between successive points along a trajectory connecting the initial and final states in the multidimensional space of elemental variables. In robotics, this method is commonly addressed as Moore–Penrose pseudoinverse (e.g., Whitney, 1969).

Consider, for example, two fingers taking part in the task of accurate change in the total force illustrated in Figure 10.6. The fingers increase the total force starting from a certain force combination (point 1 in Figure 10.6). The minimum-norm

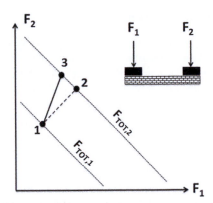

Figure 10.6 A two-finger accurate force production task. A change in the total force may be associated with a minimum-norm trajectory (from point 1 to point 2). Typically, however, the total force sharing is preserved leading to a different trajectory (from point 1 to point 3).

criterion generates a solution shown by the thin, dashed line to point 2, which is orthogonal to the solution space (shown by thin slanted lines for the initial and final total force values, $F_{TOT,1}$ and $F_{TOT,2}$). Note that this solution is observed in experiments with two-finger force production only if the initial sharing of the total force between the fingers is 50:50. Otherwise, the sharing is preserved leading to a solution that deviates from the minimum-norm one (shown by the solid, thick line to point 3, Li et al., 1998).

Within this approach, any movement of a redundant system may be represented as the superposition of two components, a movement along the optimal direction defined using the minimum-norm criterion (e.g., using the Moore–Penrose pseudoinverse), and a movement orthogonal to that direction. These two components are commonly addressed as *range-space motion* and *self-motion*. By definition, self-motion of a system includes changes in the values of the elemental variables that lead to no changes in the task-related performance variable to which the elemental variables contribute. For example, if you touch your nose with the index fingertip and then move the arm joints without losing contact with the nose, the joints will demonstrate self-motion with respect to this rather arbitrary task. The main assumption of the minimum-norm approaches in that the controller tries to minimize self-motion, which seems a reasonable assumption related to avoiding wasting energy.

Recent studies produced data both compatible with the minimum-norm approach and contradicting it. For example, minimum-norm in the space of hypothetical commands to fingers (finger modes, Danion et al., 2003) was able to account for nontrivial force distribution patterns during static multifinger tasks (Zatsiorsky et al., 2002). Note that this application addressed the state redundancy problem. On the other hand, large amounts of self-motion were observed in experiments with unexpected perturbations applied during multijoint reaching tasks (Mattos et al., 2011, 2013). Taken together, these results suggest that minimum-norm may be a useful criterion but its area of applicability is likely limited.

10.3.1 Kinematic cost functions

Several cost functions have been suggested to describe movements to a target produced in the presence of trajectory redundancy. One of them is the *minimum-time* criterion (Nelson, 1983). It can be illustrated with the example of driving a car as fast as possible to a new location and stopping there. It is intuitively clear that the optimal strategy would be to press the gas pedal to the floor, keep it pressed until some optimal time, and then press the brake pedal to the floor (also known as the teenager-driving principle). The optimal switch time from acceleration to deceleration is defined by many factors including friction, relative power of the engine, and the brakes. Assume that the task is to move a joint with one degree of freedom from an initial to a final position. The minimum-time criterion would suggest applying the highest possible acceleration, waiting until the optimal switching time, and then applying the highest possible deceleration (Figure 10.7).

For a purely inertial system with equally powerful agonist and antagonist muscle groups, the optimal switching time is exactly in the midpoint of the movement. This strategy leads to the velocity time profile illustrated in the right panel of Figure 10.7. If the opposing muscle groups have different maximal torque capabilities, the optimal time of switching from acceleration to deceleration would change. Changes in the optimal switching time could also be expected from a system that is not purely inertial; for example, high friction (damping, see Chapter 3) would help decelerate the movement, and the optimal switching time would shift toward the second part of the movement. Note that the time profile shown in Figure 10.7 is very different from the typical bell-shaped velocity profiles observed during natural movements.

Much more realistic velocity time profiles are generated by the *minimum-jerk* criterion (Hogan, 1984; Flash and Hogan, 1985), which is arguably the most commonly used cost function in studies of movement kinematics. According to this criterion, a trajectory is selected that minimizes the integral of the third time derivative of the coordinate (jerk, $J = \mathrm{d}^3x/\mathrm{d}t^3$) squared over movement time:

$$J = \frac{1}{MT} \int_0^{MT} \left(\mathrm{d}^3x/\mathrm{d}t^3\right)^2 \mathrm{d}t \qquad (10.4)$$

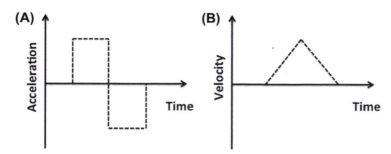

Figure 10.7 Minimum-time criterion leads to a "bang–bang" control for a purely inertial system. The acceleration and velocity time profiles do not look like those of natural trajectories.

where MT stands for the movement time. Application of this criterion leads to a solution in the form of a sixth-order polynomial function. The velocity profiles generated by the minimum-jerk criterion are bell shaped, resembling ones seen in natural movements. One has to keep in mind that application of the criterion, Eqn (10.4), shares a common feature with many other optimization approaches, namely the requirement of knowing movement time in advance. This feature makes it unlikely that this criterion can be used in planning or executing a movement.

The minimum-jerk criterion was originally associated with minimizing joint wear, but this association fails if one considers multijoint movements. While the minimum-jerk criterion remains the method of choice if one wants to solve the problem of trajectory redundancy, this criterion does not deal with the problem of state redundancy. In particular, applying the minimum-jerk criterion to generate a trajectory of the endpoint of a multijoint limb does not necessarily generate minimum-jerk trajectories in individual joints. Moreover, individual joints sometimes show large-amplitude nonmonotonic trajectories, from an initial joint position to a final joint position that may be very close to each other. Such trajectories definitely do not comply with the minimum-jerk criterion applied at the joint level and do not lead to minimization of joint wear.

10.3.2   Kinetic cost functions

Application of cost functions formulated in terms of forces and force-based computations to describe human movements implies using a mechanical model of the moving effectors and addressing, albeit implicitly, the problem of inverse kinetics. Such models are typically very much simplified, and accurate estimates of their parameters are all but absent (see Chapters 2 and 3). As a result, relatively few researchers have explored kinetic cost functions as optimization criteria.

One of the early approaches considered minimization of the impulse (time integral of force) as the optimality criterion. For the task of moving a purely inertial effector from an initial to a final position, the *minimum-impulse* criterion produces a solution illustrated in Figure 10.8—the effector is accelerated at the maximal rate to a certain velocity, kept at this velocity to a certain time, when deceleration at the maximal rate is applied to stop the movement. If other mechanical properties of the effector

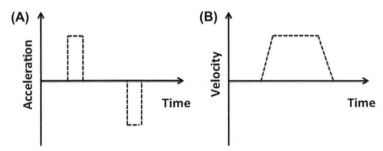

Figure 10.8 The minimum-impulse criterion leads to a trapezoidal velocity time profile when applied to a purely inertial system.

are considered, the timing and amplitude of the force pulses are adjusted. For example, in a system with large energy dissipation (due to high damping), the deceleration force pulse is much smaller than the acceleration pulse. The velocity profiles generated by this criterion are typically very different from the profiles of natural movements, for example, the profile for a purely inertial system shown in Figure 10.8.

There was an attempt to extend the minimum-jerk criterion to kinetic variables. This attempt resulted in the *minimum-torque-change* model (Uno et al., 1989):

$$C = \frac{1}{MT} \int\limits_0^{MT} (\mathrm{d}T/\mathrm{d}t)^2 \mathrm{d}t \tag{10.5}$$

where C is the cost function and T is torque. For a purely inertial one-element system, the criteria, Eqn (10.4) and (10.5), produce identical results because rotational acceleration is proportional to torque. Hence, torque change is proportional to change in acceleration, which is jerk. The two approaches start to differ when more realistic models of moving systems are considered with mechanical properties not limited to inertia, for example, when length- and velocity-dependence of muscle force production is considered (see Chapters 2 and 3). In such cases, the minimum-jerk criterion produces more realistic predictions. Just like the minimum-jerk criterion, minimum-torque change fails when movements of multijoint systems are considered.

10.3.3   Physiological cost functions

One of the relatively broadly used cost functions for the problem of sharing force among multiple muscles has been related to *minimization of fatigue*. One of the criteria was based on experimental observations that linked endurance time (T_E) (time during which a muscle can produce a certain force level) to muscle stress, that is, the ratio of force to the physiological cross-sectional area ($F/PCSA$): $T_E = (F/PCSA)^{-3}$ (Crowninshield and Brand, 1981). The inverse of T_E may be defined as muscle fatigue, and the minimum fatigue criterion can be formulated using the following cost function:

$$C = \sum_{i=1}^{n} (F_i/PCSA_i)^3 \tag{10.6}$$

where n stands for the number of muscles. This criterion is a particular example of the minimum-norm criteria.

Another physiological cost criterion is based on minimization of energy expenditure. For example, a minimum metabolic cost criterion (Alexander, 2002) was formulated as:

$$C = \sum_{i=1}^{n} \alpha_i^p F_i^{max} V_i^{max} \phi(V_i/V_i^{max}) \tag{10.7}$$

where function ϕ determines the metabolic cost; $F_i^{\max}$ and $V_i^{\max}$ are known maximum force and maximum velocity of each of the n muscles, respectively; V_i is instantaneous velocity of the i-th muscle; and α_i $(0 \leq \alpha_i \leq 1)$ is the unknown normalized activation of the i-th muscle, sought by minimizing the cost function C. Minimizing energy expenditure has an evident biological advantage. It is mainly important for actions that last for long times, such as locomotion, not for a discrete movement. Indeed, it has been suggested that, at a given speed of ambulation, people and animals select the stride length and frequency at which the energy expenditure is minimal (Ralston and Lukin, 1969).

An attempt has been made to link optimization to changes in neural and mechanical variables associated with the ideas of control with shifts of equilibrium positions (Hasan, 1986; see also Chapter 12). This *minimum-effort* criterion associated effort with the integrated squared velocities of the equilibrium positions over movement time taking into account a parameter representing joint stiffness (see Chapter 2):

$$C = \int\limits_0^{MT} k(\mathrm{d}\beta/\mathrm{d}t)^2 \mathrm{d}t \tag{10.8}$$

where β is equilibrium position, MT is movement time, and k stands for joint apparent stiffness. The application of this criterion allowed estimating optimal values of joint apparent stiffness, but its association with variables β that could not be easily measured or verified independently limited its further development.

10.3.4 Psychological cost functions

One of the optimization approaches considered discomfort as a variable minimized during voluntary movements (Cruse and Brower, 1987; Cruse et al., 1990). In those studies, subjects were asked to produce psychophysical estimates of the degree of discomfort for the whole biomechanical range of joint angles, for each joint separately. These functions typically showed low discomfort scores in the mid-range and higher scores as the joint approached its limit of rotation. In the mid-range, the scores changed very little with joint rotation and hence showed low sensitivity to joint motion. Further, the discomfort scores were summed across joints and a trajectory with the minimal overall score was considered optimal.

The application of this method showed good fit to natural human reaching movements suggesting that humans did avoid uncomfortable joint configurations. The method suffered from a few obvious drawbacks. One of them was the mentioned low sensitivity of the discomfort scores to joint rotation in the mid-range. The second is the method of obtaining discomfort scores. Discomfort ratings for positions in a joint may depend on positions in other joints, for instance, due to the links provided by biarticular muscles. For example, one and the same position in the elbow joint may be perceived as more or less comfortable for different shoulder positions. This criterion also suffers from circular logic. Indeed, during the lifetime, humans make numerous movements, most commonly in a way that they consider comfortable. So, joint rotations within that range are conditioned to be associated with low discomfort. It is of no surprise, therefore, that joint positions perceived as comfortable are close to the ones used most frequently during typical movements.

10.3.5 Complex cost functions

Many published cost functions produce movements that look realistic. This should not be very surprising; otherwise, those studies would probably not have been published. Taken together, these observations suggest that natural human movements do not violate any of those criteria by much. Whether one of those criteria dominates for specific motor tasks remains an open question. If several criteria are united into a single cost function with modifiable weights, the resultant model becomes even more powerful and able to account for movement patterns over a broader range of tasks (Seif-Naraghi and Winters, 1990).

A particular example of a complex cost function addressing both classes of the problem of motor redundancy, state redundancy, and trajectory redundancy (we imply here both selection of a trajectory from a potentially infinite set and selection of a particular time profile of motion along that trajectory) has been developed based on ideas of posture planning (Rosenbaum et al., 1993). Within this model, kinematic planning a movement to a target involves two steps. First, a search within the space of joint configurations (postures) stored in memory is performed, and the total cost is computed for each configuration as the sum of the travel cost to this configuration and the resultant spatial error cost. The travel cost is defined as:

$$C = \sum_{i=1}^{n} \xi_i(\alpha_i/MT) \tag{10.9}$$

where n is the number of joints involved in the planned movement, α is the angular displacement in a joint, MT is movement time, and ξ is a factor that defines cost of moving a joint given many factors including inertial load, apparent stiffness, and damping. Further, a single target posture is selected as a weighted average of all the postures with larger weights assigned to postures with smaller total costs. Note that averaging postures in not an obvious procedure. For instance, averaging individual joint angles may result in a posture that is anatomically impossible. So, a more complex procedure has to be used, such as finding a posture with a minimal sum of squared deviations in the joint angle space from the original set of postures. During the second step, movement into the selected target is performed based on a criterion that defines time profiles of instantaneous velocities in individual joints as functions of their distance to the target posture and also an inertia-related index.

10.3.6 Analytical inverse optimization

All the mentioned optimization methods start from accepting a cost function based on some intuitive considerations, frequently biased by personal views and preferences of individual researchers. Indeed, minimizing energy expenditure, fatigue, discomfort, joint wear, etc. all sound like reasonable assumptions with respect to natural movements. This element of subjectivity and arbitrariness has been a weak spot of the approach. Recently, a step has been made toward an objective estimation of a cost function (assuming that it exists). This method, termed *analytical inverse optimization* (ANIO), attempts to find an analytical function that is likely minimized by a system. The method requires multiple observations of the system's behavior over a range of tasks.

So far, two approaches to this problem have been used. One of them performs a two-stage search, first among classes of analytical functions and then among parameters within each class (Bottasso et al., 2006). This method requires a lot of computational power. The other method is limited to a particular class of cost functions, namely functions additive with respect to the elemental variables (Terekhov et al., 2010). This approach is based on the Lagrange principle, which allows estimating the derivatives of unknown cost functions based on observations of performance, assuming that each action obeys the same unknown optimization principle.

So far, this method has been applied successfully only to the problems of statics of parallel kinematic chains, specifically to multifinger force and moment-of-force production (a problem of state redundancy). It has shown sensitivity to fatigue, healthy aging, and neurological disorders (Park et al., 2010, 2011, 2012). On the other hand, the method failed in attempts to apply it to force production by serial kinematic chains, for example, the hand force and moment production on a fixed handle (Xu et al., 2012).

10.4   Optimal feedback control

The goal of optimal control, a branch of mathematics, is to find a way to control a system, which changes in time, in such a way that certain criteria of optimality are satisfied. Assume that the system of interest can be described with a set of variables (state variables) that change in time, $\mathbf{X}(t)$. The rate of change in these variables, $d\mathbf{X}(t)/dt$, is defined by their initial values ($\mathbf{X}_0$ at some instant of time, $t = 0$) and by the current values of the control variables, $\mathbf{C}(t)$:

$$\frac{d\mathbf{X}(t)}{dt} = f[\mathbf{X}(t), \mathbf{C}(t), t], \quad \mathbf{X}(0) = \mathbf{X}_0. \tag{10.10}$$

If one knows $\mathbf{C}(t)$ over a time interval of interest, for example, from 0 to T, the initial state of the system $\mathbf{X}_0$, and the form of the function f, Eqn (10.10) can be solved to find $\mathbf{X}(t)$. It is possible to choose a time profile $\mathbf{C}(t)$ such that a cost function J is minimized over the same time interval:

$$J = \int_0^T F[\mathbf{X}(t), \mathbf{C}(t), t]dt + S[\mathbf{X}(T), T]. \tag{10.11}$$

In this equation, F is a function reflecting the costs associated with changes in both $\mathbf{C}(t)$ and $\mathbf{X}(t)$. For example, this function may reflect any of the cost functions reviewed in the earlier sections, such as those related to energy expenditure, fatigue, effort, etc. The S-function in Eqn (10.11) gives the so-called *salvage value* of the final state $\mathbf{X}(T)$. Its purpose is to make sure that the control is not only minimizes J but also leads to a movement goal. For example, the salvage function does not allow the trivial solution of minimization of joint motion, which is not moving at all. There are constraints imposed on each component of J, such as limiting the $\mathbf{X}(t)$ and $\mathbf{C}(t)$ variables to a range compatible with the human anatomy and physiology.

A method of *optimal feedback control* was applied to address the control of a redundant system (Todorov and Jordan, 2002). A cost function, similar to the function J in Eqn (10.11), was formulated in such a way that it included a measure of neural effort related to changes in $\mathbf{C}$ variables and a measure of accuracy of performance. The model minimized the sum of two components converted to commensurate units with appropriately selected coefficients. One of them was the squared difference between a task-specific performance variable and its required value. The other was the variance of the control signals to individual effectors during task execution. Within this model, the controller recomputes a new desired trajectory in the redundant space of elemental variables at every moment in time. If there are deviations in the trajectories of elemental variables from the previously computed ones, the deviations are corrected only if they interfere with important performance characteristics that lead to an increase in the cost function. If the deviations are within the self-motion space with respect to that variable, no corrective actions are taken.

This approach offers a particular method of producing flexible behaviors leading to a desired goal, which is one of the characteristic features of voluntary movements, for example, as in the mentioned Bernstein's study of movements by professional blacksmiths described earlier (see also Chapter 11). Figure 10.9 illustrates the task of producing a constant output with two noisy effectors starting from a relaxed state. The effectors, on average, contribute equally to the output. Each point in this graph shows a combination of effector forces in an attempt to perform this task.

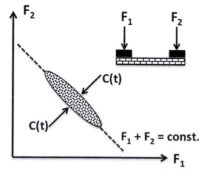

Figure 10.9 Optimal feedback control can account for nonspherical data distributions across trials, for example, in a two-finger accurate total force production task. Control signals, C(t), change only when changes in the elemental variables (finger forces) lead to changes in the total force, not when such changes leave total force unchanged (parallel to the solution space shown with the dashed line).

The controller allows relatively large deviations of the points along the slanted line corresponding to perfect performance of the task—these deviations do not add to the component of the cost function related to performance error, so no variance (changes) in control signals is spent. In contrast, deviations orthogonal to the line introduce errors in performance and lead to an increase in the function $\mathbf{J}$; then, control signals, $\mathbf{C}(t)$, are used to reduce these deviations.

The optimal feedback control approach has clear advantages as compared to more traditional optimization approaches. In particular, it can account for the patterns of natural variability observed during movements by redundant systems. On the other hand, this approach shares a few features with the earlier optimization approaches that may be viewed as drawbacks. Among them is the assumption of neural computations and the necessity to know movement time before action can be initiated, since the computation of the projected cost involves integration until movement termination, as in Eqn (10.11).

10.5   Abundance in movements

The mentioned classical experiment of Bernstein on blacksmiths (see Section 10.1) was arguably the first strong evidence for the principle of abundance. Indeed, in that experiment, the best-trained subjects, professional blacksmiths, showed large intertrial variability in the joint trajectories during their professional movement organized to keep variability of the hammer trajectory relatively invariant. This observation strongly suggested that, even after years of practice, elements within a mechanically redundant system do not converge on a single, optimal, solution but continue to show large variability organized in a task-specific way.

This feature of movements performed by mechanically redundant systems was later termed *error compensation* (Latash et al., 1998). This term implies that, during repetitive movements, if one of the elements shows a deviation from its preferred trajectory (estimated as the average across trials trajectory), other elements are also likely to show deviations from their preferred trajectories organized to keep the trajectory of a task-specific performance variable relatively unaffected. Several experiments provided support for error compensation. For example, when subjects performed a planar quick-pointing task with the shoulder, elbow, and wrist joints, intertrial variance of coordinates of the marker placed over the wrist was larger than the variance of the pointer-tip coordinates (Jaric and Latash, 1998). Note that variance of the wrist marker coordinates was affected by variability of the rotations in the shoulder and elbow joints only. Variance of the pointer-tip coordinates was expected to be larger due to two factors. First, the variability in the proximal joint rotations was expected to lead to larger spatial variability of the pointer tip because of kinematic reasons—it was farther away from the axes of rotations. Second, wrist rotation variability was expected to contribute to the pointer-tip coordinate variance. The fact that the pointer-tip coordinates showed a smaller variance than the wrist marker variability suggested strongly that there was substantial covariation of the wrist rotation with the rotations in the elbow and shoulder joints organized to reduce the pointer-tip error.

In another study, the subjects were instructed to produce constant total force while pressing with three fingers on individual force sensors (Latash et al., 1998). Then, they were asked to tap with one of the fingers. During the tap, the force of the tapping finger dropped and became zero. At the same time, the forces of the other two fingers increased in such a way that the total force remained nearly unchanged (the changes were on the order of 5%). Since those experiments, a large number of studies were performed documenting covariation among elements within redundant systems organized to stabilize (reduce variance of) task-related performance variables, to which all the elements contributed (reviewed in Latash et al., 2007; Latash, 2008).

Taken together, these results suggest that no single, optimal solution is being selected during performance of motor tasks by redundant systems. Rather, families of solutions are facilitated and organized in a way that reduces variance of potentially important performance variables. On the other hand, experiments show that while human subjects do not converge on a single solution, they use a relatively narrow range of solutions for typical tasks. For example (Figure 10.10), during two-finger accurate force production, over repetitive trials, subjects show clouds of data points elongated along the solution space (the dashed, slanted lines in Figure 10.10). These clouds are examples of covariation in the elemental variables (finger forces) that helps reduce variance of the total force. Such clouds, however, are relatively small and may be centered on different points on the solution space in different persons (compare the A and B ellipses in Figure 10.10). Moreover, when a different level of force has to be produced, corresponding to a different solution space (from $F_{TOT,1}$ to $F_{TOT,2}$), the cloud of data points typically shifts while keeping the relative contributions of the two fingers relatively unchanged (compare ellipses A1 − A2 and B1 − B2). Such observations suggest that, in addition to showing error compensation, the CNS facilitates solutions within a limited subspace, possibly selected based on an optimization principle. So, ideas of optimization and abundance are not mutually exclusive. Indeed, optimization may be used to define a preferred point in the space of elemental variables

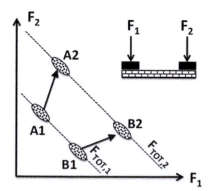

Figure 10.10 Different persons may show different average locations of data point distributions (shown by ellipses A and B) in a two-effector accurate total force production task. When total force changes, commonly, the ellipses shift to a new solution space ($F_{TOT,2}$) while preserving the average sharing pattern of force between the two effectors, A1 to A2 and B1 to B2.

and then numerous solutions may be facilitated in a vicinity of that point in accordance with the principle of abundance (Park et al., 2010).

10.6   Relations to everyday voluntary movements

As emphasized in earlier sections of this chapter, problems of redundancy are inherent to many levels of analysis of the system for movement production, kinematic, kinetic, and electrophysiological. All natural everyday movements are characterized by motor redundancy at all the mentioned levels. Only in artificial laboratory tasks, sometimes, are some of the redundancy problems avoided, for example, by giving subjects tasks of moving on a plane while using only two joints or producing force while pressing with a single finger or another effector. Note that while at the level of task formulation such tasks may avoid problems of motor redundancy, these problems are always present at other levels of analysis. For example, two-joint movements on a plane always involve more than two muscles, and activation of a single muscle always involves numerous motor units.

Are natural voluntary movements optimal? This question cannot be answered unambiguously. Many studies applied ideas of optimization and explored different cost functions in various natural movements ranging from whole-body actions such as locomotion to accurate multifinger tasks such as handwriting (reviewed in Nelson, 1983; Seif-Naraghi and Winters, 1990; Prilutsky and Zatsiorsky, 2002). The results of those studies may be summarized as follows. Natural movements do not violate in a major way many possible cost functions including those reviewed earlier (such as minimum jerk, minimum fatigue, minimum energy, etc.). Larger costs have been described for studies of populations with impaired motor function such as healthy elderly (Monaco and Micera, 2012), patients with neurological disorders (Baraduc et al., 2013), persons with atypical development (Agiovlasitis et al., 2011), and others. The application of the analytical inverse optimization method (ANIO; see Section 10.3.6) showed that young, healthy persons follow a single cost function more consistently as compared to older adults and patients with neurological disorders (Park et al., 2011, 2013). Taken together, these observations suggest that natural human movements may be optimal with respect to some cost functions. Unfortunately, the traditional methods are based on guessing cost functions, while the application of ANIO has been limited.

All the mentioned examples of the principle of abundance used comparisons of intertrial variance at the level of elemental variables to variance at the level of performance. Assuming that movements represent time evolutions of a multielement time-varying system, intertrial variance can be used as a proxy of dynamic stability (Schöner, 1995; Scholz and Schöner, 1999). Indeed, each trial starts from somewhat different initial conditions (including different states of the neuronal pools involved in the movement execution), and external forces acting during each movement differ across trials. Under such conditions, a dynamically stable system is expected to show relatively little variance in performance at the final state. Hence, the intertrial covariation among elemental variables may be viewed as a reflection of task-specific stability of performance variables.

The notion of task-specific stability in a multidimensional (redundant) space was introduced by Gregor Schöner (1995). Later, this notion led to the development of the uncontrolled manifold (UCM) hypothesis (Scholz and Schöner, 1999), which will be discussed in more detail in the next chapter. Briefly, the UCM hypothesis assumes that the CNS acts in a multidimensional space of elemental variables and organizes in that space a subspace (the UCM) corresponding to a desired value (time evolution) of an important, task-specific performance variable. Further, the CNS facilitates covariations among elemental variables that keep most variance within the UCM. As a result, the UCM hypothesis provides direct links between stability of performance and variance in different directions within a multidimensional space of elemental variables. It has been applied successfully to analysis of a variety of natural human movements, from multifinger prehensile tasks to whole-body actions (posture and locomotion) at different levels of analysis (kinematic, kinetic, and electromyographic) (reviewed in Latash et al., 2007).

Building movements on the principle of abundance offers at least two important advantages. First, task-specific stability of performance variables can be organized. Note that stability of movements is paramount for success in the environment characterized by changing external force, targets, and states of the body. On the other hand, if one wants to change a performance variable quickly, high stability of that variable may be undesirable. Within the principle of abundance, stability of a variable may be modified independently of its value or trajectory. Indeed, changes in the structure of intertrial variance have been observed 200−300 m prior to the initiation of a quick action from a steady state (Olafsdottir et al., 2005; Klous et al., 2011). These changes, termed *anticipatory synergy adjustments* (see Chapter 11 for more detail), have been linked to a decrease in stability of the variable that the person prepared to change quickly.

Another major advantage of this method of control is the opportunity to perform several tasks simultaneously with shared sets of elemental variables and without negative interference between the tasks (Zhang et al., 2008). Within the principle of abundance, performance of each task is associated not with a single optimal solution but with a subspace within the solution space. Other tasks may be performed within the same subspace without interfering with the original task. For example, if a person performs a three-finger accurate force production task (Figure 10.11), the solution space (the UCM for this task) represents a plane shown as the dotted plane in Figure 10.11. If, simultaneously, the person has to balance the frame with the sensors with respect to a pivot (see the insert in Figure 10.11), a secondary task emerges with another solution space (the dashed plane in Figure 10.11). The two planes intersect, creating a solutions space common for both tasks (the thick line). So, if solving the former task is associated with a relatively large subspace within the corresponding UCM, the second task can be solved without leaving this subspace.

A similar idea has been developed in robotics under the name of the *principle of superposition* (Arimoto et al., 2000) and confirmed in human experiments (Zatsiorsky et al., 2004). According to this principle, two or more independent controllers may be involved in solving a task, and their projections on elemental variables are summed up to define magnitudes of those variables. This principle has been applied to the control

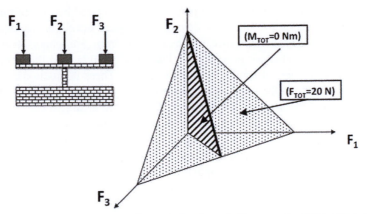

Figure 10.11 A three-finger accurate force production task (for example, 20 N) with an additional constraint of balancing the frame with the sensors on a pivot under finger 2. The solution space for the task $F_{TOT} = 20$ N is shown with the dotted plane. The solution space for the task $M_{TOT} = 0$ is shown with the dashed plane. There is a common solution space shown with the thick line.

of robotic grippers and the human hand with separate controllers for the production of the gripping action and the rotational action (Arimoto et al., 2001; Shim et al., 2005).

The principle of abundance suggests that motor variability is not necessarily a detrimental feature of motor performance but may be useful (e.g., for performance of secondary tasks) or reflect a useful feature (such as dynamic stability of task-specific performance variables). Two groups of studies provided evidence for the hypothesized usefulness of variability in natural motor tasks. A series of studies of professional butchers in chicken processing factories have shown that butchers with larger variability of joint rotations during their professional movements were less likely to develop chronic pain in the shoulder joint when compared to those who used more stereotypical movements (Madelaine et al., 2008). More experienced butchers were also more likely to show higher kinematic variability as compared to younger workers (Madelaine and Madson, 2009).

Studies of the effects of fatigue of one of the elements contributing to a multielement action have shown that variance of overall performance did not change much following fatigue while variance of the fatigued element increased significantly when this element performed as isolated task (Côté et al., 2002, 2008). A more detailed analysis has shown that an increase in intertrial variability of a fatigued element was associated with an increase in variability of other elements involved in a common task and a simultaneous increase in covariation among the elements that channeled most of the increased variance into the subspace that had no effect on performance (the UCM for the corresponding performance variable) (Singh et al., 2010; Singh and Latash, 2011). These results have suggested strongly that variability is not a consequence of unavoidable noise in the system but a reflection of task-specific control that can be tuned depending on the state of the body.

According to the principle of abundance, the ability to organize task-specific covariation among elemental variables is a crucially important feature of the neural systems involved in movement production. In populations characterized by suboptimal motor function, for example, as a result of healthy aging, atypical development, or a neurological disorder, indices of task-specific covariation drop (Shinohara et al., 2004; Latash et al., 2002; Olafsdottir et al., 2007; Park et al., 2012), and this drop shows correlations with the clinical status (Park et al., 2013). These populations also show an impaired ability to adjust patterns of covariation in preparation to a quick action known as anticipatory synergy adjustments (Olafsdottir et al., 2008; Park et al., 2012, 2013). Practice has shown relatively strong and quick effects on task-specific patterns of covariation (Kang et al., 2004; Olafsdottir et al., 2008; Wu et al., 2013; Wu and Latash, 2014). These studies carry an optimistic message that practice can lead to better coordination in those populations.

10.7   The bottom line

Currently, both notions, *redundancy* and *abundance*, coexist in the movement science literature. *Redundancy* is used more frequently, and Bernstein's formulation "problem of motor redundancy" remains classical. While the two terms are used to address the same groups of problems, they emphasize different approaches to those problems. Within the traditions of redundancy, elimination of redundant DOFs leading to a single solution is implied. This has been commonly modeled using various optimization criteria with the cost functions selected rather arbitrarily based on personal preferences and intuitive considerations of researchers. Recently, a method of computing cost functions based on experimental observations has been introduced, the analytical inverse optimization. So far, the range of applicability of this method has been relatively narrow.

In contrast, the principle of abundance emphasizes using all the available DOFs, leading to families of solutions that ensure stability of behavior under natural conditions. It also affords the possibility to perform several tasks simultaneously by the same set of elements. The concept of task-specific stability and the framework of the uncontrolled manifold hypothesis have been used to generate quantitative characteristics of intertrial variance.

The ideas of abundance and optimization are not mutually exclusive. They have been merged into a single scheme that uses a criterion of optimality to define a preferred solution and simultaneously allows individual elements to produce variable outputs about the preferred solution. The outputs of the elements are organized to reduce variability of task-specific performance variables.

References

Agiovlasitis, S., Motl, R.W., Ranadive, S.M., Fahs, C.A., Yan, H., Echols, G.H., Rossow, L., Fernhall, B., 2011. Energetic optimization during over-ground walking in people with and without down syndrome. Gait and Posture 33, 630−634.

Alexander, M.R., 2002. Energetics and optimization of human walking and running: the 2000 Raymond Pearl memorial lecture. American Journal of Human Biology 14, 641−648.

Arimoto, S., Nguyen, P.T.A., Han, H.Y., Doulgeri, Z., 2000. Dynamics and control of a set of dual fingers with soft tips. Robotica 18, 71—80.

Arimoto, S., Tahara, K., Yamaguchi, M., Nguyen, P.T.A., Han, H.Y., 2001. Principles of superposition for controlling pinch motions by means of robot fingers with soft tips. Robotica 19, 21—28.

Baraduc, P., Thobois, S., Gan, J., Broussolle, E., Desmurget, M., 2013. A common optimization principle for motor execution in healthy subjects and parkinsonian patients. Journal of Neuroscience 33, 665—677.

Bernstein, N.A., 1926. General Biomechanics. Moscow (In Russian).

Bernstein, N.A., 1930. A new method of mirror cyclographie and its application towards the study of labor movements during work on a workbench. Hygiene, Safety and Pathology of Labor. # 5, pp. 3—9, and # 6, pp. 3—11. (in Russian).

Bernstein, N.A., 1967. The Co-ordination and Regulation of Movements. Pergamon Press, Oxford.

Bottasso, C.L., Prilutsky, B.I., Croce, A., Imberti, E., Sartirana, S., 2006. A numerical procedure for inferring from experimental data the optimization cost functions using a multibody model of the neuro-musculoskeletal system. Multibody System Dynamics 16, 123—154.

Côté, J.N., Feldman, A.G., Mathieu, P.A., Levin, M.F., 2008. Effects of fatigue on intermuscular coordination during repetitive hammering. Motor Control 12, 79—92.

Côté, J.N., Mathieu, P.A., Levin, M.F., Feldman, A.G., 2002. Movement reorganization to compensate for fatigue during sawing. Experimental Brain Research 146, 394—398.

Crowninshield, R.D., Brand, R.A., 1981. A physiologically based criterion of muscle force prediction in locomotion. Journal of Biomechanics 14, 793—801.

Cruse, H., Bruwer, M., 1987. The human arm as a redundant manipulator: the control of path and joint angles. Biological Cybernetics 57, 137—144.

Cruse, H., Wischmeyer, E., Bruwer, M., Brockfield, P., Dress, A., 1990. On the cost functions for the control of the human arm movement. Biological Cybernetics 62, 519—528.

Danion, F., Schöner, G., Latash, M.L., Li, S., Scholz, J.P., Zatsiorsky, V.M., 2003. A force mode hypothesis for finger interaction during multi-finger force production tasks. Biological Cybernetics 88, 91—98.

Flash, T., Hogan, N., 1985. The coordination of arm movements: an experimentally confirmed mathematical model. Journal of Neuroscience 5, 1688—1703.

Gelfand, I.M., Latash, M.L., 1998. On the problem of adequate language in movement science. Motor Control 2, 306—313.

Gelfand, I.M., Latash, M.L., 2002. On the problem of adequate language in biology. In: Latash, M.L. (Ed.), Progress in Motor Control, Structure-Function Relations in Voluntary Movement, vol. 2. Human Kinetics, Urbana, IL, pp. 209—228.

Gelfand, I.M., Tsetlin, M.L., 1961. The non-local search principle in automatic optimization systems. Doklady Akademii Nauk SSSR (Proceedings of the USSR Academy of Sciences) 137, 295.

Gelfand, I.M., Tsetlin, M.L., 1966. On mathematical modeling of the mechanisms of the central nervous system. In: Gelfand, I.M., Gurfinkel, V.S., Fomin, S.V., Tsetlin, M.L. (Eds.), Models of the Structural-Functional Organization of Certain Biological Systems. Nauka, Moscow, pp. 9—26 (in Russian, a translation is available in 1971 edition by MIT Press: Cambridge MA).

Hasan, Z., 1986. Optimized movement trajectories and joint stiffness in unperturbed, inertially loaded movements. Biological Cybernetics 53, 373—382.

Henneman, E., Somjen, G., Carpenter, D.O., 1965. Excitability and inhibitibility of moto-neurones of different sizes. Journal of Neurophysiology 28, 599—620.

Hogan, N., 1984. An organizational principle for a class of voluntary movements. Journal of Neuroscience 4, 2745–2754.

Hughlings Jackson, J., August 17, 1889. On the comparative study of disease of the nervous system. British Medical Journal 355–362.

Jaric, S., Latash, M.L., 1998. Learning a motor task involving obstacles by a multi-joint, redundant limb: two synergies within one movement. Journal of Electromyography and Kinesiology 8, 169–176.

Kang, N., Shinohara, M., Zatsiorsky, V.M., Latash, M.L., 2004. Learning multi-finger synergies: an uncontrolled manifold analysis. Experimental Brain Research 157, 336–350.

Kilbreath, S.L., Gandevia, S.C., 1994. Limited independent flexion of the thumb and fingers in human subjects. Journal of Physiology 479, 487–497.

Kim, S.W., Shim, J.K., Zatsiorsky, V.M., Latash, M.L., 2006. Anticipatory adjustments of multi-finger synergies in preparation for self-triggered perturbations. Experimental Brain Research 174, 604–612.

Klous, M., Mikulic, P., Latash, M.L., 2011. Two aspects of feed-forward postural control: anticipatory postural adjustments and anticipatory synergy adjustments. Journal of Neurophysiology 105, 2275–2288.

Krishnamoorthy, V., Latash, M.L., Scholz, J.P., Zatsiorsky, V.M., 2003. Muscle synergies during shifts of the center of pressure by standing persons. Experimental Brain Research 152, 281–292.

Latash, M.L., 2008. Synergy. Oxford University Press, New York.

Latash, M.L., 2012. The bliss (not the problem) of motor abundance (not redundancy). Experimental Brain Research 217, 1–5.

Latash, M.L., Scholz, J.F., Danion, F., Schöner, G., 2001. Structure of motor variability in marginally redundant multi-finger force production tasks. Experimental Brain Research 141, 153–165.

Latash, M.L., Kang, N., Patterson, D., 2002. Finger coordination in persons with Down syndrome: atypical patterns of coordination and the effects of practice. Experimental Brain Research 146, 345–355.

Latash, M.L., Scholz, J.P., Schöner, G., 2007. Toward a new theory of motor synergies. Motor Control 11, 276–308.

Latash, M.L., Li, Z.-M., Zatsiorsky, V.M., 1998. A principle of error compensation studied within a task of force production by a redundant set of fingers. Experimental Brain Research 122, 131–138.

Li, Z.M., Latash, M.L., Zatsiorsky, V.M., 1998. Force sharing among fingers as a model of the redundancy problem. Experimental Brain Research 119, 276–286.

Madelaine, P., Madson, T.M.T., 2009. Changes in the amount an structure of motor variability during a deboning process: effects of work experience and neck-shoulder discomfort. Applied Ergonomics 40, 887–894.

Madelaine, P., Voigt, M., Mathiassen, S.E., 2008. Cycle to cycle variability in biomechanical exposure among butchers performing a standardized cutting task. Ergonomics 51, 1078–1095.

Mattos, D., Latash, M.L., Park, E., Kuhl, J., Scholz, J.P., 2011. Unpredictable elbow joint perturbation during reaching results in multijoint motor equivalence. Journal of Neurophysiology 106, 1424–1436.

Mattos, D., Kuhl, J., Scholz, J.P., Latash, M.L., 2013. Motor equivalence (ME) during reaching: Is ME observable at the muscle level? Motor Control 17, 145–175.

Monaco, V., Micera, S., July 2012. Age-related neuromuscular adaptation does not affect the mechanical efficiency of lower limbs during walking. Gait and Posture 36 (3), 350–355.

Nelson, W., 1983. Physical principles for economies of skilled movements. Biological Cybernetics 46, 135—147.

Nichols, T.R., 1994. A biomechanical perspective on spinal mechanisms of coordinated muscular action: an architecture principle. Acta Anatomica 151, 1—13.

Nichols, T.R., 2002. Musculoskeletal mechanics: a foundation of motor physiology. Advances in Experimental and Medical Biology 508, 473—479.

Olafsdottir, H., Yoshida, N., Zatsiorsky, V.M., Latash, M.L., 2005. Anticipatory covariation of finger forces during self-paced and reaction time force production. Neuroscience Letters 381, 92—96.

Olafsdottir, H., Zhang, W., Zatsiorsky, V.M., Latash, M.L., 2007. Age related changes in multi-finger synergies in accurate moment of force production tasks. Journal of Applied Physiology 102, 1490—1501.

Olafsdottir, H.B., Zatsiorsky, V.M., Latash, M.L., 2008. The effects of strength training on finger strength and hand dexterity in healthy elderly individuals. Journal of Applied Physiology 105, 1166—1178.

Park, J., Zatsiorsky, V.M., Latash, M.L., 2010. Optimality vs. variability: an example of multi-finger redundant tasks. Experimental Brain Research 207, 119—132.

Park, J., Sun, Y., Zatsiorsky, V.M., Latash, M.L., 2011. Age-related changes in optimality and motor variability: an example of multi-finger redundant tasks. Experimental Brain Research 212, 1—18.

Park, J., Singh, T., Zatsiorsky, V.M., Latash, M.L., 2012. Optimality vs. variability: effect of fatigue in multi-finger redundant tasks. Experimental Brain Research 216, 591—607.

Park, J., Jo, H.J., Lewis, M.M., Huang, X., Latash, M.L., 2013. Effects of Parkinson's disease on optimization and structure of variance in multi-finger tasks. Experimental Brain Research 231, 51—63.

Prilutsky, B.I., Zatsiorsky, V.M., 2002. Optimization-based models of muscle coordination. Exercise and Sport Science Reviews 30, 32—38.

Ralston, H.J., Lukin, L., 1969. Energy levels of human body segments during level walking. Ergonomics 12, 39—46.

Rosenbaum, D.A., Engelbrecht, S.E., Busje, M.M., Loukopoulos, L.D., 1993. Knowledge model for selecting and producing reaching movements. Journal of Motor Behavior 25, 217—227.

Schieber, M.H., Santello, M., 2004. Hand function: peripheral and central constraints on performance. Journal of Applied Physiology 96, 2293—2300.

Scholz, J.P., Schöner, G., 1999. The uncontrolled manifold concept: identifying control variables for a functional task. Experimental Brain Research 126, 289—306.

Scholz, J.P., Schöner, G., Latash, M.L., 2000. Identifying the control structure of multijoint coordination during pistol shooting. Experimental Brain Research 135, 382—404.

Schöner, G., 1995. Recent developments and problems in human movement science and their conceptual implications. Ecological Psychology 8, 291—314.

Seif-Naraghi, A.H., Winters, J.M., 1990. Optimized strategies for scaling goal-directed dynamic limb movements. In: Winters, J.M., Woo, S.L.-Y. (Eds.), Multiple Muscle Systems. Biomechanics and Movement Organization. Springer-Verlag, New York, pp. 312—334.

Shim, J.K., Olafsdottir, H., Zatsiorsky, V.M., Latash, M.L., 2005. The emergence and disappearance of multi-digit synergies during force production tasks. Experimental Brain Research 164, 260—270.

Shinohara, M., Scholz, J.P., Zatsiorsky, V.M., Latash, M.L., 2004. Finger interaction during accurate multi-finger force production tasks in young and elderly persons. Experimental Brain Research 156, 282—292.

Singh, T., SKM, V., Zatsiorsky, V.M., Latash, M.L., 2010. Fatigue and motor redundancy: adaptive increase in force variance in multi-finger tasks. Journal of Neurophysiology 103, 2990–3000.

Singh, T., Latash, M.L., 2011. Effects of muscle fatigue on multi-muscle synergies. Experimental Brain Research 214, 335–350.

Terekhov, A.V., Pesin, Y.B., Niu, X., Latash, M.L., Zatsiorsky, V.M., 2010. An analytical approach to the problem of inverse optimization: an application to human prehension. Journal of Mathematical Biology 61, 423–453.

Todorov, E., Jordan, M.I., 2002. Optimal feedback control as a theory of motor coordination. Nature Neuroscience 5, 1226–1235.

Tresch, M.C., Jarc, A., 2009. The case for and against muscle synergies. Current Opinions in Neurobiology 19, 601–607.

Turvey, M.T., 1990. Coordination. American Psychologist 45, 938–953.

Uno, Y., Kawato, M., Suzuki, R., 1989. Formation and control of optimal trajectory in human multijoint arm movement. Biological Cybernetics 61, 89–101.

Whitney, D.E., 1969. Resolved motion rate control of manipulators and human prostheses. IEEE Transactions on Man Machine Systems 10, 47–53.

Wu, Y.-H., Pazin, N., Zatsiorsky, V.M., Latash, M.L., 2013. Improving finger coordination in young and elderly persons. Experimental Brain Research 226, 273–283.

Wu, Y.-H., Latash, M.L., 2014. The effects of practice on coordination. Exercise and Sport Sciences Reviews 42, 37–42.

Xu, Y., Terekhov, A.V., Latash, M.L., Zatsiorsky, V.M., 2012. Forces and moments generated by the human arm: variability and control. Experimental Brain Research 223, 159–175.

Zatsiorsky, V.M., Li, Z.M., Latash, M.L., 2000. Enslaving effects in multi-finger force production. Experimental Brain Research 131, 187–195.

Zatsiorsky, V.M., Gregory, R.W., Latash, M.L., 2002. Force and torque production in static multi-finger prehension: biomechanics and control. Part II. Control. Biological Cybernetics 87, 40–49.

Zatsiorsky, V.M., Latash, M.L., Gao, F., Shim, J.K., 2004. The principle of superposition in human prehension. Robotica 22, 231–234.

Zhang, W., Scholz, J.P., Zatsiorsky, V.M., Latash, M.L., 2008. What do synergies do? Effects of secondary constraints on multi-digit synergies in accurate force-production tasks. Journal of Neurophysiology 99, 500–513.

Motor Synergy

11

Synergy means "work together" in Greek. Hence, two requirements have to be met in order for this word to be applicable to an action. First, more than one element has to take part in any synergy. Second, the elements have to do something together, act to-ward a common goal. The concept of "synergy" is used in many fields of knowledge. We limit our discussion to motor synergies, that is, to the synergetic action of elements within the system for movement production such as limbs, joints, digits, muscles, and motor units. In motor control, the first of the above-mentioned requirements links the word *synergy* to concepts such as *redundancy* and *abundance* described in the previous chapter (Chapter 10). The second requirement implies that motor synergies are always task specific or intention specific—they work toward an identifiable goal (although the term *synergy* is sometimes used in a meaning that does not imply a goal, see Synergy-A in Section 11.2.1).

There are at least three different meanings of *synergy* in the movement science literature discussed in more detail later in this chapter. They meet the two mentioned requirements to different degrees. In particular, they are based on different understandings of what "work together" may mean. One of the definitions assumes that showing simultaneous parallel changes in activity is sufficient to satisfy this requirement even if no obvious task is being accomplished. Another definition views motor synergies as linked to a specific task and characterized by a certain pattern of sharing the task among contributing elements, for example, muscles with similar mechanical actions. The third definition is also task specific; it links synergies to a crucial feature of biological movements, namely their stability. The latter two definitions may be viewed as addressing two characteristics of natural movements, which will be addressed as *sharing* and *stability* (in earlier studies, the latter feature was addressed as *error compensation*, Latash et al., 1998). Within this chapter, we will imply under stability a feature of a time-varying system to return to a trajectory in response to a small perturbation originating either within the system or in the environment. In a number of studies, the amount of variance in different directions in a multi-dimensional space of variables produced by the elements of the system (elemental variables) has been used as a proxy for stability (see Section 11.2.3).

A recent PubMed search with *synergy* and *movement* as key words produced close to 1000 entries dating back to 1951 while a Google Scholar search yielded over 120,000 results. The frequent use of this term in the movement science literature warrants a clarification of the most commonly used meanings of this term.

Biomechanics and Motor Control. http://dx.doi.org/10.1016/B978-0-12-800384-8.00011-9

11.1   Elements of history

The earliest usage of the word *synergy* dates back to at least 2000 years. The Greek Fathers of Christianity used the concept of synergy for the joint effort of man and God to help man surpass himself and to reveal God to him. In the fourteenth century, St. Gregory Palamas, an outstanding Greek philosopher, developed this concept based on observations that some people turned into true believers while others failed to do so. Palamas thought that God offered help in transforming people's minds, souls, and bodies with grace. The collaboration between man's own effort and grace (addressed as σινεργια) resulted in elevating man to a "divine state" (St Gregory Palamas, 1983; Meyendorff, 1964). Palamas thought that God offered more help to people with weak faith to achieve salvation, while those with strong faith received proportionally less help from God. This feature of "negative covariation" between efforts by two actors comes very close to one of the current definitions of motor synergy (Synergy-C in Section 11.5).

At the end of the nineteenth century, two great neurologists, John Hughlings Jackson (1835–1911) and Felix Babinski (1857–1932), introduced the notion of synergy into studies of movement disorders. In particular, Babinski (1899) used the term *asynergia* to describe the discoordinated movements of patients with cerebellar disorders. Now, this feature of disordered movements is commonly addressed as *ataxia*. Further developments within the field of motor disorders led to a somewhat different understanding of synergies (Bobath, 1978). This term started to mean stereotypical, pathological patterns of muscle activation seen in certain groups of neurological patients, in particular stroke survivors. These stereotypical patterns were assumed to interfere with natural patterns of muscle activation associated with functional movements. So, in the clinical literature, both *synergy* and *asynergia* implied negative features of muscle coordination with an emphasis on either stereotypy or excessive freedom (lack of coordinated changes), respectively.

The word *synergy* became much more broadly used in the movement science literature after the classical works of Nikolai Bernstein (1896–1966). Bernstein used this term to refer to a specific level in his multi-level scheme for the construction of movements (Bernstein, 1947). The second level of Bernstein's scheme, level B or the level of synergies, was supposed to provide coordinated involvement of large muscle groups. Synergy was used to mean "many muscles working together." This level was assumed to play a major role in the control of multi-joint kinematic chains taking into consideration reactive and motion-dependent forces. One of the examples used by Bernstein to illustrate functioning of the level of synergies was locomotion. Since locomotion can be observed in animals with the spinal cord separated surgically from the brain, the importance of the spinal cord and proprioceptive sensory signals for proper functioning of this level was emphasized.

Over the past 30 years or so, the application of correlation and matrix factorization techniques to analysis of multi-dimensional data sets led to the emergence of a number of quantitative indices of synergies. These studies form two large groups (reviewed in Latash et al., 2007). One of them searches for regularities in patterns of muscle activation (or other types of variables, kinetic and kinematic) with time during task execution or across changes in task conditions that would allow describing these patterns with fewer variables as compared to the original data set. These methods follow the original insight of Bernstein that a major problem in motor control is solving the problem of motor redundancy by reducing the number of degrees of freedom (DOFs) ("elimination of redundant DOFs", see Chapter 10). The other group searches for patterns of covariation within a redundant set of elemental variables that may suggest preferential stabilization of particular task-specific performance variables, to which the elemental variables contribute.

11.2 Current terminology

Three definitions of synergy dominate the contemporary literature. They emphasize strikingly different aspects of movements, and using the same word for such a different set of meanings has been the source of major misunderstanding and confusion. Therefore, within this chapter, we will use three different terms for the three different meanings: Synergy-A, Synergy-B, and Synergy-C.

11.2.1 Synergy-A: stereotypical muscle activation patterns

In most contemporary clinical papers, the word *synergy* has a strong negative connotation and is used to describe stereotypical patterns of muscle activation seen in some neurological patients, in particular after cortical stroke (Bobath, 1978; DeWald et al., 1995). Such patterns frequently lead to strong activation of all the major flexor muscles (flexor synergy) or extensor muscles (extensor synergy) in an affected limb interfering with voluntary movements that may require different, more flexible, patterns of muscle activation.

Figure 11.1 illustrates patterns of activation of a number of upper arm muscles during isometric tasks that required the production of force in different directions against a stop placed at the forearm level (DeWald et al., 1995). The upper panel shows data for a mildly impaired patient; similar patterns were also reported for healthy persons in earlier studies (Buchanan et al., 1989). Each muscle shows maximal level of activation for a certain, preferred, direction of force production. When the direction of force deviates from the preferred direction, the amount of muscle activation drops, and it is minimal when the force produced is in the opposite direction (180° to the preferred direction).

The bottom panel of Figure 11.1 shows the data for a severely impaired patient after stroke performing a similar set of tasks by the contralesional (strongly affected) arm. Clearly, there is much less modulation of muscle activation across the directions of force production. One may even say that muscles are activated to about the same level independently of what the subject tries to do. Such lack of specificity in muscle activation is a reflection of a pathological Synergy-A. Results similar to the one illustrated in Figure 11.1 have been observed across a variety of force production and movement tasks (Acosta et al., 2011; Sin et al., 2014).

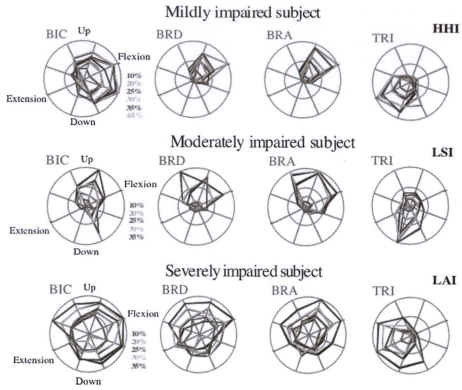

Figure 11.1 Illustrations of muscle activation levels (tuning curves) during isometric force production in different directions against a stop placed against the forearm by a severely impaired patient with a single contralateral cortical stroke, by a moderately impaired person (middle panels), and by a mildly impaired patient person (top). Note the selective, direction-specific activation of muscles in the healthy person and the lack of selectivity in the patient. BIC—biceps, BRD—brachioradialis, BRA—brachialis, TRI—triceps.
Reproduced by permission from DeWald et al. (1995), © Oxford University Press.

Breaking Synergy-A down is frequently one of the goals of motor rehabilitation. It is assumed that when a person becomes better able to activate muscles in a direction-specific way, this may lead to better performance of everyday functional tasks.

11.2.2   Synergy-B: groups of variables with parallel scaling

A number of studies explored time changes in large sets of kinematic, kinetic, and electromyographic variables during execution of natural actions and also across trials with changes in parameters of the task. Further, methods of matrix factorization were applied to correlation or covariation matrices to extract groups within the original set of variables that tended to show parallel changes in the original (elemental) variables. The commonly used methods included principal component analysis (PCA) with and without further rotation and factor extraction (factor analysis), independent component analysis (ICA), and nonnegative matrix factorization (NNMF). Several of these methods were applied to the same data sets within one of the studies (Tresch et al., 2006), resulting in comparable results across factor analysis, ICA and NNMF, while PCA without rotation performed worse. These commonly used methods belong to linear decomposition techniques. This means that they allow representing the original data set as the sum of weighted base vectors, with some residual error.

Note that this movement analysis is "objective," which may be viewed as its strength or drawback. In particular, it does not consider muscle function based on its biomechanical action, for example, whether a muscle is an agonist or an antagonist for the analyzed action. In anatomy and biomechanics, the notion of a *synergist* has been used with at least three meanings (Zatsiorsky and Prilutsky, 2012): (a) anatomical synergists or agonists (the joint agonists are the muscles that move the joint in the same direction), (b) joint moment synergists/agonists (muscles that generate moment of force in the same direction as the joint moment of force), and (c) task synergists/antagonists (depending on whether the muscle action assists or resists the task performance). For instance, the human quadriceps and hamstrings are joint antagonists in both the hip and knee joints, but they work together in many movements, for example, in the sit-to-stand movement.

Application of any of the mentioned methods resulted in reducing the original multi-dimensional data set to a lower-dimensional set of factors, principal components, independent components, etc. These have been addressed in different studies as synergies (Saltiel et al., 2001; D'Avela et al., 2003; Ivanenko et al., 2004; Ting and McPherson, 2005) or modes (Krishnamoorthy et al., 2003a; Danna-dos-Santos et al., 2007). We will address them as Synergy-B. An important difference between Synergy-B (or mode) and Synergy-A is that the composition of Synergy-B can change with a major change in the task or in the conditions of its execution, while Synergy-A is stereotypical and not task specific.

There are both advantages and disadvantages in using different methods of Synergy-B identification. In particular, PCA with factor extraction leads to a set of orthogonal eigenvectors in the space of elemental variables accounting for a certain amount of variance in the original data set. To account for 100% of the variance, the number of accepted factors has to be the same as the number of elemental variables (unless some directions have exactly zero variance, which is practically impossible), and application of the method may become meaningless. Depending on the nature of elemental variables, researchers have been satisfied with accounting for different amounts of the variance in the original data set ranging from >90% to about 60% (Ivanenko et al., 2004; Ting and McPherson, 2005; Danna-dos-Santos et al., 2007). The amount of variance accounted for by a relatively small number of PCs reflected, in particular, the magnitude of the signal-to-noise ratio in the original data. For example, the inherently noisy electromyographic data during not-very-fast actions typically show a large proportion of noise in the original data set resulting in a relatively low proportion of the original variance accounted for by a selected set of PCs or factors.

The NNMF method results in selection of a few independent eigenvectors, which are not orthogonal to each other. This method also considers only positive values of elemental variables, which may be viewed as its advantage in the analysis of electromyographic data since muscles cannot show negative activation levels. On the other hand, if changes in muscle activation levels are analyzed as compared to some baseline, the requirement of nonnegativity may be viewed as a drawback.

Figure 11.2 illustrated the application of PCA and NNMF to a data distribution within a two-dimensional set of elemental variables. To make both methods applicable, it is assumed that the original data set includes only positive values for both variables. PCA identifies two orthogonal vectors oriented along the main axes of the data cloud. In contrast, NNMF identifies two vectors, which are found through an optimization process that minimizes the deviation of the computed values from the observed ones. If the only goal of this analysis is reducing the original data set to a set of fewer variables (Synergies-B or modes) that account for a substantial amount of variance, all the mentioned methods are equally acceptable. If, however, this analysis is only the first step and is followed by further analysis in the space of the identified new variables, some methods may be viewed as offering an advantage as compared to other methods. In particular, if analysis of Synergy-C (see the next section) is planned, having the data represented as a set of orthogonal vectors has an advantage for the future variance analysis.

Here are a few examples of identifying Synergy-B within different sets of variables and different tasks. When a person is asked to perform a grasping movement, many of the finger-hand joints flex together, and they extend together when the object is released (Figure 11.3). PCA on the kinematic data across manipulative and gestural acts has shown that nearly all the variance during the parallel changes in the joint angles can be accounted for by changes in the magnitude of a couple of principal components (Braido and Zhang, 2004). That is, two Synergy-Bs accounted for the various actions. A similar analysis applied to hand motion during more sophisticated tasks, for

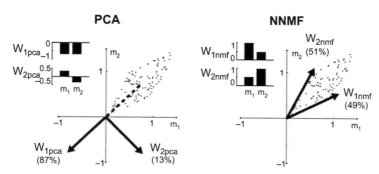

Figure 11.2 Imagine that two variables, m_1 and m_2, produce a joint output. A cloud of data points may be expected across trials, for simplicity an ellipse. The application of principal component analysis (PCA) produces two eigenvectors orthogonal to each other and corresponding to the main axes of the ellipse (W_{1PCA} and W_{2PCA}). Negative loadings for the PCA are selected for convenience, to avoid drawing the vectors over the data points. Nonnegative matrix factorization (NNMF) produces two vectors that are not orthogonal to each other (W_{1NMF} and W_{2NMF}). Both sets of vectors can be used to describe the original data set. The loadings of the W vectors in the original system of coordinates are shown in the inserts. Modified by permission from Ting and Chvatal (2011), © Oxford University Press.

Subject 1 Subject 2 Subject 3 Subject 4 Subject 5 Subject 6 Subject 7 Subject 8

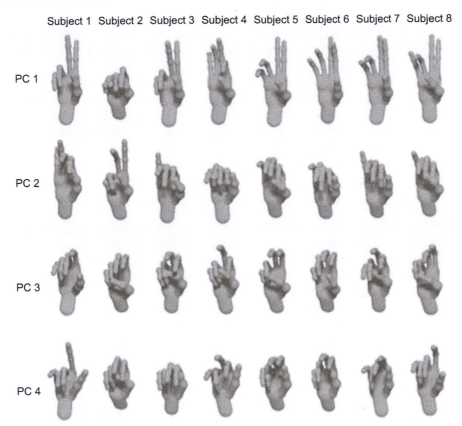

Figure 11.3 The first four principal components (PC) used by eight subjects during static performance of the American Sign Language alphabet.
Modified by permission from Jerde et al. (2003), © IEEE.

example, sign language, showed that a few Synergies-B accounted for much of the variance in the multi-dimensional joint configuration space (Jerde et al., 2003; Weiss and Flanders, 2004). When a standing person sways naturally, one PC accounts for a large amount of variance in the three-dimensional space of the ankle, knee, and hip joints in a sagittal plane (Freitas et al., 2006).

When a person is asked to hold an object in the hand, the many force and moment of force variables produced by the individual digits are correlated with changes in the external load and torque. PCA shows that the kinetic elemental variables form groups related to the control of the force vector and moment of force vector (Shim et al., 2003; Zatsiorsky et al., 2004). These results suggest that the neural control of the static hand actions obeys the principle of superposition introduced in robotics (Arimoto et al., 2001).

NNMF and PCA with factor extraction have been applied to analysis of the patterns of muscle activation observed during various human movements such as keeping balance in response to a perturbation, swaying, walking, and performing arm actions

(Ivanenko et al., 2004, 2005; Ting and McPherson, 2005; Krishnamoorthy et al., 2003a,b, 2007; Danna-dos-Santos et al., 2007). In all those studies, relatively small sets of factors (Synergies-B, modes) were identified able to account for much of the variance in the original data sets.

Note that Synergy-A may be viewed as a particular example of Synergy-B. Indeed, if a set of variables (e.g., muscle activations) shows a rigid pattern of parallel activation independently of the task, application of any of the mentioned methods would show that changes in the magnitude of a single eigenvector account for nearly all the variance in the original data set.

11.2.3   Synergy-C: task-specific stability of movement and the uncontrolled manifold concept

The notion of Synergy-C is tightly linked to the principle of abundance described in Chapter 10. To review, this principle views apparent problems of motor redundancy associated with performing actions with large sets of elements not as problems but as a rich design that allows combining certain features of performance crucial for success in natural, everyday actions in the changing environment. These involve task-specific stability of functionally important performance variables, to which all the elemental variables contribute, and possibility to perform several actions with a single set of elemental variables and without negative interference between the actions.

The idea of task-specific stability of actions was originally introduced and developed by Schöner (1995). If n elemental variables contribute to an m-dimensional performance variable $(n > m)$, there is a $(n-m)$-dimensional space within which the performance variable does not change its magnitude. This space has been addressed as the *uncontrolled manifold* (UCM, Scholz and Schöner, 1999) for that particular variable. A similar notion of *no-motion manifold* was developed in studies of the control of speech (Laboissiere et al., 1996). In robotics, a related notion of *self-motion* space has been used. Uncontrolled manifold implies that the neural controller does not have to correct deviations of the system along the UCM, since these deviations, by definition, do not lead to changes in the performance variable. Figure 11.4 illustrates a two-dimensional movement from a starting position of the endpoint (the fingertip) to a target performed by a three-joint arm. If the same movement is performed several times, intertrial variability is expected in all the kinematic variables measured at comparable phases of each movement. Figure 11.4 illustrates three such trajectories that start from the same initial joint configuration. At any given movement phase, the three trajectories differ in the joint configuration space. However, their deviations are primarily confined to the UCMs computed at those phases for the endpoint coordinate (curved, dashed lines in Figure 11.4). As a result, the relatively large intertrial variance in the joint configuration space leads to a relatively small variance in the performance variable. Solid lines in Figure 11.4 show linear approximations of the UCM at the selected movement phases. These were computed as the null spaces of the corresponding Jacobian ($\mathbf{J}$) matrices that link infinitesimally small changes in the joint angles to changes in the endpoint coordinates: $\Delta \mathbf{X} = \mathbf{J} \Delta \boldsymbol{\alpha}$, where the vectors $\mathbf{X}$ and $\boldsymbol{\alpha}$ stand for

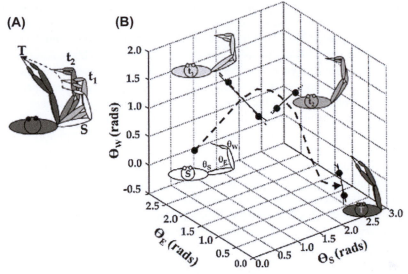

Figure 11.4 An illustration of a two-dimensional pointing task performed by a three-joint arm (A - in external space; B - in the joint configuration space). Successive trials may be expected to show variance in the joint angle space across consecutive trials. Joint configurations for two such trials are illustrated. Most of this variance is confined to the uncontrolled manifold (UCM), a space where joint motion does not affect the endpoint coordinate (solid lines, dashed lines show linear approximations of the UCM).
Modified by permission from Latash et al. (2007), © Human Kinetics.

the endpoint coordinate vector and joint configuration vector, respectively. Note that in this example, the **J** matrix changes with the joint configuration during the movement.

The illustration in Figure 11.4 suggests that, during voluntary movements, variance in the high-dimensional space of elemental variables may show a structure correspond-ing to larger variance within the UCM as compared to variance orthogonal to the UCM computed for a task-related performance variable. This conclusion fits well the classical observation by Bernstein that professional blacksmiths show relatively high intertrial joint rotation variance and relatively low variance of the hammer trajectory during repet-itive hitting movements (see Chapter 10). Note that having small or large variance within the UCM by itself has no effect on performance since errors in the performance variable are only defined by deviations orthogonal to the UCM. So, the observation of larger vari-ance within the UCM as compared to directions orthogonal to the UCM is nontrivial.

Structure of intertrial variance may be linked to different stability of the multi-element system in different directions. Indeed, if one assumes that different trials start from slightly different initial states of the system and that external forces can also show small changes across trials, trajectories in stable directions are expected to show relatively small variance across trials while trajectories in unstable directions may be ex-pected to show larger intertrial variability. Indeed, experiments with external perturba-tions have shown larger deviations of a multi-element system in directions within the UCM as compared to deviations orthogonal to the UCM (Mattos et al., 2011).

Within this framework, Synergy-C is defined as a neural organization providing task-specific covariation of elemental variables that keeps variance of a particular performance variably low. This definition has several direct implications. First, synergies are only defined with respect to a specific performance variable. This makes the phrase "human fingers form a synergy" meaningless, while the statement "human fingers form a synergy reducing variance of (stabilizing) the total pressing force" meaningful. Second, synergies are task specific. This means that if a multi-element synergy reduces variance of a performance variable in one task, it may not do so in another task. Third, synergies have a purpose, namely, reduction of variance in important performance variables, which may also be associated with higher stability of that variable.

The introduced definition of Synergy-C makes it tightly linked to the UCM hypothesis. According to the UCM hypothesis, the central nervous system facilitates covariation in a multi-dimensional space of elemental variables that keeps intertrial variance primarily limited to the UCM computed for a potentially important performance variable. In other words, the UCM hypothesis states that Synergies-C exist during natural multi-element actions.

11.3   Analysis of Synergy-C

The framework of the UCM hypothesis offers a toolbox for an operational definition of Synergy-C and its quantitative analysis. Application of this toolbox is nontrivial and involves a few steps described later. We will use a few examples of estimating Synergies-C in different spaces, different tasks, and for different performance variables. Briefly, the steps are (1) Selecting a set of elemental variables; (2) Defining a Jacobian (J) matrix that links small changes in the elemental variables to changes in a performance variable; and (3) Quantifying variance within the null space of that matrix (which is assumed to approximate the UCM) and orthogonal to that space per degree of freedom within each of the two spaces.

11.3.1   Identifying elemental variables

The first step is always to commit to a certain level of analysis, that is, to select a set of elemental variables. This set has to be redundant with respect to the analyzed tasks; otherwise, a single solution exists for each task, and the notion of synergy becomes inapplicable. Since the analysis of Synergies-C is basically analysis of variance (and covariation) across data points, the elemental variables are expected to show no covariation in the absence of a specific neural control strategy. In other words, the neural controller has to be able to change the elemental variables one at a time. Otherwise, false negative or false positive conclusions may be drawn on a synergy stabilizing or not stabilizing a performance variable.

Figure 11.5 illustrates a simple redundant task involving two elements, for example, two fingers pressing on individual force sensors. Fingers are not independent force generators—when a person tries to increase force by one finger, other fingers of the hand show an unintentional force increase (Li et al., 1998). This phenomenon, known

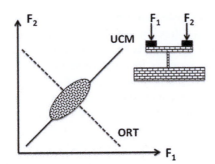

Figure 11.5 An illustration of a two-finger pressing task of producing zero total moment with respect to the pivot in between the fingers (see the inset). If commands to the fingers (finger modes) vary across trials independently of each other, the phenomenon of enslaving is expected to lead to positive covariation between the finger forces resulting in a cloud of data points elongated along the UCM for this task (the thick, solid line) This observation may lead to a false conclusion that there is a task-specific synergy.

as *lack of individuation* or *enslaving* (Kilbreath and Gandevia, 1994; Lang and Schieber, 2003; Zatsiorsky et al., 2000), is likely due to a range of factors from overlapping cortical representations of individual digits to multi-digit extrinsic hand muscles, and to passive links between the fingers provided by connective tissue (reviewed in Schieber and Santello, 2004; Zatsiorsky and Latash, 2008). Enslaving is expected to lead to positive covariation of finger forces independently of any task-specific control strategy. So, if a finger force in a trial deviates from its average across trial value, the other finger's force is expected to show a deviation in the same direction resulting in a cloud of data points elongated along an axis with a positive slope (see the dashed ellipse in Figure 11.5). If the two fingers are involved in a task of producing zero total moment of force with respect to a pivot in between the fingers, the UCM (the solution space) is a line with a positive slope (solid, thick line in Figure 11.5). Clearly, enslaving by itself leads to higher variance within the UCM than within the orthogonal space (the thin, dashed line). This observation may lead to a false conclusion that there is a task-specific synergy reducing variance of the total moment of force. On the other hand, if the task is to produce a certain value of the total force, the UCM and the orthogonal space switch. In this case, enslaving contributes to higher variance within the orthogonal space, potentially resulting in a situation when a synergy stabilizing total force may not be reflected in a difference between the two variance components.

A similar, but less obvious, interdependence of elemental variables may be expected at other levels of analysis. For example, individual joint rotations have been commonly viewed as elemental variables in studies of kinematic multi-joint Synergies-C. There are, however, several factors suggesting that an intentional small motion of one joint can lead to motion of other joints. These include the presence of two-joint and multi-joint muscles and interjoint reflexes (Nichols, 1994, 2002).

Since the classical works of Hughlings Jackson (1889), it has been commonly assumed that the brain cannot activate muscles one at a time. This makes individual muscle activation levels unlikely candidates for elemental variables. Such stable covariation patterns across tasks and across time samples during task execution have been studied as Synergies-B (see Section 11.2.2). During the analysis of Synergies-C within large multi-muscle sets, matrix factorization techniques described in the previous sections have been used to identify elemental variables. All the mentioned methods result in identifying a set of independent vectors in the original muscle activation space, which makes them adequate candidates for elemental variables.

The requirement of mutual independence of elemental variables is not easy to meet. To solve the problem, the notion of *modes* has been introduced and used for analysis of synergies. Originally, this notion was introduced for analysis of multi-finger action (Zatsiorsky et al., 1998). A *finger mode*, by definition, is a hypothetical neural command to a finger that can be changed by the person independently of commands to other fingers (Figure 11.6). A change in the magnitude of a single finger mode results in force changes by all the fingers. In a linear approximation, the space of finger forces (**F**) and modes (**m**) may be linked with the enslaving matrix **E**: $\Delta \mathbf{F} = \mathbf{E} \cdot \Delta \mathbf{m}$. In the original paper, magnitudes of modes were assumed to vary from 0 (no attempt to involve the finger in a task) to 1 (maximal involvement of the finger in the task), and the **E** matrix transformed those values to force units (newtons). In later studies of multi-finger synergies, modes were measured in newtons to facilitate linking analysis of variance in the mode space to changes in mechanical performance variables such as forces and moments of force (Latash et al., 2001; Scholz et al., 2002). Within this approach, a mode of 1 N means a force vector observed when the person tries to produce total force of exactly 1 N by pressing with one finger only.

The notion of finger modes has been used in many studies of multi-finger pressing tasks. However, in more natural multi-digit prehensile tasks, each digit produces a three-component vector of force and a three-component vector of moment of force. In general, enslaving effects may be expected to couple any of the numerous

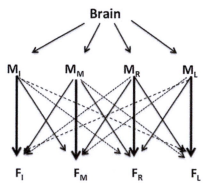

Figure 11.6 An illustration of the concept of finger modes. A finger mode is a hypothetical variable reflecting intended involvement of a finger in a task and resulting in force production (or movement) by all the fingers of the hand due to enslaving. M_I, M_M, M_R, and M_L stand for modes for the index, middle, ring, and little finger, respectively. F stands for force. The different widths of the lines that link each mode to forces produced by different fingers reflect the relative amounts of induced finger force.

force/moment vectors produced by the digits. Such effects, however, have not been studied in detail, and analysis of prehension synergies typically uses digit forces and moments of force as elemental variables (Shim et al., 2004, 2005a).

Similarly to the notion of finger modes, the notion of *muscle modes* has been introduced and used in the analysis of synergies in muscle activation spaces (Krishnamoorthy et al., 2003a,b; Danna-dos-Santos et al., 2008). Muscle modes have been typically identified using PCA with factor extraction. This approach results in identification of an orthogonal set of independent eigenvectors in the muscle activation space, which facilitates further analysis of variance as compared with methods that identify independent, but not necessarily orthogonal, eigenvectors such as non-negative matrix factorization.

11.3.2 Finding the Jacobian

After a set of elemental variables has been identified, a mapping between changes in the magnitudes of those variables and changes in a selected performance variable has to be computed or discovered. At this step, a hypothesis has to be accepted on a performance variable that may or may not be stabilized by covaried adjustments of elemental variables. In some studies, this process has been referred to as formulating a *control hypothesis* (Scholz and Schöner, 1999; Scholz et al., 2000).

Formally, this step involves defining a Jacobian matrix ($\mathbf{J}$) that links infinitesimal changes in the elemental variables ($\mathbf{EV}$) to changes in the selected performance variable ($\mathbf{PV}$): $d\mathbf{PV} = \mathbf{J} \cdot d\mathbf{EV}$. Note that the performance variable may be multidimensional and is, therefore, represented by a vector.

In some cases, finding the $\mathbf{J}$ matrix is trivial. For example, if a set of n effectors press in parallel and participate in the task of producing a certain value of total force, the $\mathbf{J}$ matrix becomes a vector with n elements, $\mathbf{J} = [1,1,\ldots,1]$. If a similar set of effectors participates in a moment of force production task, the $\mathbf{J}$ matrix is also reduced to a vector with elements representing the lever arms (d) of the effector forces, $\mathbf{J} = [d_1,d_2,\ldots,d_n]$. In analysis of multi-finger force and moment production, effects of enslaving may be taken into consideration, and the analysis is performed in the space of modes. This introduces minimal, trivial adjustments into the $\mathbf{J}$ matrix.

During the analysis of multi-joint kinematic synergies, $\mathbf{J}$ matrix that links small deviations in joint angles to a variety of performance variables such as, for example, coordinates and orientation of the end effector can be found from a geometrical model of the moving system (Scholz and Schöner, 1999). Note that $\mathbf{J}$ involves nonlinear terms (sine and cosine functions of the joint angles) and changes with joint configuration. While these nonlinearities may complicate further analysis and introduce errors for some joint configurations, defining the $\mathbf{J}$ matrix is relatively straightforward.

This step becomes less trivial when no mechanical model can be used to link elemental variables with performance variables. This is the case, for example, for analysis of synergies in muscle activation space with respect to mechanical performance variables. In such analyses, as described earlier, linear combinations of muscle activations (muscle modes) are used as elemental variables. Linking small changes in those variables (measured in millivolts integrated over small-time intervals) to changes in potentially important mechanical variables (measured in units of force, moment of force, or displacement) is nontrivial since no accurate direct biomechanical model is available. In such situations, existence of a $\mathbf{J}$ matrix is assumed, and multiple linear regression analysis is used to identify such a matrix across multiple observations with changes in the elemental variables over their whole ranges involved in the studied task (Krishnamoorthy et al., 2004; Danna-dos-Santos et al., 2008). The method of multiple linear regressions can also be used for analysis of other data sets, for which a mechanical or geometric model is available but its accuracy cannot be validated (Freitas et al., 2010).

11.3.3 Analysis of intertrial and intratrial variance

After the $\mathbf{J}$ matrix has been identified, variance indices computed for a set of data points within the UCM and orthogonal to the UCM have to be compared. In general, the UCM may be a nonlinear subspace within the original space of elemental variables. For example, the UCM for a certain position of the endpoint of a multi-joint limb in the space is a curved hypersurface in the space of individual joint rotations. This next step of analysis involves linear approximation of the UCM as the null space of the corresponding $\mathbf{J}$ matrix. This step assumes that the scattering of data points is relatively small, allowing for a linear approximation.

In most studies, analysis of variance is performed across repetitive trials. This analysis is based on an assumption that the subject is trying "to do the same" across the repetitive trials, and differences in the amounts of variance along different directions in the space of elemental variables (those spanning the UCM and orthogonal subspaces) reflect selective stabilization of particular performance variables. The assumption may be wrong if significant learning or fatigue happens across repetitive trials or if the subject's motor function is impaired.

If the task involves a change in a performance variable, with respect to which the analysis is performed, time normalization is frequently needed to ensure that intertrial variance analysis is performed over data points measured at comparable phases of the action. Figure 11.7 illustrates three trials at an action leading to a smooth change in a performance variable. The trials were performed at somewhat different speeds (panel A); after normalization by movement time, the trials show much less intertrial variance in the performance variable at all phases of the action (panel B). Errors in the time alignment of individual trials and in the estimation of movement time can potentially lead to substantial inflation of variance, particularly in movement phases characterized by high rates of change of the performance variable. Even after very accurate trial alignment, the two components of variance, within the UCM (V_{UCM}) and orthogonal to the UCM (V_{ORT}), typically show characteristic time profiles illustrated in panels C and D of Figure 11.7. Note that the V_{UCM} time profile is similar to the profile of the performance variable itself, while the V_{ORT} time profile resembles the first time derivative of that variable, suggesting that timing errors play an important role in the computed V_{ORT}. Qualitatively similar time profiles have been obtained within a model, assuming that intertrial variance is defined by variance in setting two parameters related to magnitude and timing of a performance variable, respectively (Goodman et al., 2005).

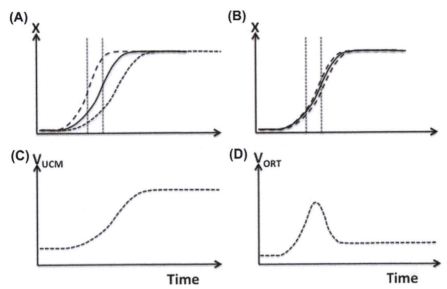

Figure 11.7 An illustration of a task associated with a smooth change in a variable X(t) performed by a redundant set of effectors: effects of time normalization on the variances along the UCM and orthogonal to it. If in different trials the task is performed at different speeds (panel A), across-trials comparison at different times may lead to large intertrial variance in X (compare the data points along the two selected times shown with dashed lines). After time normalization (panel B), variance in X at similar phase drops. The component of variance within the UCM, V_{UCM}, shows a time profile similar to that of X(t)—panel C. The component of variance orthogonal to the UCM, V_{ORT} typically shows a time profile similar to that of dX(t)/dt—panel D.

Several attempts have been made to analyze synergies that may or may not stabilize performance across samples within a time series recorded during a single trial. The idea is illustrated in Figure 11.8. If two elements produce a joint output that changes in a ramp fashion, it is possible that the elemental variables show primarily out-of-phase changes over the time series, which could be quantified using a method similar to the one involving comparison of V_{UCM} and V_{ORT}. In the example shown in Figure 11.8, the two elemental variables change primarily out of phase, and analysis of intersample variance would show $V_{UCM} > V_{ORT}$ confirming the hypothesis that the elemental variables covary to reduce variance in performance. The problem is that during actual performance, the performance variable deviates from the prescribed perfect ramp, and the subject introduces corrections leading to parallel changes in the two elemental variables. Such in-phase changes contribute to V_{ORT}. So, if both a synergy-stabilizing performance is present and there are spontaneous deviations of performance from the template corrected by the subject, the two factors would lead to opposite effects on V_{UCM} and V_{ORT}. As a result, depending on specific conditions of such an experiment, including such factors as visual resolution of the feedback and motivation of the subject to perform very accurately, analysis of variance may be expected to lead to different quantitative relations between V_{UCM} and V_{ORT}.

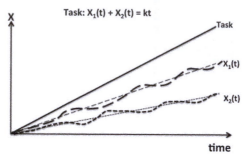

Figure 11.8 The task of accurate ramp-like increase in the sum of two variables, $X_1(t) + X_2(t) = kt$. Imagine that the two elemental variables show close to ramp changes (thick dashed lines; the corresponding ramps are shown as thin lines). If deviations from the perfect linear increase are primarily out of phase, analysis of intersample variance of the detrended data would show $V_{UCM} > V_{ORT}$, confirming the hypothesis that the elemental variables covary to reduce variance in performance. There are, however, problems with this analysis described in the text.

11.3.4 Comparisons to the goal-equivalent manifold and tolerance, noise, and covariation concepts

The idea that covariation within a set of variables may be organized in a task-specific way to reduce variance of important performance variables led to the development of a few approaches that share some features with the UCM hypothesis but also differ in important ways. One of them is the notion of goal-equivalent manifold (GEM; Cusumano and Cesari, 2006). This approach is based on a body-goal variability mapping derived from goal functions that link body variables, goal variables, and the environment needed for perfect task execution. In particular, the GEM approach considers such factors as sensitivity of solutions within the GEM (different sharing patterns) to deviations of elemental variables produced by the body. While the concepts of UCM and GEM are computationally similar, the UCM-based analysis has been developed with the goal of reflecting physiological processes within the body while the GEM approach is based on mechanical variables reflecting behavior rather than physiology.

Another related approach is based on mapping a redundant set of elemental variables on task-related variables and on considering variability of task performance as a function of three potentially independent contributing factors: tolerance, noise, and covariation (the TNC approach, Müller and Sternad, 2003, 2004). The TNC approach links elemental variables, potentially with different units of measurement, to a task-related performance variable with a formal model. *Tolerance* reflects the effects of small errors in each of the elemental variables on performance. *Noise* is assumed to represent a typical amount of variation in each elemental variable, for example, variation across repetitive trials. *Covariation* reflects whether deviations of elemental variables from their average across trial values covary in a way that reduces the effect of noise in individual elemental variables on performance.

For example, consider throwing a basketball into the basket (Figure 11.9). Success of performance in this task is defined by several factors such as ball velocity vector at release and ball coordinates at release. The elemental variables, such as ball velocity

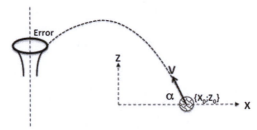

Figure 11.9 The task of accurate basketball throw requires covariation of several variables such as coordinates of release $\{X_0; Z_0\}$, angle of release α, and magnitude of velocity at release $|V|$. Perfect performance (the ball passing through the center of the basket) can be achieved by an infinite number of combinations of the variables describing the ball at the moment of release forming a solution space. Deviations from the perfect performance are related to deviations of the mentioned variables from perfect combinations in a nonlinear fashion.

magnitude, angle, and coordinates at release, have to covary to lead to success. Since this task is performed successfully even if the ball trajectory does not pass exactly through the center of the basket, some errors in the perfect combinations of the elemental variables can be tolerated. The magnitude of this tolerance can be different for different combinations of the elemental variables. The TNC method can be used, in particular, to trace the contributions of changes in the three factors—tolerance, noise, and covariation—to improve performance in the process of practice (Müller and Sternad, 2009).

The TNC method has both advantages and disadvantages as compared to the UCM-based approach. In particular, the UCM-based method of analysis requires linearization of the UCM (as the null space of the corresponding Jacobian matrix), which may lead to mistakes if the system is analyzed within a strongly nonlinear subspace. The TNC method does not use linearization. On the other hand, the UCM-based method tries to perform analysis in a space of elemental variables that can be mapped on physiological processes within the body, for instance, action of feedback loops from sensory receptors and within the central nervous system (Latash et al., 2005; Martin et al., 2009) and hierarchical control with referent body configurations (see Chapter 12) while elemental variables in the TNC approach (e.g., velocity of the ball at release in Figure 11.9) are complex and may not allow an easy mapping on such physiological variables as neural variables or muscle forces and activations. Mixing variables with different units of measurement is another complicating factor in interpreting results of the TNC-based analysis.

11.4 Synergy-C: examples of synergies

11.4.1 Kinematic synergies

The first two experimental studies of motor synergies using the framework of the UCM hypothesis were based on formal geometric models linking joint rotations to potentially important performance variables in such tasks as sit-to-stand and pistol shooting

(Scholz and Schöner, 1999; Scholz et al., 2000). In both studies, individual joint rotations were viewed as elemental variables, the Jacobian matrices were computed based on the geometrical models, and intertrial analysis of variance was performed for each time sample (phase), resulting in quantitative estimates of the two variance components, V_{UCM} and V_{ORT}. Such an analysis was performed with respect to different performance variables. For example, in the study of sit-to-stand, the variables included horizontal and vertical coordinates of the center of mass of the subject's body. In the study of pistol shooting, the variables included the angle between the pistol barrel and direction to the target, pistol coordinates in space, and coordinates of the center of mass of the arm with the pistol. Both studies confirmed the existence of multi-joint synergies (in a sense $V_{UCM} > V_{ORT}$) stabilizing some of the performance variables, but not others. In particular, the angle between the pistol barrel and direction to the target was stabilized throughout the arm motion, even when the pistol was pointing away from the target. In contrast, the pistol position in space was not stabilized close to the time of pressing the trigger. In the sit-to-stand study, only the horizontal coordinate of the center of mass was stabilized by joint covariation (not its vertical component).

Further studies explored a range of kinematic variables during a variety of tasks ranging from one-arm and two-arm pointing, to Frisbee throwing, and to whole-body voluntary sway and spontaneous sway during quiet standing (Domkin et al., 2002, 2005; Yang and Scholz, 2005; Scholz et al., 2012). The study of spontaneous sway during standing has shown, in particular, that the commonly used single-axis inverted pendulum model of the body (with rotation about the ankle joints only, Winter et al., 1996) is inadequate because sway is associated with large amounts of V_{UCM}, computed with respect to the center-of-mass trajectory and with respect to the head trajectory, with significant contributions from the small body deviations in all the joints along the body vertical axis.

11.4.2   Kinetic synergies

Studies of kinetic synergies used analysis of both parallel chains and serial chains. Multi-element parallel chains are typically considered redundant while serial chains are considered overconstrained in kinetic tasks (see Chapter 10). This is true, however, only if the complete force/moment vector produced at the endpoint of a serial chain is prescribed by the task. If only a few components of that vector are task related, the serial chain becomes redundant. For example, if a person is required to produce a certain force vector by pressing with the hand on a handle, while the moment of force is not prescribed, a solution space (UCM for the endpoint force vector) emerges with variance in that space having no effect on the task-specific performance variable.

A typical example of a set of parallel force/moment generating elements is the human hand. Studies of multi-digit synergies stabilizing various kinetic performance variables form two large groups. In pressing tasks, typically, accurate production of total force and/or total moment of force produced by a set of fingers of a hand (from two to four fingers) was studied. In prehensile tasks, multi-digit static grasping and object manipulation were studied with the thumb opposing the fingers (prismatic grasp).

As mentioned earlier, finger forces show limited independence across a variety of tasks (enslaving, Zatsiorsky et al., 2000). This suggested using not digit forces but digit modes as elemental variables in the analyses of multi-digit synergies stabilizing total force and total moment of force in pressing tasks (Latash et al., 2001). On the other hand, patterns of enslaving are complex (and not well known) if one considers not only the normal force produced by a fingertip but shear force components and moments of force (e.g., Pataky et al., 2007). In general, any of the six components of a force/moment vector produced by one digit may have enslaving effects on any of these components produced by another digit. The lack of information on these enslaving patterns is forcing researchers to study prehensile multi-digit synergies using digit forces and moments as elemental variables (Shim et al., 2003, 2005a). Synergies have been typically studied with respect to total force and total moment of force produced by a set of digits; hence, the corresponding Jacobians were defined based on straightforward mechanics.

During pressing tasks, one of the very first nontrivial results was the preferential stabilization of the moment of force in pronation-supination, which was seen even when the subjects were instructed to produce accurate force time profiles and were given visual feedback on the total force, while no feedback and no instruction regarding the moment were given. In particular, in two-finger tasks, the UCMs for the total force and for the total moment are orthogonal to each other (Figure 11.10). So, if $V_{UCM} > V_{ORT}$ for one of the variables, $V_{UCM} < V_{ORT}$ for the other variable. In such tasks, stabilizing one of the two performance variables is incompatible with stabilization of the other one. In such tasks, the subjects stabilized moment of force, not force, although the formulation of the task emphasized accurate production of the total force, not moment. Involving the thumb in pressing tasks did not change the general patterns of multi-finger interaction, suggesting that the thumb is indeed a fifth finger (Olafsdottir et al., 2005a,b).

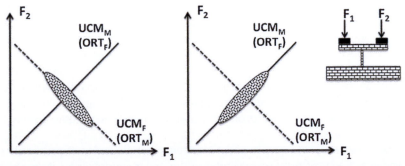

Figure 11.10 In a two-finger pressing task, the UCMs for the total force (the slanted dashed line) and for the total moment of force with respect to a horizontal line in a sagittal plane (the slanted solid line) are orthogonal to each other. Only one of the two variables can be stabilized in a sense $V_{UCM} > V_{ORT}$—total force in the left panel and total moment in the right panel. For the other variable, $V_{ORT} > V_{UCM}$.

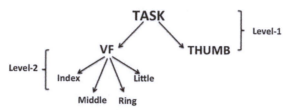

Figure 11.11 A two-level hierarchical scheme for the control of the hand during grasping actions. At the upper level (Level-1), the task (resultant force and moment vectors) is shared between the thumb (TH) and the virtual finger (VF), an imagined digit with the mechanical action equal to that of the four fingers together. At the lower level (Level-2), VF action is shared among the four fingers, I—index, M—middle, R—ring, and L—little.

Studies of prehensile synergies have been based on an idea of a hierarchical two-level control. Most of these studies used the so-called prismatic grasp with the thumb opposing the four fingers. At the upper level of the assumed hierarchy, the task is shared between the thumb and the virtual finger (VF, an imagined digit with the mechanical action equivalent to the combined action of the four fingers, Arbib et al., 1985). At the lower level, action of the virtual finger is shared among the four actual fingers (Figure 11.11). At both levels, the system is redundant. For example, for a static task of holding an object vertically, the normal forces produced by the thumb and VF have to sum up to zero (one equation with two unknowns) while the vertical shear forces plus the weight of the object have to sum up to zero as well (also, a single equation with two unknowns). The task of balancing an external torque at the upper level, as well as all the task components at the lower level of the synergy, are redundant as well (Zatsiorsky and Latash, 2008).

Synergic adjustments of digit forces and moments stabilizing the total force and moment of force have been shown in many studies (reviewed in Zatsiorsky and Latash, 2008). One of the nontrivial results was support for the principle of superposition. This principle was introduced in robotics for the control of grippers (Arimoto et al., 2001). According to the principle of superposition, two controllers organize force-related and moment-related components of any gripping action, and their outputs are superimposed at the level of elements. In human studies, changes in the external load and torque led to adjustments in all elemental variables. However, no interactions between the effects of load and torque were observed (Zatsiorsky et al., 2003). Further studies of tasks in the three-dimensional space confirmed that elemental variables formed groups that were responsible for force and moment production, and changes in elemental variables from one group had no effects on elemental variables from the other group (Shim et al., 2003, 2005a).

Another nontrivial consequence of the hierarchical organization of prehensile tasks is the apparent trade-off between synergies at the upper and lower levels of the hierarchy. Indeed, consider a simple example of the thumb opposing two fingers. If there is a strong synergy stabilizing the total force of the thumb and the virtual finger ($V_{UCM} \gg V_{ORT}$), variance of the virtual finger force is expected to be high due to the high V_{UCM} (see Figure 11.12). At the lower level, variance of the virtual finger force by

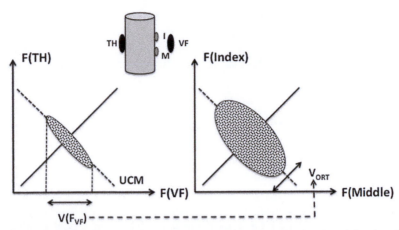

Figure 11.12 In a three-finger grasp, the task is shared between the thumb and the virtual finger (VF)—left panel. If there is a strong synergy stabilizing their total force ($V_{UCM} \gg V_{ORT}$), variance of the VF force, $V(F_{VF})$ is expected to be high. In the right level, high F_{VF}, by definition, leads to high V_{ORT}. To have a synergy at that level, V_{UCM} may have to be very high.

definition is V_{ORT}. So, to have a synergy at the lower level, variance of individual finger forces may have to be very high, possibly beyond the available force range. Synergies at both levels may coexist if the following conditions are met: $V_{ORT,H} < V_{UCM,H} = b \cdot V_{ORT,L} < V_{UCM,L}$, where the subscripts H and L stand for the high and low levels of the hierarchy, and b is a constant. During static prehensile tasks, synergies at both levels were documented for the load-resisting shear force and for the total moment of force. In contrast, for the normal force, a synergy was seen at the thumb—virtual finger level but not at the level of individual fingers (Gorniak et al., 2009).

There have been only a handful of studies of kinetic synergies within a serial chain. This is partly due to the fact that serial chains are commonly viewed as overconstrained in kinetics (Zatsiorsky, 2002). This means that, for a given chain configuration, a force/moment vector produced at the endpoint of a serial chain defines unambiguously all the joint moments independently of the number of joints. Multi-joint kinetic synergies can be studied in tasks that specify only some of the components of the force/moment vector produced at the endpoint (Xu et al., 2012) and in tasks that allow variations in the joint configuration across trials (Auyang et al., 2009; Yen et al., 2009). In the latter example, the multi-joint serial chain becomes redundant due to the kinematic redundancy inherent to serial chains that allows changing the Jacobian. When the kinematic configuration of a serial chain changes, multiple solutions emerge for kinetic tasks. Studies of the relations between joint moments and force applied by the foot on the supporting surface during cyclic locomotor tasks used this approach and confirmed intertrial covariation of joint moments that led to relatively low variance of the foot force (Auyang et al., 2009). Another study combined kinetic and kinematic variables and confirmed synergies in the space of individual joint angular momentums that led to relatively low variance of the linear momentum at the center of mass (Robert et al., 2009).

Another example of a kinetic synergy within a serial chain was presented in a study, showing that grip force and hand force could be united into a synergy in a rather artificial task of producing a constant sum of the two forces (Paclet et al., 2014). However, the formulation of the task effectively converted the serial chain into a parallel one.

11.4.3   Muscle synergies

Studies of multi-muscle Synergies-C have faced a number of problems. First, as reviewed earlier, muscle activations cannot be considered elemental variables because of the covariation among those variables that covers a range of tasks and task parameters (as in Synergy-B). Second, effects of changes in muscle activations on mechanical variables cannot be easily computed, which makes it impossible to compute the Jacobian based on a mechanical or geometric model of the multi-muscle system.

To solve the first problem, analysis of multi-muscle synergies has always included the first step of identifying elemental variables addressed as muscle modes (M-modes; Krishnamoorthy et al., 2003a). M-modes are stable muscle groups with parallel (covaried) changes of muscle activation levels common for a range of similar tasks. In other words, M-modes are Synergies-B (see Section 11.2.2). While different methods of matrix factorization have been used to explore Synergies-B, this first step in the analysis of Synergies-C favors methods resulting in a set of orthogonal M-mode vectors. Indeed, analysis of intertrial variance components within the UCM and ORT spaces is greatly simplified if the computations are performed in spaces spanned by orthogonal eigenvectors. For this reason, PCA with factor extraction has been the method of choice at this stage. In most studies of whole-body actions, a few (three to five) M-modes have been identified; two of them united leg and trunk muscles crossing several joints on the ventral ("push-forward mode") and dorsal ("push-back mode") surfaces of the body (Figure 11.13(A)). In more challenging conditions, for example, during standing on a narrow support, during vibration applied to the leg muscles, or under fatigue, the composition of M-modes changed; frequently, agonist–antagonist muscle pairs entered the same M-mode with the same signs of the loading coefficients—the so-called co-contraction M-modes (Figure 11.13(B), Krishnamoorthy et al., 2004; Danna-dos-Santos et al., 2008).

The second problem—the problem of formulating a task Jacobian— was solved assuming that small changes in the magnitudes of M-modes can be linked to changes in mechanical variables of interest (such as forces, moment of force, and displacements) via a linear model. Multiple linear regression methods were used to identify the coefficients in such linear models; the coefficients form the Jacobian of the system. A recent study compared using a straightforward geometric model and multiple regression methods in analysis of multi-joint synergies during whole-body movements (Freitas et al., 2010). The results of the two methods were basically identical supporting the use of multiple regression methods to identify the Jacobian in situations when no other method can be used.

(A)

MUSCLE	M-mode 1	M-mode 2	M-mode 3
Tibialis anterior	−0.27	0.05	**−0.73**
Gastrocnemius medialis	**0.81**	0.05	0.04
Soleus	**0.75**	−0.08	0.11
Rectus femoris	−0.25	**0.65**	−0.01
Vastus lateralis	0.07	**0.77**	0.08
Biceps femoris	**0.81**	0.20	0.06
Rectus abdominis	−0.31	0.22	**0.69**
Erector spinae	**0.74**	−0.05	−0.43

(B)

MUSCLE	M-mode 1	M-mode 2	M-mode 3
Tibialis anterior	0.02	**0.64**	−0.35
Gastrocnemius medialis	0.16	**0.70**	0.02
Soleus	−0.25	**0.72**	0.07
Rectus femoris	**0.91**	−0.05	0.13
Vastus lateralis	**0.70**	0.41	0.25
Biceps femoris	**0.61**	0.034	0.12
Rectus abdominis	−0.15	−0.04	**0.80**
Erector spinae	0.30	−0.02	**0.74**

Figure 11.13 (A) Examples of loading factors for the first three M-modes in a study of whole-body sway tasks by a standing person. Significant loading factors (with absolute value > 0.5) are shown with a larger font. Note that the first two M-modes unite muscles of the dorsal and ventral surfaces of the body. (B) Under challenging conditions (standing on a board with the narrow support beam), some of the M-modes may involve a so-called co-contraction pattern shown with bold values.
Modified by permission from Krishnamoorthy et al. (2003b, 2004), © Springer.

After the elemental variables (M-modes) and the Jacobian are identified, the rest of the analysis of multi-muscle modes follows the general scheme of the UCM analysis described earlier. Multi-muscle synergies stabilizing mechanical variables have so far been studied during arm movements (Krishnamoorthy et al., 2003b) and whole-body actions such as swaying, preparing, and reacting to perturbations, and initiating a step (Wang et al., 2005; Danna-dos-Santos et al., 2007; Krishnan et al., 2011). For example, during voluntary rhythmic whole-body sway movements, multi-muscle (more exactly, multi-M-mode) synergies were shown to stabilize the center of pressure trajectory across a broad range of movement frequencies. Such synergies have also been observed during postural preparation to a self-triggered perturbation such as those occurring during fast arm movements or load manipulations (anticipatory postural adjustments, APAs, reviewed in Massion, 1992). Other studies provided evidence in support of the principle of superposition in the control of multi-M-mode synergies stabilizing different mechanical variables (Klous et al., 2010). Studies of multi-M-mode synergies during arm action showed stabilization of the force vector applied by the arm and of the arm coordinate in conditions when the endpoint force vector was specified (Krishnamoorthy et al., 2007).

Several studies have documented changes in synergy indices in preparation to a whole-body action that can be observed 200−300 ms prior to the action initiation (Klous et al., 2011; Krishnan et al., 2011). These phenomena, called *anticipatory synergy adjustments* (ASAs), represent a relatively recently discovered mechanism of feedforward control (Olafsdottir et al., 2005a,b); they are described in more detail in the next section.

11.5   Anticipatory synergy adjustments

Changes in Synergies-C stabilizing a performance variable in preparation to an action that required a change in that performance variable were first reported in multi-finger accurate force production studies. In those studies, the subjects had to produce an accurate level of total force and then to generate a force pulse into a target (Olafsdottir et al., 2005a,b). During accurate force production, all subjects showed strong multi-finger synergies stabilizing the total force level (in a sense, $V_{UCM} > V_{ORT}$). About 200 ms prior to the force pulse initiation, a significant drop in the index of synergy was seen, an anticipatory synergy adjustment (Figure 11.14). Similar phenomena were observed during a change in the total force at a controlled rate (Shim et al., 2005b). ASAs were observed only when the subjects initiated the force change at a self-selected time or knew the required timing of the force change in advance. Changes in the synergy index moved to the action initiation time in conditions when the subjects had to initiate the action as quickly as possible to an imperative signal under the typical simple reaction time conditions. In later studies, ASAs have been documented in prehensile tasks and in whole-body postural tasks with analysis at the level of digit forces/moments and M-modes, respectively (Shim et al., 2006; Klous et al., 2011; Krishnan et al., 2011).

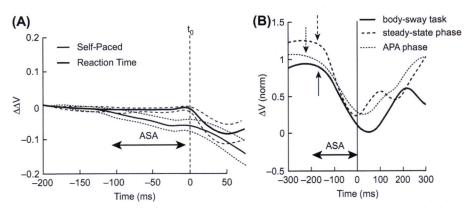

Figure 11.14 (A) An illustration of anticipatory synergy adjustments (ASA, changes in the synergy index ΔV, $\Delta\Delta V$) in a multi-finger task of producing an accurate force pulse from a steady-state force level in a self-paced manner (this lines with standard error lines—dashed) and under the simple reaction time instruction (thick lines). (B) An illustration of ASAs in a multi-muscle synergy stabilizing center of pressure in preparation to a quick arm movement. ASA is characterized by the drop of ΔV. The three lines correspond to the three methods of estimation of the synergy index used in the study (using a different task and different phases within the main task).
Modified with permission from Olafsdottir et al. (2007a), © Pergamon Press, and Klous et al. (2011), © American Physiological Society.

There are both similarities and important differences in two mechanisms of preparation to an action, APAs and ASAs. APAs are typically seen about 100 ms prior to action initiation, for example, when a standing person prepares to perform a fast arm action or to catch/drop a load; they lead to changes in the net forces and moments applied to the supporting surface (reviewed in Massion, 1992). ASAs are seen significantly earlier, 200−300 ms prior to action initiation; they are not accompanied by net changes in the forces and moments, only in covariation in the space of elemental variables, such as M-modes, stabilizing those variables. Both ASAs and APAs move toward the movement initiation time when the person is required to perform the same action in simple reaction time conditions. Both ASAs and APAs are reduced and delayed in older individuals (Olafsdottir et al., 2007a; Woollacott et al., 1988).

ASAs have been assumed to reflect a gradual loss of stability of a variable prior to its planned quick change. Indeed, trying to change a variable that is strongly stabilized by the body would require overcoming the stabilizing mechanisms and, in a way, fighting one's own synergies. This makes ASAs a universal mechanism in preparation to a quick action, which can be used even in conditions when action direction is unknown in advance (Zhou et al., 2013). In contrast, APAs can only be used when direction of the action or perturbation is known in advance; otherwise, the net changes in forces and moments produced during APAs would not counteract the perturbation but add to its effects on the body. Within one of the current schemes of synergy formation, APAs and ASAs reflect feed-forward changes in two groups of neural variables that define patterns of performance variables and their stability properties, respectively (see Section 11.8.3).

11.6 Atypical synergies

By definition, Synergies-A are signs of impaired neural control of voluntary movements, which are not expected in healthy persons and, hence, are always atypical.
Synergies-B and Synergies-C show changes with development, aging, and neurological disorders. In particular, muscle activation patterns in toddlers show poor organization into a few Synergies-B (modes) typical of those observed during grown-up
walking (Dominici et al., 2011). In older persons, the composition of Synergies-B
changes with more frequent occurrence of co-contraction patterns (i.e., agonist–
antagonist muscle pairs entering a single synergy with loading factors of the same
sign, Wang et al., 2013). In the following few sections, we focus primarily on atypical
Synergies-C.

11.6.1 Synergies in persons with atypical development

Persons with Down syndrome have a spectrum of differences from persons without
Down syndrome ranging from characteristics of the body anatomy to functioning of
internal organs, and to relative size of brain structures defined by the extra chromo-
some in the twenty-first pair. Movements of persons with Down syndrome are
commonly described as *clumsy* although no exact definition of clumsiness exists
(reviewed in Latash, 1992). Behavioral characteristics of movements in Down syndrome include slowness in movement initiation and performance, use of excessive
forces (e.g., when gripping objects), and predominance of muscle co-contraction patterns across a variety of actions and reactions to sensory signals.

Studies of multi-finger synergies during simple accurate force-production tasks in
Down syndrome showed in these persons predominance of positive covariation of
finger forces (and modes) across trials. In other words, these persons used their fingers
as prongs of a fork turned upside down—when one finger pressed stronger, all the fingers showed larger forces. Estimation of variance components within the UCM
computed for the total force and orthogonal to the UCM confirmed large amounts
of variance orthogonal to the UCM, that is, lack of synergies stabilizing total force.
After three days of practice, however, these persons showed more typical patterns
of variance, with more variance within the UCM suggesting the emergence of total
force stabilizing synergies. These observations could be interpreted as switching
from the "fork strategy" to a more typical flexible use of the fingers that allowed these
persons to show negative covariation among finger forces.

11.6.2 Changes in synergies with advanced age

Changes in several aspects of synergic control of fingers have been documented in
older persons. One of the least expected results is the drop in enslaving (*better* finger
individuation) with advanced age. The drop in enslaving is associated with significantly weaker synergies in multi-digit tasks documented across a variety of tasks
such as multi-finger pressing and static prehension (Shinohara et al., 2004; Olafsdottir
et al., 2007b). Taken together, these observations led to a hypothesis on two stages in

the development of synergies. First, during early development, anatomical elements (e.g., muscles, joints, and digits) are united into Synergies-B (modes) based on their involvement in typical everyday tasks. This is accomplished by making and strengthening neural projections in the brain that unite the elements into groups (Synergies-B) (cf. the notion of cortical piano, Schieber, 2001). Second, during aging, the progressive death of neurons at all levels of the central nervous system leads to destruction of some of those projections. As a consequence, the central nervous system has to switch to the preexistent element-based method of movement control. This "back-to-elements" hypothesis accounts for the better individual control of elements and worse synergic control in older persons.

Another consequence of aging is the impairment of feed-forward mechanisms of the neural control of movements. These include changes in both anticipatory postural adjustments (Woollacott et al., 1988) and anticipatory synergy adjustments (Olafsdottir et al., 2007a) seen in preparation to a quick action. Both APAs and ASAs are delayed in time and reduced in magnitude in older persons.

11.6.3   Changes in synergies with neurological disorders

There are only a few studies exploring changes in Synergies-C in patients with neurological disorders. The very first study of patients after a single mild cortical stroke (Reisman and Scholz, 2003) led to an unexpected finding. In this study, the patients performed reaching movements to a target by the ipsilesional (relatively unimpaired) and contralesional (impaired) arms. The general movement patterns were rather different on the two sides of the body, with atypical kinematic profiles during reaching by the contralesional arm. However, when the relative amount of variance in the joint configuration space within the UCM for the endpoint trajectory was quantified, no difference was observed between the reaching movements performed by the two arms. In other words, movement patterns were dramatically affected by the stroke while multi-joint synergies were not.

These observations are in stark contrast to findings in studies of multi-digit synergies in patients with subcortical disorders such as Parkinson's disease and olivo-ponto-cerebellar atrophy (a multi-system brain disorder affecting loops involving both the cerebellum and the basal ganglia) (Park et al., 2012, 2013). In those studies, significantly lower indices of the multi-finger synergies stabilizing total force and moment of force were observed accompanied by significantly reduced ASAs (Figure 11.15). Such changes were seen in both hands, even in patients with stage I of Parkinson's disease. This stage is defined as including clinical symptoms that can be detected in only one side of the body. The changes were seen when the patients were on their prescribed medications, and the changes became larger when the patients were tested off their medications. Note that the general performance of patients with early-stage Parkinson's disease on the medications was only mildly different from the performance of age-matched healthy subjects.

Taken together, these observations suggest that, while general movement patterns are defined with significant participation of cortical structures, synergies stabilizing these patterns seem to be much more sensitive to functioning of subcortical loops.

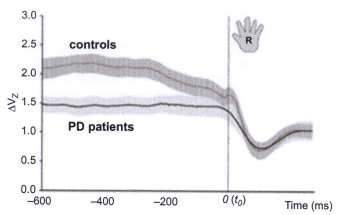

Figure 11.15 The index of synergy computed across trials for a multi-finger task or producing an accurate force pulse from a steady-state force level. Averages across patients with early-stage Parkinson's disease (dashed line, darker shade) and healthy control subjects (solid line, lighter shade) are shown with standard error bars. Note the lower synergy index at steady state and smaller anticipatory synergy adjustment (ASA) in the patients.
Reproduced by permission from Park et al. (2012), © American Physiological Society.

11.7   Changes in synergies with practice

Practice has been shown to lead to changes in both Synergies-B and Synergies-C. Changes in Synergies-B have been documented in studies of both multi-muscle and multi-finger coordination. In particular, when humans perform quick actions while standing in unstable conditions (e.g., on a board with a narrow supporting surface), the composition of muscle modes (synonymous with Synergies-B) changes as compared to similar actions performed while standing on a stable board. These changes involve the emergence of muscle modes with pronounced co-contraction patterns of agonist—antagonist muscles crossing major leg joints (Krishnamoorthy et al., 2004). Practicing while standing on an unstable board for a couple of days led to progressive disappearance of the atypical co-contraction patterns, and muscle modes with a more typical composition have emerged (Asaka et al., 2008). Practicing multi-finger accurate force production tasks in challenging conditions has also been shown to lead to a change in the composition of finger modes reflected in changed indices of unintentional finger force production, that is, enslaving (Wu et al., 2012).

Complex effects of practice have been reported for Synergies-C (reviewed in Latash, 2010a). To describe these effects, we will accept here an operational definition of Synergy-C based on the uncontrolled manifold hypothesis (see Section 11.2.3). Namely, a synergy with respect to a particular performance variable is associated with an inequality $V_{UCM} > V_{ORT}$, where V_{UCM} stands for the intertrial variance that is confined to the UCM for that variable, and V_{ORT} is variance orthogonal to the UCM.

Commonly, the primary goal of practice is to lead to more accurate performance with respect to task-specific salient variables. This means that V_{ORT} computed for those variables is expected to drop after practice. Changes in V_{UCM}, by definition, have no effect on these variables. This means that lower variability of task-related variables may be associated with higher, unchanged, proportionally lower, and even disproportionally lower V_{UCM} as compared to the pre-practice performance. These four scenarios are illustrated in Figure 11.16 using the task of constant total force production with two effectors acting in parallel (e.g., pressing with two fingers). Assume that prior to practice, there was already a synergy between the two effector's outputs. This means that intertrial variance was primarily confined to the UCM for the total force (shown with the slanted solid line).

In the four "after-practice" panels of Figure 11.16, V_{ORT} is the same and it is smaller than V_{ORT} before practice. So, practice led to improved accuracy of performance in all four cases. If V_{UCM} changed in proportion to V_{ORT}, one may conclude that improved accuracy of performance was associated with unchanged Synergy-C stabilizing the total force (panel A). If V_{UCM} decreased less than V_{ORT} ($\Delta V_{UCM} < \Delta V_{ORT}$), or did not change (panel B), or even increased (panel C) as a result of practice, a stronger corresponding synergy may be claimed. However, V_{UCM} may decrease more than V_{ORT} ($\Delta V_{UCM} > \Delta V_{ORT}$, panel D). In that case, improved accuracy of performance is associated with weaker synergy stabilizing the corresponding performance variable.

Experiments have confirmed all the scenarios shown in Figure 11.16. During multi-joint pointing studies, a drop in the synergy index ($\Delta V_{UCM} > \Delta V_{ORT}$, Domkin et al., 2002) and no changes in the synergy index ($\Delta V_{UCM} = \Delta V_{ORT}$, Domkin et al., 2005)

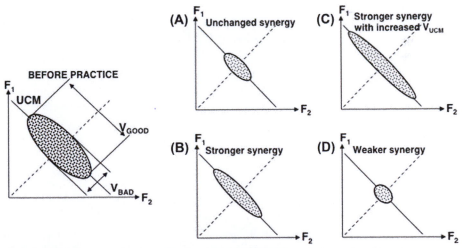

Figure 11.16 Four scenarios of changes in the intertrial data distributions and variance components, V_{UCM} and V_{ORT}, for the task of accurate force production with two effectors (E_1 and E_2). A synergy ($V_{UCM} > V_{ORT}$) is assumed prior to practice. Practice is expected to lead to a drop in V_{ORT}, while V_{UCM} can drop proportionally to V_{ORT} change (A), decrease less than V_{ORT} or stay unchanged (B), increase (C), or decrease more than V_{ORT} (D). Modified by permission from Wu and Latash. (2014), © Wolters Kluwer Health.

were reported. Practicing a very unusual multi-finger task resulted in an increase in the synergy index (as in panel B, Kang et al., 2004). Recently, one of the least probable scenarios has been observed—an increase in the total amount of variance accompanied with a drop in V_{ORT} (Wu et al., 2012). Clearly, this was only possible if V_{UCM} increased after practice (as in panel C). This was observed when the subjects practiced a two-finger accurate force production task with the visual feedback adjusted to simulate higher instability of the task with improved performance. The subjects produced accurate total force profiles as if moving over a ridge with modifiable slopes (see Figure 11.17).

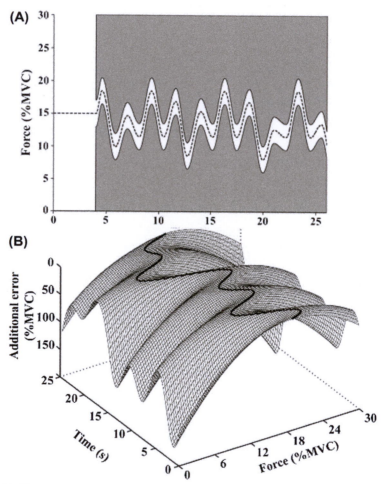

Figure 11.17 An illustration of a multi-finger accurate force production task designed to encourage higher V_{UCM}. The subjects pressed to match the center (dashed) line of the "Tube" (A) while visual feedback was adjusted to amplify their deviations from the center line as if moving over a ridge with variable slopes (B). Standard half-cycles were "hidden" within each "Tube" to facilitate UCM-based analysis of variance.
Reproduced by permission from Wu et al. (2012), © Taylor and Francis.

The various changes in the synergy index with practice suggest that depending on the task and person's proficiency, different effects of practice on Synergy-C can be expected. A hypothesis has been offered on two stages on effects of practice on synergies (Latash, 2010a). According to this hypothesis, when a person practices a novel task (or in challenging conditions), V_{ORT} drops and the relative amount of V_{UCM} increases, which may be associated with strengthening of the corresponding synergy. At some point, the person becomes very precise, V_{ORT} reaches a very low value, and cannot be reduced more. Further practice may lead to a drop in V_{UCM} reflecting a search for better solutions with respect to factors that may be outside the explicit task formulation. The corresponding synergy becomes weaker (smaller difference between V_{UCM} and V_{ORT}). Both stages of practice were observed in a study of accurate multi-finger force production with complicating factors such as a constraint related to the total moment of force and transcranial magnetic stimuli applied at random times, which challenged stability of the performance. Practice-related changes in Synergy-C have been shown to be associated with plastic changes within the central nervous system.

11.8   Origins of synergies

11.8.1   Synergies and optimal feedback control

One of the approaches to solving the problem of motor redundancy, reviewed briefly in the previous chapter, is optimization. This approach searches for solutions that optimize (minimize or maximize) a cost function, which may reflect biologically relevant task goals. Examples of cost functions include energy, fatigue, an integrated measure of jerk, and an integrated measure of joint torque change. This approach produces a single optimal solution for a given task that may be characterized by certain patterns of sharing the task among elements. So, optimization may form the basis of Synergies-B and also may be used to define one of the major characteristics of Synergies-C, namely *sharing*.

The *optimal feedback control* approach (reviewed in Diedrichsen et al., 2010) allows addressing the other major feature of Synergies-C, namely *error compensation* (or structure of variance), by including consideration of the role of sensory feedback. This method was applied to address the control of a redundant motor system (Todorov and Jordan, 2002). Within this approach, a cost function was formulated including a measure of internal effort spent on control and a measure of accuracy of performance with respect to the task-specific variables. Namely, the model minimized the weighted sum where the first summand was the squared difference between a function of effector outputs and its required value, and the second summand was defined as the variance of the control signals during task execution.

The controller was assumed to recompute a new desired trajectory at every moment in time, making no effort to correct deviations away from the previously planned behavior unless those deviations interfered with important performance characteristics. In the presence of noise, this method leads to families of trajectories in successive trials that show variance primarily in directions not associated with changes in the task-specific performance variable (along the UCM for that variable). As a result, analysis of intertrial variance is expected to show ellipses of data points similar to those illustrated in Figure 11.10 and characterized by the inequality $V_{UCM} > V_{ORT}$, which is a signature of Synergy-C stabilizing the performance variable.

While optimal feedback control is able to reproduce both characteristics of synergies, a sharing pattern and more variance within the UCM, its formulation is not well compatible with the knowledge on the neural control of movement. In particular, motor commands have been conceptualized as the neural drive to the muscles (measured as electromyographic signals, EMGs). Since muscle activation reflects the action of both descending and reflex pathways, a hypothetical neural computational device cannot, in principle, predict the total neural drive to a muscle. Using sensory signals as the primary source of signals used in feedback loops is another questionable approach because of the unavoidable conduction time delays, which may be comparable with movement times during very fast actions. Finally, the basic assumption of symbolic computations performed by the central nervous system (with such operations as computing squares of variables, rectifying EMG, and integrating variables over predicted movement time intervals) has no support in studies of the central nervous system. The next two approaches do not assume neural computations, use physical (physiological), rather than computational, feedback loops, and avoid long delays associated with transmission of sensory signals.

11.8.2   Back-coupling as a mechanism of synergies

A particular scheme involving fast-acting feedback loops within the central nervous system has been proposed (Latash et al. 2005) to account for the experimentally observed patterns of variance within redundant motor systems such as, for example, a set of fingers involved in a pressing task. The task-specific input in this scheme encodes a desired profile of total force. This input is shared among a redundant set of N neurons (e.g., $N = 4$ for the example of four-finger force production involved in the task of accurate total force production) with some noise added at that stage. Further, the output of each of those neurons has two effects. First, it leads to force production by the corresponding target finger. Second, it excites a neuron from a different group that makes feedback projections on all the N neurons, somewhat similar to the action of the well-known system of Renshaw cells. This part of the design was addressed as *central back-coupling*. The action of the feedback projections can be described with an $N \times N$ gain matrix **G**. Computer simulations have shown that this system can reduce variance of (stabilize) the total force output of four fingers when all the entries of the G matrix are negative (all back-coupling projections are negative feedbacks). Changing the elements of the G matrix allowed the four fingers to stabilize total moment of force in pronation-supination or both total force and total moment.

A similar notion of back-coupling has also been invoked in theoretical analysis of synergies within the UCM hypothesis (Martin et al. 2009). Within this scheme, feedback projections from sensory receptors are organized to adjust signals encoding referent trajectories of the elements (see Chapter 12) primarily within the UCM, i.e., in a way that does not lead to changes in task-specific performance variables. This is done by a neurophysiological mechanism that remains hypothetical. Assuming an important role of signals from sensory receptors introduces time delays that may create a problem for stability of action by such a mechanism, in particular during fast movements.

11.8.3   Synergies and control with referent configurations

A general scheme uniting the ideas of control with referent body configurations (see Chapter 12), hierarchical control, and principle of abundance has been developed (Latash, 2010b). Within this scheme (Figure 11.18), at the upper level of a hypothetical hierarchy, a few neural variables specify referent values for a handful of salient task-specific performance variables, RC_{TASK}. Further, a chain of few-to-many transformations leads to the involvement of lower levels of the hierarchy. At each of these levels, referent values for corresponding performance variables are specified. Ultimately, at the lowest level, neural control signals to individual muscles are generated associated with setting values of the threshold of the tonic stretch reflex (λ, as in the classical version of the equilibrium-point hypothesis, Feldman, 1986; see Chapter 12).

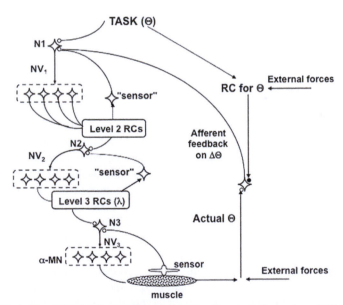

Figure 11.18 A hierarchical scheme of control with referent configurations (RCs). The task specifies an RC for a low-dimensional variable Θ. Further, a sequence of few-to-many transformations leads to changes in tonic stretch reflex values (λ) for numerous muscles and muscle activations that drive the actual Θ value to its referent value. This is achieved by a feedback loop on the difference between actual and referent Θ. At each level of the hierarchy, the few-to-many transformation is organized in a synergic way with low-latency back-coupling loops.

For example, if the task is to move the hand to a target location in space, RC_{TASK} may be associated with hand referent coordinates specified at the highest level. They lead to referent coordinates for individual joint rotations (RC_{JOINT}), which lead to referent coordinates for the muscles (RC_{MUSCLE} equivalent to setting λ).

Each of the few-to-many mappings is redundant. Its output may show variations in the corresponding RC as long as these variations are within the UCM for the RC specified at the hierarchically higher level. This can be achieved with a back-coupling scheme, for example, similar to the one described in Section 11.8.2. Neurons at the highest level of the hierarchy also receive excitation from sensors measuring the difference between RC_{TASK} and actual values of the corresponding variables—actual body configuration (AC). Neural processes within the scheme in Figure 11.18 continue until $RC_{TASK} = AC$. This result may be achieved with different changes at the lower levels of the hierarchy leading to different time profiles and final magnitudes of the corresponding performance variables. For the earlier example of hand movement to a target, different attempts are expected to show different trajectories in the joint configuration space and different patterns of muscle activation, all leading to more or less the same AC in the space of task-specific variables. In other words, the scheme in Figure 11.18 is expected to lead to one of the signatures of Synergy-C, $V_{UCM} > V_{ORT}$, when across-trials variance is analyzed using a redundant set of elemental variables.

The scheme in Figure 11.18 offers a particular example how stability for a handful, task-specific variables can be organized by means of varying contributions from elemental variables at lower levels of the hierarchy—at each of those levels, within the high-dimensional space of elemental variables, stability within the UCM is expected to be lower than orthogonal to the UCM. In particular, transient changes in the external conditions are not expected to violate the final state of the system with respect to the variables specified by RC_{TASK}, that is, transient perturbations are expected to lead to equifinality at the level of task-specific performance variables. In contrast, at any of the intermediate levels, perturbations are expected to lead to violations of equifinality along the directions that span the corresponding UCMs—a prediction that has been confirmed experimentally (Wilhelm et al., 2013; Zhou et al., 2014).

A change in RC_{TASK}, for example, associated with correction of the ongoing movement, is expected to lead to motion of the system to a new body configuration. Within a redundant set of elemental variables, motion may be viewed as superposition of two components. One of them changes the performance variables in a way prescribed by the change in RC_{TASK}. In other words, this component is orthogonal to the corresponding UCM. The other component is confined to the UCM—it changes the configuration of the system without a change in the corresponding performance variables. In robotics, these two components have been addressed as range motion and self-motion of the system. In motor control, the two components have been addressed as motor equivalent (ME) and non-motor equivalent (nME) (Mattos et al., 2011; Scholz et al., 2011). Experimental studies have shown that external perturbations of an ongoing movement produce deviations of the redundant system (multi-joint, multi-finger, or multi-muscle) with a large ME component that typically is larger than the nME component. Corrections of errors produced by such perturbations are also characterized by a very large ME component. In other words, most of the joint/finger/ muscle motion produced in the course of a correction happens within the corresponding UCM. As a result, much of the overall motion does not correct the deviation of the task-specific performance variable.

These observations fit naturally the scheme in Figure 11.18. Indeed, any change in any of the inputs in this scheme is expected to lead to its deviations primarily in less stable directions, i.e., within the corresponding UCM. This is true for effects of changes in the external force field (perturbations) and for the effects of changes in the task-specific input (corrections).

11.8.4   Possible relations between Synergies-B and Synergies-C

The multi-level hierarchical scheme in Figure 11.18 suggests that elemental variables at one level of the hierarchy may be viewed as performance variables at the next level. For example, studies of muscle activation in large muscle groups commonly involve two steps (see Section 11.4.3). At the first step, a smaller set of muscle groups is identified, addressed sometimes as muscle modes or muscle synergies (as in Synergy-B). These are typically identified as eigenvectors in the muscle activation space using one of the matrix factorization techniques. At the next step, muscle modes are viewed as elemental variables, and intertrial variance in that space is analyzed within the framework of the UCM hypothesis to identify and quantify Synergy-C. The set of muscle modes is still considered redundant with respect to salient mechanical variables related to the task formulation.

While many studies emphasized consistency of the composition of muscle modes across conditions (Ivanenko et al., 2004, 2005; Torres-Oviedo et al., 2006; Chvatal et al., 2011), other studies showed that muscle modes can show major changes when the task becomes challenging and also as a result of practice (Asaka et al., 2008; Danna-dos-Santos et al., 2008). These changes were primarily associated with the emergence and disappearance of co-contraction patterns. Changes in the composition of finger modes (elemental variables considered in studies of multi-finger action, Danion et al., 2003) have also been reported with practice (Wu et al., 2012). These observations suggest that elemental variables at one level of analysis may be viewed as performance variables stabilized by a hierarchically lower set of elemental variables at another level.

11.9   The bottom line

The term *synergy* (or *motor synergy*) used in movement science has at least three very different meanings. These include "stereotypical patterns of muscle activation observed in certain patient populations" (Synergy-A), "groups of variables that change in parallel with time or with changes in task parameters" (Synergy-B), and "patterns of covariation in a redundant set of elemental variables that stabilize (reduce variance of) a performance variable" (Synergy-C). To avoid confusion, the word *synergy* has to be always defined explicitly.

Synergy-B has been studied using a variety of matrix factorization techniques, particularly in muscle activation spaces. Performance-stabilizing Synergy-C has been extensively studied in kinematic, kinetic, and muscle activation space using the framework of the uncontrolled manifold hypothesis. It has been linked to the ideas of optimal feedback control, hierarchical control, and control with referent body configurations. Changes in Synergy-C with practice, atypical development, aging, fatigue, and a range of neurological disorders have been documented. These studies promise practical applications of the methods of synergy studies and also suggest links of the hypothetical synergic mechanisms to brain structures.

References

Acosta, A.M., Dewald, H.A., Dewald, J.P., 2011. Pilot study to test effectiveness of video game on reaching performance in stroke. Journal of Rehabilitation Research and Development 48, 431−444.

Arbib, M.A., Iberall, T., Lyons, D., 1985. Coordinated control programs for movements of the hand. In: Goodwin, A.W., Darian-Smith, I. (Eds.), Hand Function and the Neocortex. Springer Verlag, Berlin, pp. 111−129.

Arimoto, S., Tahara, K., Yamaguchi, M., Nguyen, P.T.A., Han, H.Y., 2001. Principles of superposition for controlling pinch motions by means of robot fingers with soft tips. Robotica 19, 21−28.

Asaka, T., Wang, Y., Fukushima, J., Latash, M.L., 2008. Learning effects on muscle modes and multi-mode synergies. Experimental Brain Research 184, 323−338.

Auyang, A.G., Yen, J.T., Chang, Y.H., 2009. Neuromechanical stabilization of leg length and orientation through interjoint compensation during human hopping. Experimental Brain Research 192, 253−264.

Babinski, F., 1899. De l'asynergie cerebelleuse. Revue Neurologique 7, 806−816.

Bernstein, N.A., 1947. On the Construction of Movements. Moscow, Medgiz (in Russian).

Bobath, B., 1978. Adult Hemiplegia: Evaluation and Treatment. William Heinemann, London.

Buchanan, T.S., Rovai, G.P., Rymer, W.Z., 1989. Strategies for muscle activation during isometric torque generation at the human elbow. Journal of Neurophysiology 62, 1201−1212.

Braido, P., Zhang, X., 2004. Quantitative analysis of finger motion coordination in hand manipulative and gestic acts. Human Movement Science 22, 661−678.

Chvatal, S.A., Torres-Oviedo, G., Safavynia, S.A., Ting, L.H., 2011. Common muscle synergies for control of center of mass and force in nonstepping and stepping postural behaviors. Journal of Neurophysiology 106, 999−1015.

Cusumano, J.P., Cesari, P., 2006. Body-goal variability mapping in an aiming task. Biological Cybernetics 94, 367−379.

Danion, F., Schöner, G., Latash, M.L., Li, S., Scholz, J.P., Zatsiorsky, V.M., 2003. A force mode hypothesis for finger interaction during multi-finger force production tasks. Biological Cybernetics 88, 91−98.

Danna-Dos-Santos, A., Slomka, K., Zatsiorsky, V.M., Latash, M.L., 2007. Muscle modes and synergies during voluntary body sway. Experimental Brain Research 179, 533−550.

Danna-Dos-Santos, A., Degani, A.M., Latash, M.L., 2008. Flexible muscle modes and synergies in challenging whole-body tasks. Experimental Brain Research 189, 171−187.

d'Avella, A., Saltiel, P., Bizzi, E., 2003. Combinations of muscle synergies in the construction of a natural motor behavior. Nature Neuroscience 6, 300−308.

DeWald, J.P., Pope, P.S., Given, J.D., Buchanan, T.S., Rymer, W.Z., 1995. Abnormal muscle coactivation patterns during isometric torque generation at the elbow and shoulder in hemiparetic subjects. Brain 118, 495−510.

Diedrichsen, J., Shadmehr, R., Ivry, R.B., 2010. The coordination of movement: optimal feedback control and beyond. Trends in Cognitive Science 14, 31−39.

Dominici, N., Ivanenko, Y.P., Cappellini, G., d'Avella, A., Mondì, V., Cicchese, M., Fabiano, A., Silei, T., Di Paolo, A., Giannini, C., Poppele, R.E., Lacquaniti, F., 2011. Locomotor primitives in newborn babies and their development. Science 334, 997−999.

Domkin, D., Laczko, J., Jaric, S., Johansson, H., Latash, M.L., 2002. Structure of joint variability in bimanual pointing tasks. Experimental Brain Research 143, 11−23.

Domkin, D., Laczko, J., Djupsjöbacka, M., Jaric, S., Latash, M.L., 2005. Joint angle variability in 3D bimanual pointing: uncontrolled manifold analysis. Experimental Brain Research 163, 44−57.

Feldman, A.G., 1986. Once more on the equilibrium-point hypothesis (λ-model) for motor control. Journal of Motor Behavior 18, 17−54.

Freitas, S.M.S.F., Duarte, M., Latash, M.L., 2006. Two kinematic synergies in voluntary whole-body movements during standing. Journal of Neurophysiology 95, 636−645.

Freitas, S.M.S.F., Scholz, J.P., Latash, M.L., 2010. Analyses of joint variance related to voluntary whole-body movements performed in standing. Journal of Neuroscience Methods 188, 89−96.

Goodman, S.R., Shim, J.K., Zatsiorsky, V.M., Latash, M.L., 2005. Motor variability within a multi-effector system: experimental and analytical studies of multi-finger production of quick force pulses. Experimental Brain Research 163, 75−85.

Gorniak, S.L., Zatsiorsky, V.M., Latash, M.L., 2009. Hierarchical control of static prehension: II. Multi-digit synergies. Experimental Brain Research 194, 1−15.

Ivanenko, Y.P., Cappellini, G., Dominici, N., Poppele, R.E., Lacquaniti, F., 2005. Coordination of locomotion with voluntary movements in humans. Journal of Neuroscience 25, 7238−7253.

Ivanenko, Y.P., Poppele, R.E., Lacquaniti, F., 2004. Five basic muscle activation patterns account for muscle activity during human locomotion. Journal of Physiology 556, 267−282.

Jackson, J.H., Aug. 17, 1889. On the comparative stuy of disease of the nervous system. British Medical Journal 355−362.

Jerde, T.E., Soechting, J.F., Flanders, M., 2003. Coarticulation in fluent fingerspelling. Journal of Neuroscience 23, 2383−2393.

Kang, N., Shinohara, M., Zatsiorsky, V.M., Latash, M.L., 2004. Learning multi-finger synergies: an uncontrolled manifold analysis. Experimental Brain Research 157, 336−350.

Kilbreath, S.L., Gandevia, S.C., 1994. Limited independent flexion of the thumb and fingers in human subjects. Journal of Physiology 479, 487−497.

Klous, M., Danna-dos-Santos, A., Latash, M.L., 2010. Multi-muscle synergies in a dual postural task: evidence for the principle of superposition. Experimental Brain Research 202, 457−471.

Klous, M., Mikulic, P., Latash, M.L., 2011. Two aspects of feed-forward postural control: anticipatory postural adjustments and anticipatory synergy adjustments. Journal of Neurophysiology 105, 2275−2288.

Krishnamoorthy, V., Goodman, S.R., Latash, M.L., Zatsiorsky, V.M., 2003a. Muscle synergies during shifts of the center of pressure by standing persons: identification of muscle modes. Biological Cybernetics 89, 152−161.

Krishnamoorthy, V., Latash, M.L., Scholz, J.P., Zatsiorsky, V.M., 2003b. Muscle synergies during shifts of the center of pressure by standing persons. Experimental Brain Research 152, 281–292.

Krishnamoorthy, V., Latash, M.L., Scholz, J.P., Zatsiorsky, V.M., 2004. Muscle modes during shifts of the center of pressure by standing persons: effects of instability and additional support. Experimental Brain Research 157, 18–31.

Krishnamoorthy, V., Scholz, J.P., Latash, M.L., 2007. The use of flexible arm muscle synergies to perform an isometric stabilization task. Clinical Neurophysiology 118, 525–537.

Krishnan, V., Aruin, A.S., Latash, M.L., 2011. Two stages and three components of postural preparation to action. Experimental Brain Research 212, 47–63.

Laboissiere, R., Ostry, D.J., Feldman, A.G., 1996. The control of multi-muscle systems: human jaw and hyoid movements. Biological Cybernetics 74, 373–384.

Lang, C.E., Schieber, M.H., 2003. Differential impairment of individuated finger movements in humans after damage to the motor cortex or the corticospinal tract. Journal of Neurophysiology 90, 1160–1170.

Latash, M.L., 1992. Motor control in down syndrome: the role of adaptation and practice. Journal of Developmental and Physical Disabilities 4, 227–261.

Latash, M.L., 2007. Learning motor synergies by persons with down syndrome. Journal of Intellectual Disability Research 51, 962–971.

Latash, M.L., 2010a. Stages in learning motor synergies: a view based on the equilibrium-point hypothesis. Human Movement Science 29, 642–654.

Latash, M.L., 2010b. Motor synergies and the equilibrium-point hypothesis. Motor Control 14, 294–322.

Latash, M.L., Li, Z.-M., Zatsiorsky, V.M., 1998. A principle of error compensation studied within a task of force production by a redundant set of fingers. Experimental Brain Research 122, 131–138.

Latash, M.L., Scholz, J.F., Danion, F., Schöner, G., 2001. Structure of motor variability in marginally redundant multi-finger force production tasks. Experimental Brain Research 141, 153–165.

Latash, M.L., Scholz, J.P., Schöner, G., 2007. Toward a new theory of motor synergies. Motor Control 11, 276–308.

Latash, M.L., Shim, J.K., Smilga, A.V., Zatsiorsky, V.M., 2005. A central back-coupling hypothesis on the organization of motor synergies: a physical metaphor and a neural model. Biological Cybernetics 92, 186–191.

Li, Z.M., Latash, M.L., Zatsiorsky, V.M., 1998. Force sharing among fingers as a model of the redundancy problem. Experimental Brain Research 119, 276–286.

Martin, V., Scholz, J.P., Schöner, G., 2009. Redundancy, self-motion, and motor control. Neural Computation 21, 1371–1414.

Massion, J., 1992. Movement, posture and equilibrium – interaction and coordination. Progress in Neurobiology 38, 35–56.

Mattos, D., Latash, M.L., Park, E., Kuhl, J., Scholz, J.P., 2011. Unpredictable elbow joint perturbation during reaching results in multijoint motor equivalence. Journal of Neurophysiology 106, 1424–1436.

Meyendorff, J., 1964. A Study of Gregory Palamas. The Faith Press, London, UK.

Müller, H., Sternad, D., 2003. A randomization method for the calculation of covariation in multiple nonlinear relations: illustrated with the example of goal-directed movements. Biological Cybernetics 89, 22–33.

Müller, H., Sternad, D., 2004. Decomposition of variability in the execution of goal-oriented tasks: three components of skill improvement. Journal of Experimental Psychology: Human Perception and Performance 30, 212–233.

Müller, H., Sternad, D., 2009. Motor learning: changes in the structure of variability in a redundant task. Advances in Experimental Medicine and Biology 629, 439−456.

Nichols, T.R., 1994. A biomechanical perspective on spinal mechanisms of coordinated muscular action: an architecture principle. Acta Anatomica 151, 1−13.

Nichols, T.R., 2002. Musculoskeletal mechanics: a foundation of motor physiology. Advances in Experimental and Medical Biology 508, 473−479.

Olafsdottir, H., Yoshida, N., Zatsiorsky, V.M., Latash, M.L., 2005a. Anticipatory covariation of finger forces during self-paced and reaction time force production. Neuroscience Letters 381, 92−96.

Olafsdottir, H., Yoshida, N., Zatsiorsky, V.M., Latash, M.L., 2007a. Elderly show decreased adjustments of motor synergies in preparation to action. Clinical Biomechanics 22, 44−51.

Olafsdottir, H., Zhang, W., Zatsiorsky, V.M., Latash, M.L., 2007b. Age related changes in multi-finger synergies in accurate moment of force production tasks. Journal of Applied Physiology 102, 1490−1501.

Olafsdottir, H., Zatsiorsky, V.M., Latash, M.L., 2005b. Is the thumb a fifth finger? A study of digit interaction during force production tasks. Experimental Brain Research 160, 203−213.

Paclet, F., Ambike, S., Zatsiorsky, V.M., Latash, M.L., 2014. Enslaving in a serial chain: interactions between grip force and hand force in isometric tasks. Experimental Brain Research 232, 775−787.

Park, J., Wu, Y.-H., Lewis, M.M., Huang, X., Latash, M.L., 2012. Changes in multi-finger interaction and coordination in Parkinson's disease. Journal of Neurophysiology 108, 915−924.

Park, J., Lewis, M.M., Huang, X., Latash, M.L., 2013. Effects of olivo-ponto-cerebellar atrophy (OPCA) on finger interaction and coordination. Clinical Neurophysiology 124, 991−998.

Pataky, T.C., Latash, M.L., Zatsiorsky, V.M., 2007. Finger interaction during radial and ulnar deviation: experimental data and neural network modeling. Experimental Brain Research 179, 301−312.

Reisman, D., Scholz, J.P., 2003. Aspects of joint coordination are preserved during pointing in persons with post-stroke hemiparesis. Brain 126, 2510−2527.

Robert, T., Bennett, B.C., Russell, S.D., Zirker, C.A., Abel, M.F., 2009. Angular momentum synergies during walking. Experimental Brain Research 197, 185−197.

Saltiel, P., Wyler-Duda, K., D'Avella, A., Tresch, M.C., Bizzi, E., 2001. Muscle synergies encoded within the spinal cord: evidence from focal intraspinal NMDA iontophoresis in the frog. Journal of Neurophysiology 85, 605−619.

Schieber, M.H., 2001. Constraints on somatotopic organization in the primary motor cortex. Journal of Neurophysiology 86, 2125−2143.

Schieber, M.H., Santello, M., 2004. Hand function: peripheral and central constraints on performance. Journal of Applied Physiology 96, 2293−2300.

Scholz, J.P., Danion, F., Latash, M.L., Schöner, G., 2002. Understanding finger coordination through analysis of the structure of force variability. Biological Cybernetics 86, 29−39.

Scholz, J.P., Schöner, G., 1999. The uncontrolled manifold concept: identifying control variables for a functional task. Experimental Brain Research 126, 289−306.

Scholz, J.P., Schöner, G., Latash, M.L., 2000. Identifying the control structure of multijoint coordination during pistol shooting. Experimental Brain Research 135, 382−404.

Scholz, J.P., Park, E., Jeka, J.J., Schöner, G., Kiemel, T., 2012. How visual information links to multijoint coordination during quiet standing. Experimental Brain Research 222, 229−239.

Scholz, J.P., Dwight-Higgin, T., Lynch, J.E., Tseng, Y.W., Martin, V., Schöner, G., 2011. Motor equivalence and self-motion induced by different movement speeds. Experimental Brain Research 209, 319–332.

Schöner, G., 1995. Recent developments and problems in human movement science and their conceptual implications. Ecological Psychology 8, 291–314.

Shim, J.K., Latash, M.L., Zatsiorsky, V.M., 2003. Prehension synergies: trial-to-trial variability and hierarchical organization of stable performance. Experimental Brain Research 152, 173–184.

Shim, J.K., Lay, B., Zatsiorsky, V.M., Latash, M.L., 2004. Age-related changes in finger co-ordination in static prehension tasks. Joural of Applied Physiology 97, 213–224.

Shim, J.K., Latash, M.L., Zatsiorsky, V.M., 2005a. Prehension synergies in three dimensions. Journal of Neurophysiology 93, 766–776.

Shim, J.K., Olafsdottir, H., Zatsiorsky, V.M., Latash, M.L., 2005b. The emergence and disappearance of multi-digit synergies during force production tasks. Experimental Brain Research 164, 260–270.

Shim, J.K., Park, J., Zatsiorsky, V.M., Latash, M.L., 2006. Adjustments of prehension synergies in response to self-triggered and experimenter-triggered load and torque perturbations. Experimental Brain Research 175, 641–653.

Shinohara, M., Scholz, J.P., Zatsiorsky, V.M., Latash, M.L., 2004. Finger interaction during accurate multi-finger force production tasks in young and elderly persons. Experimental Brain Research 156, 282–292.

Sin, M., Kim, W.S., Park, D., Min, Y.S., Kim, W.J., Cho, K., Paik, N.J., 2014. Electromyographic analysis of upper limb muscles during standardized isotonic and isokinetic robotic exercise of spastic elbow in patients with stroke. Journal of Electromyography and Kinesiology 24, 11–17.

St Gregory Palamas, 1983. The Triads. Classics of Western Spirituality. Paulist Press, Mahwah, NJ.

Ting, L.H., Chvatal, S.A., 2011. Decomposing muscle activity in motor tasks: methods and interpretation. In: Danion, F., Latash, M.L. (Eds.), Motor Control: Theories, Experiments, and Applications. Oxford University Press, New York, pp. 102–138.

Ting, L.H., Macpherson, J.M., 2005. A limited set of muscle synergies for force control during a postural task. Journal of Neurophysiology 93, 609–613.

Todorov, E., Jordan, M.I., 2002. Optimal feedback control as a theory of motor coordination. Nature Neuroscience 5, 1226–1235.

Torres-Oviedo, G., Macpherson, J.M., Ting, L.H., 2006. Muscle synergy organization is robust across a variety of postural perturbations. Journal of Neurophysiology 96, 1530–1546.

Tresch, M.C., Cheung, V.C., d'Avella, A., 2006. Matrix factorization algorithms for the identification of muscle synergies: evaluation on simulated and experimental data sets. Journal of Neurophysiology 95, 2199–2212.

Wang, Y., Asaka, T., Watanabe, K., 2013. Multi-muscle synergies in elderly individuals: preparation to a step made under the self-paced and reaction time instructions. Experimental Brain Research 226, 463–472.

Wang, Y., Zatsiorsky, V.M., Latash, M.L., 2005. Muscle synergies involved in shifting center of pressure during making a first step. Experimental Brain Research 167, 196–210.

Wilhelm, L., Zatsiorsky, V.M., Latash, M.L., 2013. Equifinality and its violations in a redundant system: multi-finger accurate force production. Journal of Neurophysiology 110, 1965–1973.

Winter, D.A., Prince, F., Frank, J.S., Powell, C., Zabjek, K.F., 1996. Unified theory regarding A/P and M/L balance in quiet stance. Journal of Neurophysiology 75, 2334–2343.

Weiss, E.J., Flanders, M., 2004. Muscular and postural synergies of the human hand. Journal of Neurophysiology 92, 523–535.

Woollacott, M., Inglin, B., Manchester, D., 1988. Response preparation and posture control. Neuromuscular changes in the older adult. Annals of the New York Academy of Sciences 515, 42–53.

Wu, Y.-H., Latash, M.L., 2014. The effects of practice on coordination. Exercise and Sport Sciences Reviews 42, 37–42.

Wu, Y.-H., Pazin, N., Zatsiorsky, V.M., Latash, M.L., 2012. Practicing elements vs. practicing coordination: changes in the structure of variance. Journal of Motor Behavior 44, 471–478.

Xu, Y., Terekhov, A.V., Latash, M.L., Zatsiorsky, V.M., 2012. Forces and moments generated by the human arm: variability and control. Experimental Brain Research 223, 159–175.

Yang, J.F., Scholz, J.P., 2005. Learning a throwing task is associated with differential changes in the use of motor abundance. Experimental Brain Research 163, 137–158.

Yen, J.T., Auyang, A.G., Chang, Y.H., 2009. Joint-level kinetic redundancy is exploited to control limb-level forces during human hopping. Experimental Brain Research 196, 439–451.

Zatsiorsky, V.M., 2002. Kinetics of Human Motion. Human Kinetics, Urbana, IL.

Zatsiorsky, V.M., Gao, F., Latash, M.L., 2003. Prehension synergies: effects of object geometry and prescribed torques. Experimental Brain Research 148, 77–87.

Zatsiorsky, V.M., Latash, M.L., 2008. Multi-finger prehension: an overview. Journal of Motor Behavior 40, 446–476.

Zatsiorsky, V.M., Latash, M.L., Gao, F., Shim, J.K., 2004. The principle of superposition in human prehension. Robotica 22, 231–234.

Zatsiorsky, V.M., Li, Z.M., Latash, M.L., 1998. Coordinated force production in multi-finger tasks: finger interaction and neural network modeling. Biological Cybernetics 79, 139–150.

Zatsiorsky, V.M., Li, Z.M., Latash, M.L., 2000. Enslaving effects in multi-finger force production. Experimental Brain Research 131, 187–195.

Zatsiorsky, V.M., Prilutsky, B.I., 2012. Biomechanics of Skeletal Muscles. Human Kinetics, Urbana, IL.

Zhou, T., Wu, Y.-H., Bartsch, A., Cuadra, C., Zatsiorsky, V.M., Latash, M.L., 2013. Anticipatory synergy adjustments: preparing a quick action in an unknown direction. Experimental Brain Research 226, 565–573.

Zhou, T., Solnik, S., Wu, Y.-H., Latash, M.L., 2014. Equifinality and its violations in a redundant system: control with referent configurations in a multi-joint positional task. Motor Control 18, 405–424.

Equilibrium-Point Hypothesis

The equilibrium point (EP) hypothesis was introduced by Anatol Feldman about 50 years ago (Feldman, 1966). Its fate is very unusual. It has been the target of waves of criticism and attempts to disprove it, while only a handful of researchers have tried to work with this hypothesis and develop it. Despite this imbalance between the numbers of opponents and champions, the EP hypothesis has survived and is currently considered one of the main hypotheses in the field of motor control. The community of scientists in the field of motor control owes a great deal of gratitude to Anatol Feldman who introduced the EP hypothesis, defended it from the waves of skepticism, educated generations of followers, and developed the EP hypothesis to its current state.

The EP hypothesis is built on two pillars, physics and physiology. It originated from a question: What laws of nature define interactions between the central nervous system, the rest of the body, and the environment that lead to biological movements? This formulation does not leave room for approaches that assume computations performed by the central nervous system to solve the problems related to the complex mechanics typical of human movements, in particular motor redundancy (see Chapter 10) and complexity of the interactions within the central nervous system. While laws of nature are commonly described with computational means (equations), these laws are never assumed to reflect computations performed by the objects to which they apply.

Over the time of its development, the EP hypothesis went through a few stages when it was renamed to emphasize certain aspects of the underlying mechanisms. In particular, it has been addressed as the *referent configuration hypothesis* (Feldman and Levin, 1995; Feldman et al., 2007) and the *threshold control hypothesis* (Feldman, 2011). To avoid confusion, we are going to use a single term, *EP hypothesis*, to refer to all its versions described in detail in the following sections.

12.1 Elements of history

The EP hypothesis had several main roots. The first is the argument on the role of reflexes in voluntary movements (see Chapter 6 and Prochazka et al., 2000). According to Sir Charles Sherrington, voluntary movements represented results of modulation of reflexes. This view was pushed even further by another Nobel Prize winner, Ivan Pavlov, who developed a theory according to which all behaviors represented combinations of inborn and conditioned reflexes. The alternative view considered voluntary movements as results of processes within the central nervous system, not necessarily driven, or even triggered, by sensory stimuli. Champions of this approach emphasized active rather than reactive nature of movements, their feed-forward nature. They developed such concepts as *central pattern generators* (Chapter 9) and *physiology of*

Biomechanics and Motor Control. http://dx.doi.org/10.1016/B978-0-12-800384-8.00012-0

activity (Bernstein, 1965, 1966). More recently, the importance of feedback loops and other coupling mechanisms within the body has been developed within the *dynamic systems* and *ecological psychology* approaches (Kugler and Turvey, 1987; Turvey, 1990; Kelso, 1995), while the role of central neural mechanisms has been emphasized in such concepts as *motor program* and *internal model* (Schmidt, 1975; Kawato, 1999; Shadmehr and Wise, 2005). The EP hypothesis accepted the idea of Sherrington that natural voluntary movements are produced by modulation of parameters of reflexes (for a discussion on *reflexes*, see Chapter 6). The physical and physiological ideas of the EP hypothesis are close in spirit to the dynamic systems approach. On the other hand, within the EP hypothesis, parameters of reflexes (feedback-based mechanisms) are produced by hierarchically higher structures with unknown physiology in a feed-forward manner. At some level, time profiles of these parameters may be viewed as Bernstein's *engrams* (Bernstein, 1935), a precursor of the notion of *generalized motor program* (Schmidt, 1975; see Chapter 13).

Another major argument in the field of motor control focused on the role of length (and velocity) dependence of muscle force (Chapters 2 and 3). Bernstein (1935) emphasized that this dependence made it impossible for the central nervous system to predict the mechanical consequences of neural signals given the changing and not perfectly predictable environment. This argument may be traced back to classical papers by Wachholder and Altenburger (1927) who asked a seemingly naïve question: How can a person relax at different joint positions given the well-known length dependence of muscle forces (including forces produced by relaxed muscles; see Chapter 2)? The experimental analysis did confirm the ability of humans to relax muscles crossing a joint at different joint angles, which led the authors to conclude that, during joint motion, the central nervous system resets the length-dependent properties of muscles. In his classical book *On the Construction of Movement* published in Russian in 1947 and never translated into English, Nikolai Bernstein drew a graph with a family of parallel force—length muscle characteristics (Figure 12.1) and suggested that these characteristics could be moved by the central nervous system parallel to the length axis to produce movements (in the original book, the lines were supposed to correspond to different levels of muscle excitation, in contrast to the EP hypothesis, cf. Figure 12.3). Bernstein implied that the central nervous system should not fight the natural properties of muscle force generation trying to predict length-dependent force changes and compensate for their effects on ongoing movements (as it is done in some theories based on the idea of internal models, Shadmehr and Mussa-Ivaldi, 1994; Shadmehr and Wise, 2005). Instead, he thought that these properties should form the basis for the neural control of movements.

One more root of the EP hypothesis related to the mentioned length dependence of muscle force is the *posture-movement paradox* (see Chapter 8) formulated by von Holst and Mittelstaedt (1950): How can a person produce a movement without triggering the resistance of involuntary posture-stabilizing mechanisms? Indeed, if voluntary and involuntary (reflex) mechanisms are viewed as two independent, separate contributors to muscle activation, the problem has no simple solution. The EP hypothesis united the reflex and voluntary mechanisms into a single scheme thus avoiding the posture-movement paradox (see further details later in this chapter).

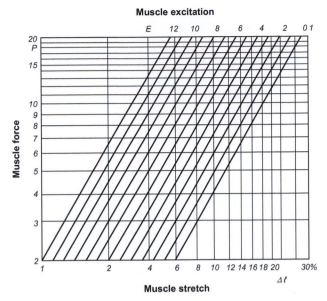

Figure 12.1 An illustration of muscle force—length characteristics from Bernstein (1947). The slanted lines correspond to different levels of muscle excitation (shown on the top). Note that higher excitation levels result in shifts of the force—length line to the left along the length axis.

One of the precursors of the equilibrium-point hypothesis was the *servo-control hypothesis* suggested by Merton in 1953. The main idea of Merton's hypothesis was to use descending neural signals to gamma-motoneurons in order to change the set point of muscle spindle endings and thus initiate movement to a new equilibrium value of muscle length (Figure 12.2(A)). Further, the stretch reflex would act to move the muscle to the specified length—to an equilibrium point (Figure 12.2(B)). The servo-control hypothesis faced two major problems. First, according to this hypothesis, voluntary movements had to be initiated by signals to gamma-motoneurons, and activation of alpha-motoneurons was expected after a substantial conduction time delay. Second, gain of the stretch reflex had to be very high (note the nearly vertical force—length lines in Figure 12.2(B)) to ensure movement to a specified value of muscle length in different external loading conditions. Both predictions were proven wrong (Matthews, 1959; Vallbo, 1970, 1974). In particular, in 1959, a paper by Peter Matthews described characteristics of the tonic stretch reflex (which had been described earlier by Liddell and Sherrington, 1924) in a decerebrate cat preparation. This study documented limited gain of the stretch reflex. While the simultaneous activation of alpha- and gamma-motoneurons (alpha-gamma co-activation, Section 6.5.1) was incorporated into a later version of the servo hypothesis, the problem with the low gain in the tonic stretch reflex arc resulted in its refutation.

The first papers describing the EP hypothesis were published in Russian in 1965—1966 (Asatryan and Feldman, 1965; Feldman, 1966) and promptly translated into English. The first papers addressed the control of a single muscle and emphasized

(A)   **(B)**

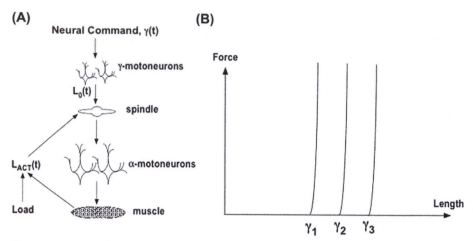

Figure 12.2 According to the servo model, movements start from a signal to gamma-motoneurons (A), which define a target value of muscle length. The stretch reflex loops act as a perfect servo stabilizing the length value independently of the external force. (B) As a result, for a given command (γ), the length–force characteristic is nearly vertical. Modified by permission from Latash (2012).

parallels between muscle control and motion of a mass-spring system. This was justified for didactic purposes but also led to numerous misinterpretations (reviewed in Feldman and Latash, 2005). A very important paper was published by Feldman and Orlovsky in 1972, which provided much needed links between the EP hypothesis and neurophysiological mechanisms. It was also the first paper on the EP hypothesis written in English and published in a respected international journal. By the mid-1970s, the EP hypothesis attracted the interest of several researchers outside the Soviet Union (Schmidt and McGown, 1980; Nichols and Houk, 1976; Bizzi et al., 1982; Kugler and Turvey, 1987). However, the broad neurophysiological community did not pay much attention to the hypothesis; clearly, the EP hypothesis was not understood at the time by most researchers in the field.

By the end of the 1970s, a series of ingenious experiments were performed on monkeys by Emilio Bizzi's research group (Polit and Bizzi, 1978, 1979; Bizzi et al., 1982). These experiments emphasized equifinality of movements under transient perturbations—one of the predictions of the EP hypothesis—and also the smooth transition from an initial to a final equilibrium point—the equilibrium trajectory. The experiments were performed on deafferented animals. As a result, the tonic stretch reflex, which was a fundamental component of the EP hypothesis, was absent. This led to the emergence of another model of equilibrium-point control addressed as the *alpha-model* (see the next section).

The collapse of the Soviet Union in the late 1980s and the subsequent emigration of Anatol Feldman and some of his colleagues (Sergei Adamovich, Olga Fookson, Mark Latash, Grigory Orlovsky, to name a few) to Western countries triggered an important step in the development of the EP hypothesis. The experimental foundation of the EP hypothesis started to grow, its appreciation among colleagues started to spread, and the hypothesis itself started to be developed much more rapidly to address various aspects of motor control such as motor variability, electromyographic patterns, patterns of equilibrium trajectories, multimuscle actions, kinesthetic perception, and links to neurophysiological mechanisms within the human body.

12.2 Current terminology

The name *equilibrium-point hypothesis* emphasizes the idea that movements of physical objects may be viewed as transitions between equilibrium states (equilibrium points). An equilibrium state, by definition, is a state in which the resultant force acting on the object is zero. For example, a muscle acting against an external load may be in an equilibrium state when muscle force is equal in magnitude and directed against the load force. So, if a muscle is in equilibrium with external forces, its state may be characterized with two variables, length and force. These two variables form the equilibrium point for the muscle on the length–force plane (Figure 12.3).

There is considerable confusion due to the existence of two versions of the EP hypothesis addressed as the α-model and the λ-model. The original Feldman's hypothesis (the λ-model) assumes that the neural control of a muscle can be adequately described as setting only one variable, *threshold of the tonic stretch reflex* (λ). Figure 12.3 illustrates the control of a muscle according to the λ-model. Setting a value of λ defines a range of muscle length values (those larger than λ) associated with muscle activation—to the right of λ in Figure 12.3. If muscle length is smaller than λ, no activation is seen. The muscle shows larger activation levels and larger forces for larger deviations of its length from λ. The muscle shows larger activation levels and larger forces for larger deviations of its length from λ. The dependence of muscle force on the difference between its length and λ is addressed as the muscle *invariant characteristic*

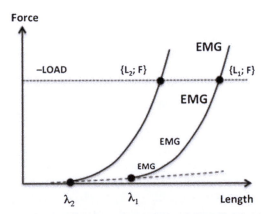

Figure 12.3 Setting a value of λ defines a range of muscle length values (those larger than λ) associated with muscle activation. Note that along the force–length characteristic, muscle activation increases (in contrast to Figure 12.1). The shallow dashed line illustrates the force–length dependence of a nonactivated muscle. A change in λ (from λ_1 to λ_2) may result in a movement (length change from L_1 to L_2), a change in force (in isometric conditions), or in both.

or *tonic stretch reflex characteristic*. If the muscle acts against a certain constant external load, it will be in equilibrium at a length corresponding to the muscle force equal in magnitude to the load force. This combination of muscle force and length is addressed as the *equilibrium point* of the system (EP in Figure 12.3).

Note that, in the absence of activation, when the muscle is relaxed, muscle—tendon complexes behave in a spring-like fashion, that is, they show an increase in force with slow stretch, at least within a certain range of muscle length values (see Chapter 2). This is reflected by the shallow dashed line in Figure 12.3. So, λ is the muscle length value at which muscle force starts to deviate from the level shown with the dashed line due to muscle activation via the reflex loops. Within the λ-model, voluntary movements are peripheral consequences of shifts of λ (cf. $λ_1$ and $λ_2$ in Figure 12.3) leading to a change in muscle length (as illustrated in the figure, cf. L_1 and L_2), force, or both depending on the external load characteristic (see Section 12.4 for more details).

The α-model was originally based on experimental observations in deafferented monkeys (Polit and Bizzi, 1978, 1979). Deafferentation led to the lack of somatosensation in those animals accompanied by the lack of muscle reflexes including the stretch reflex. Deafferented monkeys could only control their movements by specifying different levels of muscle activation directly by descending signals from the brain. It is known that, for a given level of activation, muscle force depends on muscle length and this dependence is stronger for greater activation levels (Chapter 2, see Figure 12.4).

Several important predictions of equilibrium-point control were confirmed in the mentioned experiment in deafferented monkeys. The monkeys performed movements to targets without seeing their moving limb. Since the limb was deafferented, the monkeys had no information on the actual course of its motion. In some trials, an external force perturbed the limb. In cases of quick, transient perturbations, the limb landed in the target despite moving along a different trajectory. This confirmed the

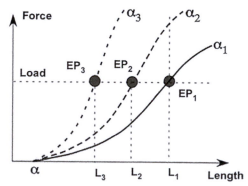

Figure 12.4 According to the alpha-model, the neural control of movement could be adequately described as specifying different levels of muscle activation (α). A shift in the value of α is expected to lead to a shift in the equilibrium point depending on the external load ($α_1 < α_2 < α_3$).
Modified by permission from Latash (2012).

prediction of equifinality, which is expected from equilibrium-point control under certain circumstances (see Section 12.7 for details). In other trials, as soon as the monkey initiated the movement, the limb was brought quickly by the external force into the target and released. In such cases, motion of the limb back toward the initial position was seen; this movement reversed and then moved back into the target. These results confirmed the gradual shift of the equilibrium point during the quick movement production.

Based on these observations, a hypothesis was offered that the neural control of movement could be adequately described as specifying different levels of muscle activation (α). A shift in the value of α was expected to lead to a shift in the equilibrium point depending on the external load (cf. Figures 12.4 and 12.3).

A few obvious differences between the λ-model and α-model can be seen immediately from Figures 12.3 and 12.4. The muscle is always active within the same length range in the α-model while the range of muscle length values within which the muscle shows nonzero activation varies with λ in the λ-model. For a given value of the assumed control signal, muscle activation varies in the λ-model (due to the varying contribution of the reflexes) while the level of muscle activation is constant within the α-model (since no reflex contribution to muscle activation is possible). Clearly, the mentioned features of the α-model contradict experimental observations in intact animals and humans. Therefore, it is fair to conclude that the α-model may be an adequate description of the neural control of movements in animals without reflexes while the λ-model describes the neural control of movements performed by intact animals. Further in this chapter, we always imply the λ-model under the EP hypothesis.

12.3 Control with threshold elements

One of the basic features of neurons is their threshold property. Indeed, a neuron (including motoneurons) does not produce output signals (action potentials) unless the membrane potential reaches a particular threshold value. So, if a neuron receives a combination of excitatory and inhibitory inputs that lead to changes in its membrane potential, its output will be zero as long as the induced variation of the membrane potential is below the threshold. If the potential reaches the threshold, a standard action potential is generated characterized by a standard amplitude and time course, which are independent of the magnitude of the input. This behavior is similar to the one of a typical toggle switch—if one presses on the switch with a force below its threshold, no apparent consequence will be seen; if the force is above the threshold, the switch will flip and lead, for example, to the light being turned on. This effect will be independent of the actual force applied to the switch.

Figure 12.5(A) illustrates a very much simplified cartoon neuron with three inputs: two inputs from other parts of the central nervous system (C1 and C2) and the third one (A) carried by afferent fibers from peripheral receptors. Note that if the central inputs, C1 and C2, lead to depolarization of the neuron above its threshold, the neuron will generate action potentials at its highest possible frequency (defined, in particular, by the duration of its *refractory period*—the state of lack of excitability following an

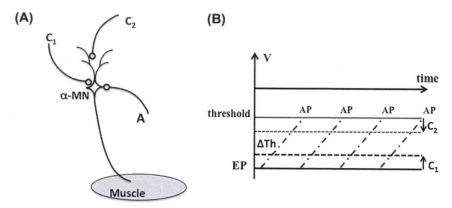

Figure 12.5 (A) A scheme of three inputs into an alpha-motoneuron (A — afferent; C_1 and C_2 — central). (B) An illustration of the three methods of control of a neuron. C_1 and C_2 define the effective distance to the threshold (ΔTh). The afferent input (A) is assumed to define the ramp-like change in the membrane potential (V). AP show the moments of action potential generation.

action potential). In this case, the frequency of firing of the neuron will not depend on the afferent input. In contrast, if the central inputs depolarize the neuronal membrane below the threshold, activity of the neuron will depend on the A input. In other words, the neuron would behave as an adaptive element, changing its activation as a function of the state of peripheral structures, from which it receives the afferent input.

Figure 12.5(B) illustrates the membrane potential and its changes under the action of the three inputs shown in Figure 12.5(A). For simplicity, assume that the afferent input leads to a ramp-like change in the membrane potential (dashed ramp lines). One of the central inputs (C1) induces a steady shift of the membrane potential (V) toward depolarization. The other input (C2) can change the threshold value itself. Clearly the frequency of firing of the neuron will be a function of all three inputs: C1 and C2 define the effective distance from the steady-state level of depolarization to the threshold (ΔTh), while A defines the speed at which the membrane potential changes (S = dV/dt). The rate of firing can be approximated as: $R_f = \Delta Th/S$. This is obviously a crude approximation because it is based on cartoon assumptions and does not consider, for example, such important features of neuronal activation as the refractory period and the action of the sodium—potassium pump that may not allow the A input to reach the activation threshold.

According to the EP hypothesis, neural control of the muscle may be adequately described as subthreshold depolarization of motoneurons in the corresponding pool. While two central inputs are illustrated in Figure 12.5(A), we will primarily focus on the one that leads to membrane depolarization (C1). The other input (C2) is much less well understood. It may be related to the phenomenon of persistent inward currents (PICs; Heckman et al., 2003, 2005) that are pronounced on dendrites; PICs lead to an effective change in the threshold for neuronal activation and can potentially even turn a neuron into an autogenerator of action potentials. While the potential role of this input may be very important in animal movement (Heckman et al., 2005),

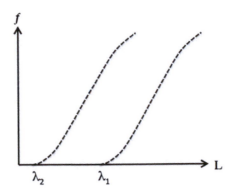

Figure 12.6 Two dependences of frequency of firing of a neuron on muscle length for two values of the threshold of the stretch reflex (λ).

we will limit the discussion to C1 only assuming that it changes the effective distance to the threshold for neuronal activation.

Note that the effects of C1 are measured in millivolts, an adequate unit for the description of neuronal activity. On the other hand, these effects can be expressed in units of muscle length given the action of the length-sensitive reflex input—tonic stretch reflex. This is illustrated in Figure 12.6, which shows two dependences of the frequency of firing of a neuron (or neurons within a pool) as a function of the length of the innervated muscle. Setting up a value of the central input defines a threshold level of sensory feedback activity (A), at which the muscle starts to show signs of activation. Assuming that much of the excitatory sensory feedback comes from length-sensitive receptors in muscle spindles, a value of C1 defines the muscle length, at which the muscle starts to show signs of activation during very slow stretch. This is the definition of the threshold of the tonic stretch reflex, which is measured in units of muscle length (meters). Following the original papers by Feldman (1966), we will use the Greek letter lambda (λ) for this threshold value in length units. Hence, the mechanism of length-sensitive feedback effectively converts millivolts (C1) into meters (λ). A change in the central input leads to a change in λ resulting in a nearly parallel shift of the dependence of the frequency of firing of the neuron on muscle length (see the two curves in Figure 12.6).

12.4   Control of a single muscle

It is common to represent the tonic stretch reflex characteristics of a muscle as curves on the force–length plane (Figure 12.7). If a muscle is stretched slowly starting from a very short length, at first it may show no signs of electrical activation. Nevertheless, it will resist the stretch due to the passive properties of the muscle and tendon tissues (see Chapter 2). This dependence is shown in Figure 12.3 as a shallow dashed curve; we do not show this line in Figure 12.7 and assume that only the active force (due to muscle activation) is plotted as a function of muscle length. Assume that at some threshold

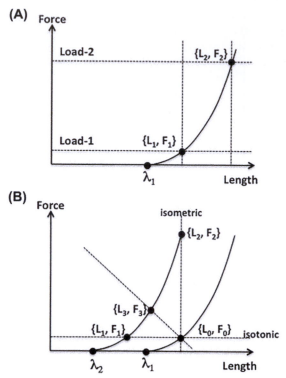

Figure 12.7 During slow stretch, the muscle will show signs of electrical activity at some value of muscle length (λ). (A) Passive movement. A change in the external load (e.g., from Load-1 to Load-2) is expected to produce a movement (a change in muscle length) to a new EP (from $\{L_1; F_1\}$ to $\{L_2; F_2\}$) accompanied by a change in muscle force and EMG. (B) Active movement. A change in λ (from λ_1 to λ_2) can lead to different peripheral consequences depending on the external load (dashed lines). If the load is constant (isotonic conditions), a movement happens to a new value of muscle length. If the muscle acts against a stop (isometric conditions), the same shift of λ leads to a change in muscle force. In the most general case of acting against a position-dependent external load, both force and length change.

value of muscle length, the muscle starts to show signs of electrical activation (λ_1 in Figure 12.7(A)). Starting from that point, further stretch will induce an increase in muscle force illustrated by the solid curve deviating from the dashed one. Along this curve, both muscle force and its level of activation (EMG) increase with muscle length. If a muscle acts against a constant external load, the system "muscle + reflexes + load" would reach an equilibrium when the active force of the muscle will be equal in magnitude to and directed against the load. This combination of muscle force and length is referred to as the equilibrium point (EP) of the system.

   Two types of movement are possible within the scheme in Figure 12.7. First, a change in the external load (e.g., from Load-1 to Load-2) is expected to produce a movement (a change in muscle length) to a new EP (from $\{L_1; F_1\}$ to $\{L_2; F_2\}$) accompanied by a change in muscle force and EMG. This movement may be called passive despite the changes in muscle activation levels, which happen only due to the changed input from peripheral receptors. An active movement is produced by a change in the descending signals to the corresponding motoneuronal pool (C1 in Figure 12.5) resulting in a change in λ (from λ_1 to λ_2). This could potentially lead to different peripheral consequences depending on the external load characteristic. If the load is constant (isotonic conditions), a movement will happen to a new value of muscle length. If the muscle acts against a stop (isometric conditions), the same shift of λ would lead to a change in muscle force. In the most general case of acting against a position-dependent external load, both force and length are expected to change. So, within the EP hypothesis, there are no qualitative differences in the control of movements and force production. Certainly, depending on the task, humans can use different time profiles $\lambda(t)$ for the involved muscles. However, the principle of control with shifts of the threshold of the tonic stretch reflex remains the same across tasks and external loading conditions.

12.4.1 Three basic trajectories

Control of a muscle may be associated with three different trajectories illustrated in Figure 12.8. First, there is a control trajectory, which may be adequately expressed as $\lambda(t)$. In some studies, $\lambda(t)$ and corresponding variables for the control of a joint (see Section 12.5) have been addressed as the *virtual trajectory* of the movement. The term *control trajectory* seems more appropriate and less ambiguous. For each moment of time, there exists an instantaneous EP defined by the current value of λ and the external load. Depending on the external load characteristic, the same control trajectory may lead to different trajectories of the EP (as in Figure 12.7). Note that the EP trajectory is a combination of the time profiles of two variables, muscle force and muscle length, $\{F_{EP}(t); L_{EP}(t)\}$. Sometimes, the term *equilibrium trajectory* is

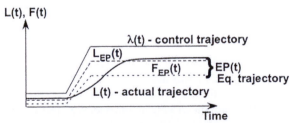

Figure 12.8 Three basic trajectories within the EP hypothesis. A control trajectory may be adequately expressed as $\lambda(t)$. Depending on the external load characteristic, the same control trajectory may lead to different trajectories of the EP $\{F_{EP}(t); L_{EP}(t)\}$. $L_{EP}(t)$ is sometimes addressed as the equilibrium trajectory. The actual trajectory, $\{L(t); F(t)\}$, is expected to reflect such factors as the inertial load, damping, and muscle force-generating capabilities. Modified by permission from Latash (2012).

used to address $L_{EP}(t)$, while the accompanying changes in $F_{EP}(t)$ are commonly ignored. Finally, an actual movement takes places. Its kinetic and kinematic characteristics, $L(t)$ and $F(t)$, are expected to reflect a host of factors including, for example, the inertial load, damping, and muscle force-generating capabilities. Only $L(t)$ and $F(t)$ are relatively easily measured in experiments, while the control and equilibrium trajectories require complex experimental procedures and data processing under certain assumed mechanical properties of the muscle (Latash and Gottlieb, 1991; Gribble et al., 1998).

12.4.2   Factors affecting λ

In the original papers published about 50 years ago, Anatol Feldman assumed that changes in λ were under exclusive control of neural mechanisms associated with the production of voluntary movements. This simplification ignored a number of factors that affect membrane depolarization of a motoneuronal pool. These factors include, in particular, reflex effects from sensory receptors located in remote muscles, those within the limb and in other limbs of the body, history dependence of the processes on the neuronal membrane, and the dependence of the length-sensitive receptors on other variables such as, for example, muscle velocity (reviewed in Feldman, 1986). As described in detail in a relatively recent paper (Feldman and Latash, 2005), the effective λ on a neuronal membrane can be viewed as the sum of the mentioned four factors:

$$\lambda = \lambda^* + \mu V + \rho + f(t), \tag{12.1}$$

where λ^* is the central contribution to λ, μ is a constant reflecting velocity sensitivity of muscle spindle endings, ρ is a reflection of proprioceptive signals from other muscles, and $f(t)$ reflects the history dependence on the membrane potential. The fact that the central input defines only a portion of the overall λ is not a major problem. In fact, when we control any device, for example, drive a car, our action is only one component that defines the car's response. In particular, effects of hand effort applied to the steering wheel on the actual wheel rotation depend on the presence of the power-assisted steering mechanism, friction between the wheels and the pavement, and a few other factors.

The sensitivity of muscle spindle endings to muscle velocity is an important factor defining the mechanical behavior of a muscle during voluntary movements. This dependence may be illustrated by a line with a negative slope on the velocity−length plane (phase plane; Figure 12.9). This line divides the phase plane into zones of muscle activation (to the right of the line) and muscle silence (to the left of the line). The point of intersection of this line with the muscle length axis is λ. The slope of the line reflects the sensitivity of the reflex effects to muscle velocity (μ in Eqn (12.1)). This dependence has been discussed as a major factor that defines the typical electromyographic (EMG) patterns observed during fast movements (the so-called *tri-phasic EMG pattern*, reviewed in Gottlieb et al., 1989) and the ability of quick human movements to stop without major terminal oscillations.

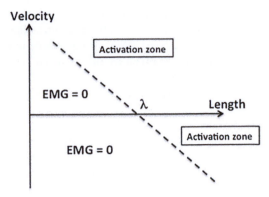

Figure 12.9 On the length–velocity plane, a line with a negative slope (dashed) divides the plane into zones of muscle activation (to the right of the line) and muscle silence (to the left of the line, EMG = 0). The slope of the line reflects the sensitivity of the reflex effects to muscle velocity.

12.5 Control of a joint

Generalization of the EP hypothesis to the control of a joint with one kinematic degree of freedom is relatively straightforward for a simple joint crossed by two opposing muscles, which we will address as *agonist* and *antagonist*. If we assume that the agonist and antagonist produce positive and negative moments of force, respectively, the tonic stretch reflex characteristics of the two muscles can be illustrated with two dashed curves shown in Figure 12.10. Note that the X-axis corresponds to joint angle,

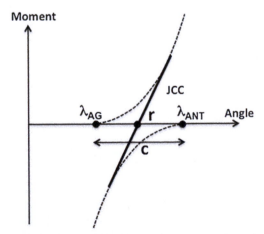

Figure 12.10 Control of a joint crossed by two muscles, an agonist and an antagonist, can be described with two variables, λ_{AG} and λ_{ANT}, respectively, or an equivalent pair, r-command and c-command corresponding to the midpoint between λ_{AG} and λ_{ANT} and to the range between λ_{AG} and λ_{ANT}.

and the length of one of the two muscles increases with α, while the length of the opposing muscle decreases with α. The overall mechanical behavior of the joint (the total moment of force or joint torque) at steady states is defined by the combined action of the two muscles, the joint compliant characteristic (JCC) shown by the thick, solid line.

The control of the two muscles can be described with two variables, the corresponding values of the threshold of the tonic stretch reflex, λ_{AG} and λ_{ANT} for the agonist and antagonist, respectively. Alternatively, another pair of variables can be used referred to as reciprocal command (r-command) and co-activation command (c-command) (Feldman, 1980). The former defines the midpoint of the angular range where both muscles are active, while the latter defines the width of this range:

$$r = (\lambda_{AG} + \lambda_{ANT})/2$$

$$c = \lambda_{ANT} - \lambda_{AG}$$

(12.2)

Note that, just like λ, the r-command and c-command are measured in spatial units. A change in the r-command leads to an effective motion of the joint (assuming a constant external torque), while changing the c-command results in an effective change in the slope of the JCC due to the nonlinearity of the tonic stretch reflex characteristic. If a joint acts against a nonzero external torque, changes in both r- and c-commands result in joint motion as illustrated in Figure 12.11. In isometric conditions, changes in both r- and c-commands lead to a change in the net joint moment of force.

12.5.1 Measuring control and equilibrium trajectories

The main control variables assumed within the EP hypothesis, such as λ, r-command, and c-command, are not directly observable. They can be, however, reconstructed

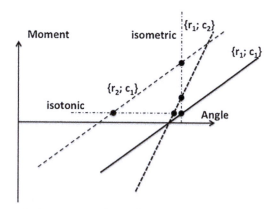

Figure 12.11 Changes in the r-command, e.g., from $\{r_1; c_1\}$ to $\{r_2; c_1\}$, lead to translation of the joint characteristic. Changes in the c-command lead to rotation of the joint characteristic, e.g., from $\{r_1; c_1\}$ to $\{r_1; c_2\}$. Peripheral effects of such changes depend on the load characteristic leading to joint motion in isotonic conditions and joint moment change in isometric conditions.

based on a number of assumptions. The main idea of this method is to ask a person to perform a well-learned movement many times and change smoothly the external loading conditions during the execution of some of the movements. The subject of this experiment is instructed not to react to possible errors in performance that may occur due to changes in external forces (perturbations). Assuming that the instruction is strictly followed by the subjects means that, across all trials, similar time profiles $\lambda(t)$, $r(t)$, and $c(t)$ are being repeated. Actual trajectories, however, are expected to differ because of the different external forces. Figure 12.12 illustrates three such single-joint trials performed during an unchanged external load (Unchanged Load), against a smoothly increasing load (Loading), and against a smoothly decreasing load (Unloading). If for a specific phase of the action, for example, shown with the dash-dotted vertical line, the matched values of joint muscle torque and angle are plotted on a torque–angle plane, they are expected to belong to a single straight line—the JCC (as in Figure 12.10). The intercept of this line reflects the r-command, while its slope reflects the c-command.

The application of this method is not trivial, however, because it requires accepting a mechanical model of joint motion. Typically, second-order linear models have been used for this purpose with an inertial term, a damping term, and a term reflecting the JCC (a spring-like linear term). The adequacy of such models has been questioned a number of times (Feldman et al., 1998; Gribble et al., 1998), but nothing better has been offered yet. In addition, there are no reliable estimates of the damping coefficient for intact muscles. Note that the velocity sensitivity of the primary muscle spindle endings makes their reflex effects major contributors to the damping properties of a joint, although at a time delay. No reliable estimates of those effects are available.

The application of this method (under all the mentioned assumptions) to fast elbow movements led to the reconstruction of so-called N-shaped equilibrium trajectories,

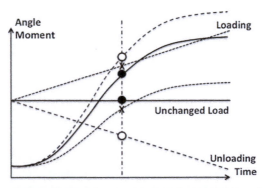

Figure 12.12 Three trials performed by a person against a constant load (Unchanged), a smoothly increasing load (Loading), and a smoothly decreasing load (Unloading). For a specific phase (dash-dotted vertical line), the matched values of load and trajectory (shown by similar symbols) are expected to fit a single JCC. The intercept of this characteristic reflects the r-command, while its slope reflects the c-command.

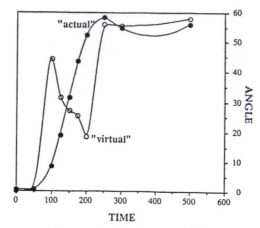

Figure 12.13 An example of the so-called N-shaped equilibrium trajectory (open circles, "virtual") reconstructed during a series of fast elbow flexion movements (closed circles, "actual").
Reproduced by permission from Latash and Gottlieb (1991).

which were in front of the actual joint trajectories during the first half of the movement, then reversed, and then reversed again and led to the desired final joint location (Figure 12.13; Latash and Gottlieb, 1991, 1992). Note that the N-shaped equilibrium trajectories were reconstructed assuming values of the damping term corresponding to an underdamped joint. If joints were overdamped, straight equilibrium trajectories leading from the initial to the final joint angle would be sufficient (Gribble et al., 1998). Similar nonmonotonic equilibrium trajectories were also reconstructed during fast movements of one joint of the two-joint, wrist-elbow, arm segment (Latash et al., 1999). Similar time patterns of equilibrium trajectories were seen for the joint that was instructed to move and for the other joint that showed only small flapping under the action of joint coupling moments. These observations suggest that nonmonotonic equilibrium trajectories may be used during natural fast movements.

12.6   Referent configuration hypothesis

The principle of control accepted within the EP hypothesis has been developed for movements involving large groups of muscles and several joints, including whole-body movements (Feldman and Levin, 1995). In particular, task-specific neural command is assumed to represent subthreshold depolarization of a group of neurons (similar to C1 in Figure 12.5); it is also expected to correspond to a set of referent spatial coordinates for task-specific performance variables (analogous to λ for the control of a single muscle, see Figure 12.6). Feedback loops are assumed to act similarly to the tonic stretch reflex loop: They generate changes in muscle activation that move the actual values of the task-specific variables to their referent values.

The set of referent coordinates for salient variables has been addressed as the body referent configuration (RC) at the task level, resulting in the name *referent configuration hypothesis*. The difference between the actual body configuration and RC drives muscle activation moving the body toward RC where it would attain a state of minimal potential energy and minimal muscle activation compatible with the anatomical and external constraints. Similarly to the illustration in Figure 12.7, if motion of a salient variable toward its referent value is impossible (blocked), a state with nonzero active forces produced on the environment will be reached.

Control with RCs may be illustrated using an intuitive example of controlling movement of a donkey with the help of a carrot, which is moved to a desired location in space. The discrepancy between the donkey's head coordinate and carrot coordinate (the carrot coordinate is the RC) leads to the generation of changes in muscle activation moving the donkey's head to the carrot. Since the head motion requires coordinated action of other body parts including legs (a baby learns early in childhood that moving to an object of interest placed outside the immediate reach without stepping leads to a fall), problems of motor redundancy emerge (see Chapter 10), which are considered in more detail in Section 12.8.

Consider as an example a person gripping an object with the thumb opposing the four digits and holding it vertically in the air (Figure 12.14)—if analysis is limited to a plane where all the digits make contacts with the object (the grasp plane, see Chapter 15), RC may be viewed as consisting of three components: referent aperture between the lines of thumb and finger contact (AP_{REF}), referent vertical coordinate (Z_{REF}), and referent orientation (α_{REF}). The rigid walls of the object do not allow the digits to move toward each other resulting in the gripping force, normal to the

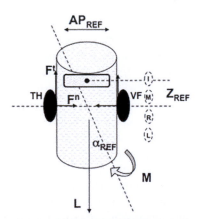

Figure 12.14 RC components for a static prehensile task. The four actual fingers (I — index, M — middle, R — ring, and L — little) are represented by an imagined digit (virtual finger, VF). The external load (L) is resisted by tangential forces (F^t) produced by a referent vertical coordinate (Z_{REF}) above the finger contact level. Referent aperture, AP_{REF}, smaller than the actual grip aperture, results in nonzero normal (grip) forces, F^n. The difference between the actual object orientation and referent angle (α_{REF}) results in a nonzero moment of force produced by the hand that acts against the external torque.

surface of contact (F^n), if AP_{REF} is smaller than the actual grip aperture. The difference between the actual vertical coordinate of the hand and Z_{REF} results is nonzero vertical forces acting against the weight of the object (tangential forces, F^t). The difference between the actual object orientation and α_{REF} results in a nonzero moment of force produced by the hand that acts against the external torque. If the external forces change, motion of the actual hand and digit coordinates toward the RC is expected (and was observed experimentally, Latash et al., 2010).

Along similar lines, movement of the tip of a finger during a pointing movement may be associated with a shift of the three-dimensional finger referent coordinate. An important component of RC is analogous to the c-command introduced earlier (Section 12.5). Indeed, one can occupy a position of the fingertip in space and then vary muscle co-contraction without moving the fingertip. Varying the co-contraction level results in scaling of the ellipsoid of apparent stiffness (see Chapter 2) of the endpoint (Flash, 1987). Hence, RC may be viewed as a superposition of two commands. One of them (similarly to the r-command) defines an equilibrium point in space for the effector, while the other command (similar to the c-command) defines a range of spatial deviations from that point when both shortened and stretched muscles continue to show nonzero activation. Mechanisms leading to unintentional changes in muscle activation during externally imposed deviations of the endpoint from its equilibrium state represent a superposition of the numerous reflex loops (including tonic stretch reflex loops for all the involved muscles). The term *generalized displacement reflex* has been suggested for this mechanism (Latash, 1998).

12.7   Equifinality and its violations

In early papers, the λ-model was associated with changing zero length of the "muscle spring," and the EP hypothesis was rather directly associated with the control of a mass-spring system. This simplification led to a prediction of *equifinality* of natural movements. This term implies that a transient perturbation applied in the course of a motor task should not affect the final state of the system, which is expected to depend only on the parameters of the assumed mass-spring system and external forces in the final state. Equifinality is a natural consequence of the described method of control with setting the tonic stretch reflex threshold for the muscles. Indeed, for a given λ value and the same external force field, there is a single equilibrium point, that is, a combination of muscle length and force. Transient forces acting during the transition of the system from an initial equilibrium state to a final equilibrium state can change the trajectory, along which the system moves, but not the final equilibrium state. Indeed, a number of studies reported equifinality of movements and postural tasks under the action of a small, brief transient perturbation (Kelso and Holt, 1980; Schmidt and McGown, 1980; Bizzi et al., 1982; Latash and Gottlieb, 1990).

A number of more recent studies reported violations of equifinality in conditions when the subjects moved in a destabilizing force field. In one series of studies (Lackner and DiZio, 1994), subjects were sitting in the center of a large centrifuge that was accelerated very slowly so that the subjects were unaware of the rotation. The arm

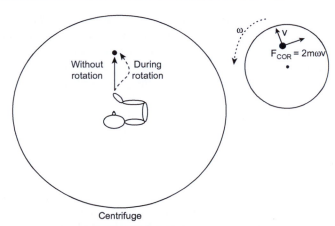

Figure 12.15 During pointing to a remembered target in a stationary environment, movements of the fingertip are nearly straight (without rotation). During rotation in a centrifuge, Coriolis force (F_{COR}) acts on the moving hand (the right insert) resulting in hand deviation from the straight trajectory. During the second half of the trajectory, the hand tends to correct the deviation but shows residual errors.
Modified by permission from Latash (2012).

reaching movements were practiced without rotation of the centrifuge and then performed in total darkness to a light target. Without rotation, the movement trajectories were nearly straight. During rotation, the Coriolis force acted on the hand during its motion. If the hand moves in a plane orthogonal to the axis of rotation of the room, Coriolis force, $\mathbf{F}_{COR} = 2m\omega\mathbf{V}$, where ω is the angular velocity of rotation of the room, $\mathbf{V}$ is hand velocity, and m is mass. As a result, the trajectories became curved, deviated strongly from the straight line, and then compensated the lateral deviation during the second half of the trajectory, but this compensation was only partial, and the hand landed off the target (Figure 12.15). Since Coriolis force acts only during the movement but not at the final steady state, the residual errors in the final position were described as violations of equifinality.

Another example of violations of equifinality was reported in experiments where subjects practiced for a long time a fast movement in a simulated force field with negative damping, that is, force acting proportional to the velocity in the direction of the velocity vector (impossible in nature!) (Hinder and Milner, 2003). After practice, the subjects consistently undershot the target when the force field was turned off. In both examples, violations of equifinality were discussed as incompatible with and, therefore, falsifying the EP hypothesis.

While the described experiments report violations of equifinality, the conclusions were based on misunderstanding of the EP hypothesis. Equifinality is not a rule but rather a rare exception possible within the EP hypothesis under certain assumptions. These include, in particular, no reaction of the central nervous system to the perturbation, no changes in muscle force-generating capabilities, and a few other assumptions reviewed in more detail in Feldman and Latash (2005). In fact, new examples of violations of equifinality have been observed in recent studies and interpreted within the framework of the EP hypothesis. For example, a transient perturbation applied to a redundant motor system may result in equifinality at the level of salient task-specific variables but not at the level of elemental variables. In particular, a transient force perturbation applied to the hand may lead to hand motion to the initial position and orientation achieved with significantly different joint configurations (Zhou et al., 2014a). Moreover, if a transient perturbation lasts for several seconds, even the salient task-specific variables may show violations of equifinality (Zhou et al., 2014b). These observations led to a hypothesis that two mechanisms define motion of a system. The first mechanism, addressed as *direct coupling*, leads to the body motion (or force production on the environment if body motion is blocked) following a shift of the RC. This process is supposed to be quick and proceed at typical times of action of spinal reflex loops, on the order of 100 ms. The second mechanism, *back-coupling*, leads to a shift of RC if the actual body configuration is kept away from the RC for a relatively long time, at least a few seconds. Note that both processes contribute to motion of the two body configurations, actual and referent, toward each other thus moving the body toward a state with minimal potential energy.

12.8   Relation of the EP hypothesis to the notion of synergies

There are several types of relations between the EP hypothesis and the notion of synergies (Chapter 11). To remind, three definitions of synergy were discussed in the previous chapter. The first definition (Synergy-A in Chapter 11) defined synergy as a stereotypical motor pattern seen in some neurological disorders characterized by spasticity. Recently, the ideas of EP hypothesis were applied to the control of spastic muscles (Jobin and Levin, 2000; Levin et al., 2000; Musampa et al., 2007). Within these ideas, the two aspects of spasticity, uncontrolled contractions and poor voluntary control, are linked via a single mechanism—an inability to shift the tonic stretch reflex threshold (λ) within its whole range used during unimpaired movements. This is illustrated in Figure 12.16. Healthy persons can relax muscles even at their longest length values and can produce large muscle forces even at their shortest length values. This means that the range of λ (from λ^- to λ^+ in the top panel of Figure 12.16) is larger than the biomechanical range of muscle length (shown with two dashed vertical lines). If the range of intentional λ changes in severely limited (e.g., between λ_S and λ_L, the bottom panel), two consequences may be expected. First, the muscle will show unintentional contractions at lengths larger than λ_L. Second, it would be unable to produce voluntary activation at lengths shorter than λ_S. A flexion synergy (or extension synergy) seen, for example, in patients after stroke, may be a consequence of λ_L shifting unintentionally to values shorter than the shortest muscle length for a group of flexor (or extensor) muscles. This is expected to move the limb into full flexion (extension) with nonzero activation of flexors (extensors) in the final posture. In terms of referent configurations, this mechanism corresponds to an unintentional shift of RC beyond the biomechanically accessible full flexion (full extension) state.

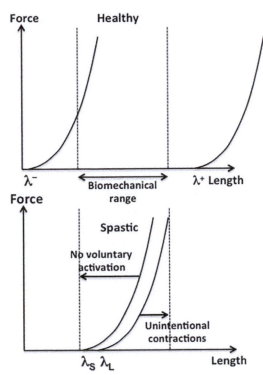

Figure 12.16 Spasticity within the EP hypothesis. In a healthy muscle, the range of λ (from λ^- to λ^+, the top panel) is larger than the biomechanical range of muscle length (shown with two dashed vertical lines). If the range of intentional λ changes is severely limited (between λ_S and λ_L, the bottom panel), the muscle will show unintentional contractions at lengths larger than λ_L, and it would be unable to produce voluntary activation at lengths shorter than λ_S.

The second definition (Synergy-B) links synergies to groups of performance variables—kinetic, kinematic, or electromyographic—that show parallel scaling over time or over task variations. Within the language of the EP hypothesis, synergies of this type are associated with preferred patterns of shifts of the body RC, which are combined to produce specific movements. Indeed, some of the typical patterns of this kind of synergies (modes) seen in whole-body postural tasks suggest muscle activation that would be observed during basic shifts of the RC of the body (Robert et al., 2008). Such basic shifts may be viewed as coordinates in a space of higher-order variables that unite muscles into stable groups. Sets of such RC-related variables are still likely redundant with respect to typical motor tasks, and hence, synergies of the third type (Synergy-C in Chapter 11) have been quantified within the space of muscle modes (Sections 11.3.1 and 11.4.3).

The third definition of synergy links it to stability of performance reflected, in particular, in covaried across trials adjustments of elemental variables that keep a salient performance variable relatively invariant (Synergy-C, reviewed in Latash et al., 2007). Within this definition, there are two links between the EP hypothesis and synergies. First, the basic mechanism of the control of a muscle (tonic stretch reflex) may be viewed as a synergy stabilizing the equilibrium state of the muscle. Second, the control with referent body configurations organized in a hierarchical way is expected by itself to lead to synergies stabilizing task-specific variables.

12.8.1   Tonic stretch reflex as a synergy

Within the EP hypothesis, tonic stretch reflex is a major mechanism of the control of a muscle. In a simplified scheme, consider only two inputs into a pool of alpha-motoneurons innervating the muscle (Figure 12.17), a central input that defines the threshold of the tonic stretch reflex (λ) and a peripheral input that provides a length-sensitive input. Imagine that for a certain value of λ, the muscle is in equilibrium acting against an external load. Imagine now that one of the alpha-motoneurons stopped producing action potentials for an unknown reason. This perturbation would lead to a drop in the total activation of the muscle. As a result, muscle force will decrease below the external load magnitude, and the resultant force will act to stretch the muscle (Figure 12.17). Muscle stretch will be sensed by the length-sensitive receptors, which will increase their level of firing and produce additional activation of the

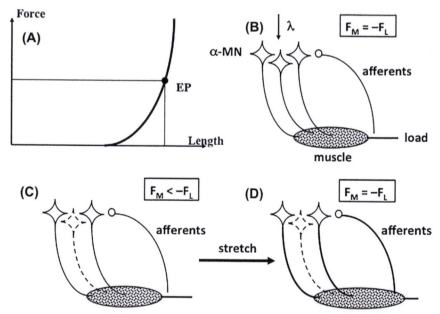

Figure 12.17 Tonic stretch reflex as a synergy. A muscle acting against a load reaches an equilibrium point (EP, panels A and B) when its force equals in magnitude the load force, $F_M = -F_L$. If one of the α-motoneurons stops producing action potentials, muscle force will drop (panel C), $F_M < -F_L$, the muscle will stretch, the length-sensitive receptors will increase their level of firing and produce additional activation of the α-motoneuronal pool. This will lead to additional force that will (partly) compensate for the original change in the equilibrium state induced by the perturbation (panel D).

alpha-motoneuronal pool. This will lead to an increase in muscle activation and development of additional force that will shorten the muscle and (partly) compensate for the original change in its equilibrium state induced by the perturbation. In other words, the mechanism of the tonic stretch reflex acts as a feedback loop providing for negative covariation of the firing rates of individual motoneurons. This qualifies this mechanism as a synergy stabilizing the equilibrium state of the muscle acting against an external load (stabilizing its EP). The EP hypothesis presents an example of how a large set of elements (motor units) can be united by a physiological mechanism (the tonic stretch reflex) to stabilize an important feature of performance—the equilibrium point characterized by values of muscle force and length.

12.8.2 Synergies created by control with referent configurations

Within the RC hypothesis, action starts with specifying a time profile of referent spatial coordinates for a handful of task-specific salient variables, $RC_{TASK}(t)$ different from actual values of those variables (actual body configuration at the task level, AC_{TASK}). Further, a sequence of few-to-many transformations leads to RC time profiles at hierarchically lower levels. Each of these transformations is redundant, and an RC at a higher hierarchical level does not specify unambiguously all the RCs at a lower level.

For example, consider a reaching movement performed by a kinematically redundant multijoint limb. Assume, for simplicity, that state of the endpoint of the limb may be described with two commands comprising the RC; one of them (R) defines the equilibrium position of the endpoint while the other (C) defines its stability about the equilibrium position. This {R; C} pair is equivalent to the {r; c} pair of commands introduced earlier for the control of a single joint (Section 12.5). An {R, C} pair maps on {r; c} commands sent to individual joints (Figure 12.18); this is likely to be a redundant mapping meaning that an infinite number of {r; c} pairs can satisfy the {R; C} constraint for the endpoint. Note that a single {r; c} pair describes the control of a simple joint with only one rotational degree of freedom. More complex joints, such as, for example, the shoulder and the wrist joints, require several {r; c} pairs to describe their control.

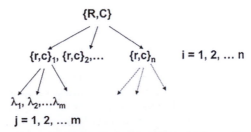

Figure 12.18 For a reaching movement, setting the referent body configuration at the task level can be described with two commands defining the equilibrium position of the endpoint (R) and its stability about the equilibrium position (C). The {R, C} pair maps on a redundant set of {r; c} commands sent to individual joints. Mapping an {r; c} pair onto a set of λs for the muscles crossing the joint is also likely redundant.

Earlier, a simple mapping between the {r; c} pair and two λs was suggested (Section 12.5). Such a mapping is possible only for a cartoon joint crossed by only two muscles. Most human joints, including relatively simple ones (such as the elbow joint), are crossed by more than two muscles. Hence, mapping an {r; c} pair onto a set of λs for the muscles crossing the joint is a problem of redundancy. For each muscle, a value of λ results in variable patterns of recruitment of alpha-motoneurons—another problem of redundancy. Ultimately, muscles interact with external forces leading to changes in body configuration. Assume that a higher-order sensory signal reflects AC_{TASK}. The difference between AC_{TASK} and RC_{TASK} drives muscle activations within the suggested scheme. When AC_{TASK} becomes equal to RC_{TASK}, the process stops. If AC_{TASK} is unable to reach RC_{TASK} because of external and/or anatomical constraints, a body configuration is achieved with nonzero forces acting on the environment.

At each of the mentioned few-to-many transformations, emergence of particular lower-level RC(t) may be based on a feedback mechanism (e.g., based on central back-coupling loops; Latash et al., 2005), leading to synergic adjustments of lower-level RC(t) functions in such a way that their combined action is compatible with the RC(t) at the higher level. This is not necessary, however, as shown in several studies demonstrating a trade-off between synergies at different hierarchical levels (Section 11.4.2 and Gorniak et al., 2007, 2009). In general, control with referent body configurations organized into a hierarchy can by itself lead to synergies revealed at different levels of analysis (for more details, see Section 11.8.3).

12.9   The bottom line

The EP hypothesis remains the only hypothesis in the field of motor control with a solid foundation in physics and physiology. Its recent developments to the control of multi-muscle and whole-body movements have provided links to such diverse phenomena as motor variability, muscle activation patterns during fast and slow movements, certain movement disorders, structure of motor variance, and motor synergies. Several of the previous chapters covered links of the EP hypothesis to muscle tone (see Chapter 5), reflexes (Chapter 6), and kinesthetic perception (see Chapter 8).

References

Asatryan, D.G., Feldman, A.G., 1965. Functional tuning of the nervous system with control of movements or maintenance of a steady posture. I. Mechanographic analysis of the work of the limb on execution of a postural task. Biophysics 10, 925—935.

Bernstein, N.A., 1935. The problem of interrelation between coordination and localization. Archives of Biological Science 38, 1—35 (in Russian).

Bernstein, N.A., 1947. On the Construction of Movements. Medgiz, Moscow (in Russian).

Bernstein, N.A., 1965. On the routes to the biology of activity. Problems of Philosophy 10, 65—78 (in Russian).

Bernstein, N.A., 1966. Essays on the Physiology of Movements and Physiology of Activity. Meditsina, Moscow (in Russian).

Bizzi, E., Accornero, N., Chapple, W., Hogan, N., 1982. Arm trajectory formation in monkeys. Experimental Brain Research 46, 139–143.

Feldman, A.G., 1966. Functional tuning of the nervous system with control of movement or maintenance of a steady posture. II. Controllable parameters of the muscle. Biophysics 11, 565–578.

Feldman, A.G., 1980. Superposition of motor programs. I. Rhythmic forearm movements in man. Neuroscience 5, 81–90.

Feldman, A.G., 1986. Once more on the equilibrium-point hypothesis (λ-model) for motor control. Journal of Motor Behavior 18, 17–54.

Feldman, A.G., 2011. Space and time in the context of equilibrium-point theory. Wiley Interdisciplinary Reviews: Cognitive Science 2, 287–304.

Feldman, A.G., Goussev, V., Sangole, A., Levin, M.F., 2007. Threshold position control and the principle of minimal interaction in motor actions. Progress in Brain Research 165, 267–281.

Feldman, A.G., Latash, M.L., 2005. Testing hypotheses and the advancement of science: recent attempts to falsify the equilibrium-point hypothesis. Experimental Brain Research 161, 91–103.

Feldman, A.G., Levin, M.F., 1995. Positional frames of reference in motor control: their origin and use. Behavioral and Brain Sciences 18, 723–806.

Feldman, A.G., Levin, M.F., Mitnitski, A.M., Archambault, P., 1998. 1998 ISEK Congress Keynote Lecture: multi-muscle control in human movements. Journal of Electromyography and Kinesiology 8, 383–390.

Feldman, A.G., Orlovsky, G.N., 1972. The influence of different descending systems on the tonic stretch reflex in the cat. Experimental Neurology 37, 481–494.

Feldman, A.G., Ostry, D.J., Levin, M.F., Gribble, P.L., Mitnitski, A.B., 1998. Recent tests of the equilibrium-point hypothesis (λ model). Motor Control 2, 189–205.

Flash, T., 1987. The control of hand equilibrium trajectories in multi-joint arm movements. Biological Cybernetics 57, 257–274.

Gorniak, S.L., Zatsiorsky, V.M., Latash, M.L., 2007. Emerging and disappearing synergies in a hierarchically controlled system. Experimental Brain Research 183, 259–270.

Gorniak, S.L., Zatsiorsky, V.M., Latash, M.L., 2009. Hierarchical control of static prehension: II. Multi-digit synergies. Experimental Brain Research 194, 1–15.

Gottlieb, G.L., Corcos, D.M., Agarwal, G.C., 1989. Strategies for the control of voluntary movements with one mechanical degree of freedom. Behavioral and Brain Sciences 12, 189–250.

Gribble, P.L., Ostry, D.J., Sanguineti, V., Laboissiere, R., 1998. Are complex control signals required for human arm movements? Journal of Neurophysiology 79, 1409–1424.

Heckman, C.J., Gorassini, M.A., Bennett, D.J., 2005. Persistent inward currents in motoneuron dendrites: implications for motor output. Muscle and Nerve 31, 135–156.

Heckman, C.J., Lee, R.H., Brownstone, R.M., 2003. Hyperexcitable dendrites in motoneurons and their neuromodulatory control during motor behavior. Trends in Neuroscience 26, 688–695.

Hinder, M.R., Milner, T.E., 2003. The case for an internal dynamics model versus equilibrium point control in human movement. Journal of Physiology 549, 953–963.

Jobin, A., Levin, M.F., 2000. Regulation of stretch reflex threshold in elbow flexors in children with cerebral palsy: a new measure of spasticity. Developmental Medicine and Child Neurology 42, 531–540.

Kawato, M., 1999. Internal models for motor control and trajectory planning. Current Opinions in Neurobiology 9, 718−727.

Kelso, J.A.S., 1995. Dynamic Patterns: The Self-organization of Brain and Behavior. MIT Press, Cambridge.

Kelso, J.A., Holt, K.G., 1980. Exploring a vibratory systems analysis of human movement production. Journal of Neurophysiology 43, 1183−1196.

Kugler, P.N., Turvey, M.T., 1987. Information, Natural Law, and the Self-assembly of Rhythmic Movement. Erlbaum, Hillsdale, NJ.

Lackner, J.R., DiZio, P., 1994. Rapid adaptation to Coriolis force perturbations of arm trajectory. Journal of Neurophysiology 72, 1−15.

Latash, M.L., 1998. Control of multi-joint reaching movement: the elastic membrane metaphor. In: Latash, M.L. (Ed.), Progress in Motor Control: vol. 1: Bernstein's Traditions in Movement Studies. Human Kinetics, Urbana, IL, pp. 315−328.

Latash, M.L., 2012. Fundamentals of Motor Control. Academic Press, New York, NY.

Latash, M.L., Aruin, A.S., Zatsiorsky, V.M., 1999. The basis of a simple synergy: reconstruction of joint equilibrium trajectories during unrestrained arm movements. Human Movement Science 18, 3−30.

Latash, M.L., Friedman, J., Kim, S.W., Feldman, A.G., Zatsiorsky, V.M., 2010. Prehension synergies and control with referent hand configurations. Experimental Brain Research 202, 213−229.

Latash, M.L., Gottlieb, G.L., 1990. Compliant characteristics of single joints: preservation of equifinality with phasic reactions. Biological Cybernetics 62, 331−336.

Latash, M.L., Gottlieb, G.L., 1991. Reconstruction of elbow joint compliant characteristics during fast and slow voluntary movements. Neuroscience 43, 697−712.

Latash, M.L., Gottlieb, G.L., 1992. Virtual trajectories of single-joint movements performed under two basic strategies. Neuroscience 47, 357−365.

Latash, M.L., Scholz, J.P., Schöner, G., 2007. Toward a new theory of motor synergies. Motor Control 11, 276−308.

Latash, M.L., Shim, J.K., Smilga, A.V., Zatsiorsky, V., 2005. A central back-coupling hypothesis on the organization of motor synergies: a physical metaphor and a neural model. Biological Cybernetics 92, 186−191.

Levin, M.F., Selles, R.W., Verheul, M.H., Meijer, O.G., 2000. Deficits in the coordination of agonist and antagonist muscles in stroke patients: implications for normal motor control. Brain Research 853, 352−369.

Liddell, E.G.T., Sherrington, C.S., 1924. Reflexes in response to stretch (myotatic reflexes). In: Proceedings of the Royal Society of London, Series B, vol. 96, pp. 212−242.

Matthews, P.B.C., 1959. The dependence of tension upon extension in the stretch reflex of the soleus of the decerebrate cat. Journal of Physiology 47, 521−546.

Merton, P.A., 1953. Speculations on the servo-control of movements. In: Malcolm, J.L., Gray, J.A.B., Wolstenholm, G.E.W. (Eds.), The Spinal Cord. Little, Brown, Boston, pp. 183−198.

Musampa, N.K., Mathieu, P.A., Levin, M.F., 2007. Relationship between stretch reflex thresholds and voluntary arm muscle activation in patients with spasticity. Experimental Brain Research 181, 579−593.

Nichols, T.R., Houk, J.C., 1976. Improvement in linearity and regulation of stiffness that results from actions of stretch reflex. Journal of Neurophysiology 39, 119−142.

Polit, A., Bizzi, E., 1978. Processes controlling arm movements in monkey. Science 201, 1235−1237.

Polit, A., Bizzi, E., 1979. Characteristics of motor programs underlying arm movement in monkey. Journal of Neurophysiology 42, 183–194.

Prochazka, A., Clarac, F., Loeb, G.E., Rothwell, J.C., Wolpaw, J.R., 2000. What do reflex and voluntary mean? Modern views on an ancient debate. Experimental Brain Research 130, 417–432.

Robert, T., Zatsiorsky, V.M., Latash, M.L., 2008. Multi-muscle synergies in an unusual postural task: quick shear force production. Experimental Brain Research 187, 237–253.

Schmidt, R.A., 1975. A schema theory of discrete motor skill learning. Psychological Reviews 82, 225–260.

Schmidt, R.A., McGown, C., 1980. Terminal accuracy of unexpected loaded rapid movements: evidence for a mass-spring mechanism in programming. J Mot Behav 12, 149–161.

Shadmehr, R., Mussa-Ivaldi, F.A., 1994. Adaptive representation of dynamics during learning of a motor task. Journal of Neuroscience 14, 3208–3224.

Shadmehr, R., Wise, S.P., 2005. The Computational Neurobiology of Reaching and Pointing. MIT Press, Cambridge, MA.

Turvey, M.T., 1990. Coordination. American Psychologist 45, 938–953.

Vallbo, A., 1970. Discharge patterns in human muscle spindle afferents during isometric voluntary contractions. Acta Physiologica Scandinavica 80, 552–566.

Vallbo, A., 1974. Human muscle spindle discharge during isometric voluntary contractions. Amplitude relations between spindle frequency and torque. Acta Physiological Scandinavica 90, 310–336.

Von Holst, E., Mittelstaedt, H., 1950. Daz reafferezprincip. Wechselwirkungen zwischen Zentralnerven-system und Peripherie (1950/1973). Naturwiss 37, 467–476. The reafference principle. In: The behavioral physiology of animals and man. The collected papers of Erich von Holst. Martin R (translator), 1 pp. 139–173, University of Miami Press: Coral Gables, FL.

Wachholder, K., Altenburger, H., 1927. Do our limbs have only one rest length? Simultaneously a contribution to the measurement of elastic forces in active and passive movements. Pflügers Archiv für die gesammte Physiologie des Menschen und der Thiere 215, 627–640. Cited after: Sternad D (2002) Foundational experiments for current hypotheses on equilibrium point control in voluntary movements. Motor Control 6: 299–318.

Zhou, T., Solnik, S., Wu, Y.-H., Latash, M.L., 2014a. Equifinality and its violations in a redundant system: control with referent configurations in a multi-joint positional task. Motor Control 18, 405–424.

Zhou, T., Solnik, S., Wu, Y.-H., Latash, M.L., 2014b. Unintentional movements produced by back-coupling between the actual and referent body configurations: violations of equifinality in multi-joint positional tasks. Experimental Brain Research 232, 3847–3859.

Motor Program

13

It is intuitively obvious that when we learn a novel movement (e.g., riding a bicycle), at first we approximate the desired motor action with suboptimal means, leading frequently to failure. With practice, we develop a way to perform the movement successfully, but the success requires constant attention and correction of the ongoing movement if it deviates from the desired pattern. When we become proficient, however, we acquire an ability to perform the movement *automatically*—an intuitive but imprecise term. This means that we do not have to monitor the movement, that is, mentally concentrate on the movement execution, anymore, and can perform other tasks while performing the learned movement with minimal or no negative interference. This qualitative step in learning has been addressed by some researchers as developing a *motor program* for the movement (other researchers prefer to describe it as developing a *neural representation* of the learned motion). We have to admit right away that contemporary scientists, including the authors of this book, do not know what exactly is stored in the brain when a performer learns a certain movement and what processes are involved in creating such representations. The purpose of this chapter is to discuss the notion of motor program, decide whether it is useful for scientific research, and, if so, offer an answer to arguably the central question: What do motor programs program?

It is possible that this question can only be answered for some classes of movements. We are not ready yet to offer a noncontroversial taxonomy of movements, although such attempts were made in the past by classics of movement science. For example, Nikolai Bernstein suggested a four-level (sometimes, five-level) account for all natural movements (Bernstein, 1947, 1996), from movements controlled at the level of muscle tone (see Chapter 5) to highly elaborated symbolic movements, such as handwriting and piano playing. It also seems that this concept may not be relevant for highly variable movements requiring constant adjustments to the quickly changing environment, such as those seen in tennis or basketball. However, since at the end of this chapter we come to a conclusion that the notion of motor program has no identifiable physical or physiological meaning, it does not seem necessary to specify how the aforementioned central question relates to particular movement types.

The notion of motor program is arguably one of the least precisely defined and most commonly used in the field of movement studies. A recent search on PubMed yielded over 5000 papers on this topic. The word *program* biases the term very strongly and is understood differently by different researchers. In different contexts, this word has a range of meanings from a general goal of a person, group of people, or society (cf. "Program of the Communist Party of the Soviet Union" or "Program of Economic Development") to a detailed schedule of events (cf. concert program) or a series of operations solving a problem (cf. computer program). The definitions of motor program

Biomechanics and Motor Control. http://dx.doi.org/10.1016/B978-0-12-800384-8.00013-2

in contemporary sources are vague. In our opinion, one of the least controversial definitions can be found in Wikipedia, which defines motor program as an abstract representation of movement that centrally organizes and controls the many degrees of freedom involved in performing an action. We will analyze aspects of this definition in more detail later. For now, compare this definition with the results of numerous studies, starting from the classical study of Bernstein (1930), showing that movements by very well-trained subjects show variable involvement of elements (degrees of freedom) compatible with accurate performance with respect to the task-specific salient variables (reviewed in Chapters 10 and 11).

The notion of motor program is tightly linked to a range of issues central to the field of movement studies. These include, in particular, the role of memory and of feedback loops in movement production. Over time, this notion went through a number of stages associated with shifts of the emphasis from the idea of reflex-based movement production to central specification of peripheral motor patterns (such as kinetic, kinematic, and electromyographic patterns), and to complex interactions between feedback and feedforward processes. While the notion of motor program is rather old, it is still used broadly and has even experienced a recent revival under the label of *internal models*.

13.1   Elements of history

The general idea that the brain controls the body as a puppeteer controls a marionette based on the available information about the body, the environment, and motor goals is very old. The great ancient Greek philosopher Plato compared human voluntary movements to a chariot moved by horses controlled by the ultimate charioteer, the soul. If one places the soul (mind) into the brain, this statement is rather similar to the aforementioned definition of motor program from Wikipedia. The discovery of muscle reflexes led in the nineteenth century to an emphasis on the reflex nature of movements. A number of scientists, including Ivan Sechenov (1829–1905) and his student Ivan Pavlov (1849–1936), developed a theory that human movements (as well as mental actions) represented sequences of reflexes, which were viewed as fundamental units of movement production. William James (1842–1910), a great American philosopher and psychologist, proposed a reflex-chaining hypothesis, within which movements represented properly selected and ordered sequences of reflexes.

The discussion on the role of reflexes and centrally initiated neural processes has never ended. It resembles the swinging of a giant pendulum with the period of tens of years. The reflex theory dominated in the early twentieth century supported, in particular, by the classical experimental studies of Sir Charles Sherrington (1857–1952) and his students, which led, in particular, to the theory that locomotion represented a sequence of flexor and crossed extensor reflexes in the extremities. Later, the reflex-based theory of locomotion was refuted in favor of the notion of central pattern generators for a range of rhythmical movements including locomotion (see Chapter 9).

In the 1930s, Nikolai Bernstein (1896–1966) published a highly influential paper (Bernstein, 1935) in which he laid the foundation for the alternative view that emphasized the active, rather than reactive, nature of biological movements. In that paper, Bernstein introduced the notion of *engrams* as abstract neural representations of movements. He emphasized that engrams were built on prior experience; they encoded topological features of movement of salient points on the body (such as the fingertip during a pointing movement) but not their metric features, which could be adjusted by the actor for specific actions performed in particular external conditions. The idea that all movements are characterized by two groups of parameters, essential (topological, encoded in engrams) and nonessential (metric, scaled for particular actions), was later developed by Israel Gelfand (1913−2009) and Michael Tsetlin (1924−1966) (Gelfand and Tsetlin, 1966). Unfortunately, our current knowledge of physiological mechanisms involved in the creation and storage of neural signals associated with learned motor skills (engrams) remains all but nonexistent.

The idea of engrams remained more philosophical than physiological for many years. It avoided the central issue of physiological (physical) processes that could be involved in storing and implementing the hypothetical topological representations of movements. Arguably, the first attempt to link descending motor commands for an action to neurophysiological processes was made by Merton who introduced the so-called *servo-model* (Merton, 1953; see Chapter 12). This model associated motor commands with descending signals to gamma-motoneuronal pools. Later studies, however, falsified Merton's gamma model, as well as its extension based on the phenomenon of co-activation of gamma- and alpha-motoneurons (Vallbo, 1970, 1974).

The next step was made in the mid-1960s, when Anatol Feldman introduced the lambda-model of the equilibrium-point hypothesis (Feldman, 1966, 1986; see Chapter 12). This model associated descending signals with time profiles of the threshold of the tonic stretch reflex (designated with the Greek λ, hence λ-model) for participating muscles. The lambda-model accepted the idea of active movement production and incorporated an important role of feedback reflex loops in defining patterns of muscle activation. Later, we discuss whether time patterns of neural variables, such as those introduced within the equilibrium-point hypothesis, for example, λ(t), can be called motor programs.

The next important step was made in the 1970s when Richard Schmidt introduced the notion of *generalized motor program* incorporated into the *schema theory* (reviewed in Schmidt, 1975). According to the schema theory, generalized motor programs represent rules that define the spatial and temporal muscle activation patterns to produce a class of movements. To produce specific actions, appropriate generalized motor programs were used with parameters defining the desired speed and effort. Defining generalized motor programs as direct precursors of patterns of muscle activation (also see seminal studies of Keele, 1968) made them relatively directly accessible to measurement in experiments and, hence, highly attractive for researchers in the field of motor behavior. This step also made generalized motor programs different from engrams. Note that engrams define topological properties of movements of salient, task-specific points of the body, while motor programs were supposed to define involvement of individual elements (e.g., muscles). Indeed, Bernstein appreciated the fact that descending neural signals could not prescribe muscle activation patterns (and their mechanical consequences such as forces and trajectories) because of the unavoidable reflex contributions to those patterns. Hence, engrams remained abstract representations of movements that could not be associated with patterns of muscle activation. The appreciation of the effects of reflexes on patterns of muscle activation led Adams (1971) to suggest that motor programs define only the first 30 ms of muscle activation patterns, before reflex-mediated changes in muscle activation could take place.

In recent years, the notion of motor programs as neural representations that prescribe desired patterns of muscle activation and their mechanical consequences (forces, trajectories, etc.) has been developed in a much more sophisticated way under the label of internal models (Wolpert et al., 1998; Kawato, 1999). The similarities and differences between generalized motor programs and internal models will be discussed later.

13.2   Current definitions for motor program

Two definitions of motor programs dominate the field. Both assume that at some, relatively high, level of the hierarchical system for movement production the central nervous system represents (prescribes) the relative timing and magnitude of involvement of the elements involved in a planned action. The first, more abstract, definition of motor program does not specify variables that are prescribed by motor programs (as in the aforementioned definition from Wikipedia). The second, more specific, definition associates the prescribed variables with patterns of muscle activation or with patterns of muscle forces. Both definitions emphasize the open-loop nature of well-learned movements. Feedback is viewed as a means of informing the central nervous system on the initial state of the body, as well as a source of signals that are used to issue corrections in cases of unexpected events, and changing/aborting the action altogether if needed.

Both definitions seem to suffer from a common flaw. They assume that planning of an action is associated with planning of involvement of all the involved output elements (degrees of freedom). This seems highly unlikely given the motor redundancy/abundance typical of all natural actions (see Chapter 10) and numerous studies documenting variability of the trajectories of elements over repeated movements, even during movements performed by the best-trained subjects (Bernstein, 1930, 1996). Note that this variability does not represent "noise" but shows structure compatible with accurate performance of task-specific (intention-specific) performance variables (Schöner, 1995, Chapter 11). As described in detail in Chapters 10 and 11, natural actions are associated with varying contributions of elements that keep the time profiles of task-specific performance variables relatively invariant (reviewed in Latash et al., 2007; Latash, 2008). This has been shown to be true for both mechanical (kinematic and kinetic) and muscle activation (electromyographic, EMG) variables.

Another major problem is the assumption that a generalized motor program can be implemented at different speeds and efforts by scaling the involvement of the elements (it does not matter what variables produced by the elements are prescribed). This is clearly impossible given the nonlinearities involved in the mechanics of multijoint movements. For example, drawing a small figure eight and a big one cannot be achieved by proportional scaling of muscle activations or muscle forces. The same is true for drawing a figure eight slowly and at a very high speed. If the requirement of scaling the involvement of elements is removed from the definition of generalized motor program, this definition becomes synonymous with Bernstein's engrams.

The second, more specific, definition also faces another problem. Patterns of muscle activation depend on both descending signals to respective alpha-motoneuronal pools and reflex inputs from sensory endings sensitive to movement mechanics, in particular, signals from muscle spindles and Golgi tendon organs. This means that descending signals cannot in principle prescribe patterns of muscle activation since this would mean predicting movement kinematics and kinetics perfectly. Since motor variability is seen at all levels during all natural movements, movement mechanics cannot be predicted perfectly. Muscle forces depend on muscle activation patterns and muscle kinematics. Therefore, prescribing patterns of muscle forces is even less feasible than prescribing patterns of muscle activation. Nikolai Bernstein (1935) appreciated and emphasized this factor and, therefore, rejected the ideas of muscle activation control and muscle force control, and purposefully kept the notion of engrams abstract.

The mentioned idea of Adams that motor programs specify the first 30 ms of muscle activations is also flawed because of similar problems. Indeed, the initial state of any motor system is time varying and cannot be assumed perfectly predictable. As shown in experiments with unexpected perturbations applied prior to the quick movement initiation (Adamovich et al., 1997), even the timing of the initial agonist muscle activation burst could be modified by unexpected mechanical perturbations applied prior to the action initiation. Only in artificial experiments with accurately controlled initial steady state may the first 30 ms of movement be viewed as being relatively immune to reflex feedback. For example, a series of studies by Gottlieb and colleagues (Gottlieb et al., 1989, 1990) selected the first 30 ms to produce integral measures of muscle activation and showed their consistent and predictable behavior across task manipulations (Figure 13.1). During natural movements, however, initial states for individual motor acts are variable and unpredictable making even the very first time segments of muscle activation dependent on reflex feedback.

Can descending signals specify relative timing of the involvement of elements as assumed in both definitions of motor program? A classical study of professional typists by Viviani and Terzuolo (1980) provided a positive answer. In that study, during repetitive typing of a standard phrase, the relative timing of individual key presses was preserved despite the variability of the actual key press timing—the time intervals between successive key pressings varied across repetitions of typing a standard phrase, but they could all be reduced to a single time series with a single scaling parameter. Later studies have shown, however, that even in simple two-element tasks, such as rhythmical flexion/extension of two index fingers or two joints of a limb, relative timing can change dramatically with a change in parameters of the task such as movement frequency (Kelso, 1984). Such shifts in the relative phase, typically from out-of-phase to in-phase, happened without the subject's intention (Figure 13.2). Later, phase shift and phase transition phenomena have been described in numerous other actions (reviewed in Kelso, 1995; Schöner, 2002). Phase shifts, however, do not happen in all

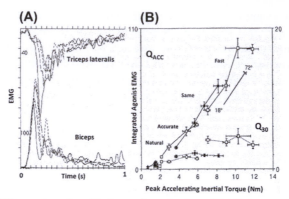

Figure 13.1 (A) Time series of rectified muscle activations (EMG) for the biceps and lateral head of triceps during elbow flexion movements performed as fast as possible over different distances. (B) The amount of agonist activation increases with movement distance while the amount of activation within the first 30 ms (Q30) remains relatively invariant. It changes with a change in the instruction related to movement speed.
Reproduced by permission from Gottlieb et al. (1990), © American Physiological Society.

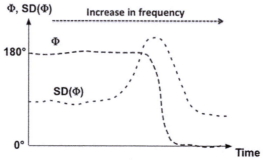

Figure 13.2 Changes in the relative phase (Φ) and its standard deviation with an increase in the frequency of rhythmic movement of two fingers that started out of phase. Note that at some critical frequency, the standard deviation of Φ increased and then the movement pattern shifted to in phase (relative phase close to zero).
Modified by permission from Latash (2012).

actions; for example, when humans walk faster and faster, they do not switch into hopping. So, at least in some movements a hypothetical central controller was unable to prescribe even such a general characteristic of an action as relative timing of the involvement of elements.

We come to a conclusion that both mentioned definitions are inadequate. However, the current meager knowledge of the neural processes associated with voluntary movements makes the notion of motor program highly attractive. This notion may be used as a poor man's substitute for the currently unknown physics and physiology of processes associated with such vaguely defined concepts as *motor representations, motor memories, decision making, target identification*, etc. This makes motor programs nearly synonymous with engrams. An important aspect is that engrams (motor programs) are assumed to specify only a relatively low-dimensional set of task-specific variables, while similar variables for individual elements emerge as a result of a hierarchy of few-to-many mappings, which is under the influence of feedback, afferent signals (Latash, 2010; see Chapter 11). This makes engrams a more attractive term.

A possible place of motor programs in a hypothetical hierarchical system for the production of voluntary movements is illustrated in Figure 13.3. We purposefully avoid the question: Where does the formulation of the task come from? It is assumed to emerge from currently unknown neurophysiological processes. A promising approach to this problem, the *dynamical neural field theory*, has been developed recently by the group of Gregor Schöner (Erlhagen and Schöner, 2002; Schöner and Thelen, 2006). However, mapping of the assumed neural processes within the dynamical neural field theory and actual neurophysiological structures remains unknown. Further in the scheme, it is assumed that the task-related signals are transformed into time patterns of neural signals (N1 in Figure 13.3), which define the time evolution of referent spatial coordinates for effectors. Referent coordinates for individual effectors may show large intertrial variability as long as their combined effects are compatible with the desired time profiles of salient, task-specific performance variables. The expression motor program can be used to address the time profiles of neural

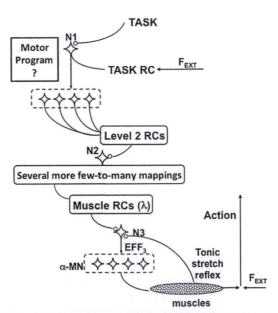

Figure 13.3 A hypothetical scheme of the generation of a voluntary action based on the ideas of control with referent body configurations (RC). Task-related signals are transformed into time patterns of neural signals (N1), which define the time evolution of referent spatial coordinates (RC) for effectors that ultimately generate salient, task-specific performance variables (Action). The expression "motor program" can be used to address the time profiles of those neural signals (or the referent spatial coordinates, RC) with minimal controversy.

signals generated by N1 with minimal controversy. These signals also seem an adequate example of Bernstein's engrams. However, associating motor programs with any of the variables downstream of N1 is highly questionable because all those variables emerge with contribution of sensory signals reflecting the current external force field. These issues are discussed in more detail in the next section.

13.3   What can be encoded by signals from the brain?

13.3.1   Relative timing: a clock in the brain versus emergent timing

The issue of relative timing was one of the central ones in the argument between the champions of the motor programming approach and of the dynamical systems approach to movements (Schmidt, 1975, 2003; Kugler and Turvey, 1987; Kelso, 1995). The former assumed that a hypothetical motor program prescribed the relative timing of involvement of effectors, while the actual magnitudes of action could vary depending on the external conditions of movement execution. The latter claimed that all characteristics of movement patterns, including relative timing, emerged during the movement as a consequence of interactions both within the body and between the body and the environment. Those interactions were assumed to be defined by laws of nature and not involve computational processes within the central nervous system.

The idea that the brain contains an internal clock is relatively old. In the middle of the twentieth century, the clock was tentatively placed into the cerebellum (Braitenberg, 1967), which remains a favorite brain structure for models of computational processes in the brain (Wolpert et al., 1998). The aforementioned study of typists (Viviani and Terzuolo, 1980) provided important experimental evidence in favor of an internal timing structure of well-learned actions. On the other hand, numerous studies of unintentional phase transitions during movements (Kelso, 1995; Mechsner et al., 2001) have provided abundant evidence for relative timing being an emergent property of interactions within the body and between the body and the environment.

The question in the title of this subsection, however, seems to be poorly formulated in the first place. Indeed, all the actual clocks that we are aware of function based not on some abstract computational processes but on laws of nature, starting from the motion of the Sun and planets and ending up with the grandfather's clock with a pendulum. Hence, the timing provided by a clock is by its very nature emergent given the environment. In a brilliant chapter, Onno Meijer (2001) wrote about Emperor Charles V (1500—1558) of the Holy Roman Empire who was obsessed with clocks in his palace and frustrated by the fact that he could not ensure that they all always showed exactly the same time. When he placed the clocks into the same room, however, their pendulum swings synchronized due to the weak mechanical coupling provided by the air vibrations. So, the timing of even the most proverbial clock is defined by laws of nature and depends on interactions of the clock parts with the environment. In our opinion, this invalidates the argument of whether the brain contains a clock or relative timing of events is emerging given the laws of nature.

13.3.2   Internal models—does the brain prescribe movement mechanics?

The notion of *internal models* is used broadly in the contemporary motor control and motor learning literature (reviewed in Wolpert et al., 1998; Kawato, 1999; Shadmehr and Wise, 2005). There are two definitions of internal models in the literature, which can be addressed as specific and nonspecific. A common feature of both definitions is that they assume computational processes within the central nervous system. The nonspecific definition assumes that internal models are functional units within the central nervous system that emulate (compute, model, represent, etc.) input—output relationships between groups of variables within the body. The specific definition assumes that internal models are stored in structures within the brain and used to compute patterns of muscle activation and/or forces needed to perform a desired movement. The latter definition makes internal models very similar to generalized motor programs. It is most commonly implied in experimental studies that explore movement mechanics under manipulations of the external force field or sensory feedback and during adaptation to these new conditions.

The applications of the specific version of internal models to the field of motor control are driven by the success of classical mechanics and control theory. The following two axiomatic statements are accepted: (1) To perform a movement, one has to ensure that a requisite pattern of resultant force acts on the object; and (2) Predictive mechanisms have to be used to compensate for the unavoidable time delays introduced by the relatively slow neural signal transmission in the body and the relatively slow force production by the skeletal muscles. The former statement follows classical mechanics while the latter one allows application of tools developed within the control theory.

Two types of internal models are commonly used in the literature, inverse and direct (Figure 13.4). The former try to answer the question: What signals from the brain are needed to implement a desired action? For example, if one wants to move the hand with the cup of tea to the mouth, a pattern of joint rotations has to be realized compatible with the required hand movement. To produce joint rotation, a time profile of the moment of force acting at the joints has to be produced. To generate desired time changes in the moment of force, a change in the forces of muscles crossing the joints has to happen. Muscle forces are functions of excitation and muscle kinematics. This means that muscle kinematics has to be predicted perfectly to enable computing time profiles of action potentials that have to be sent from the corresponding alpha-motoneuronal pools. Alpha-motoneurons receive signals from many sources including descending signals from the brain and those from other neurons reflecting the activity of receptors sensitive to a variety of mechanical variables including muscle length, velocity, and force (including variables characterizing other muscles). So, to compute a required descending input into an alpha-motoneuronal pool, the reflex input has to be predicted perfectly. This chain of problems is addressed as *inverse* in Figure 13.4 because it tries to compute an input based on a desired output, while in the natural course of events outputs are produced by inputs.

It is already clear that the described process involves several nontrivial steps. First, accurate prediction of movement kinematics, including length changes of all muscles,

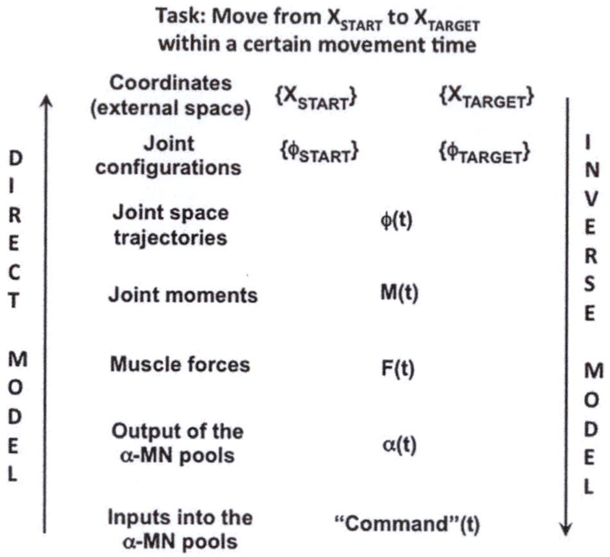

**Task: Move from X$_{START}$ to X$_{TARGET}$
within a certain movement time**

Coordinates (external space)	{X$_{START}$}		{X$_{TARGET}$}
Joint configurations	{ϕ_{START}}		{ϕ_{TARGET}}
Joint space trajectories		$\phi(t)$	
Joint moments		M(t)	
Muscle forces		F(t)	
Output of the α-MN pools		$\alpha(t)$	
Inputs into the α-MN pools		"Command"(t)	

(Left vertical label: DIRECT MODEL; right vertical label: INVERSE MODEL)

Figure 13.4 A simplified scheme of the main steps involved in the production of a voluntary movement to a spatial target by a multijoint limb. The inverse model tries to predict what descending inputs into alpha-motoneuronal pools would be necessary to produce the desired mechanical effect. The direct model tries to predict changes in the peripheral system expected given its current state and the recent descending signals. Direct models help solve problems associated with the relatively low speed of information transmission within the human body (see also Figure 13.5).
Modified by permission from Latash (2012).

is necessary prior to movement initiation. Given the natural variability of even the best-learned actions, this is impossible. Second, many steps require computation (determination) of signals based on a smaller set of inputs, that is, solving problems of motor redundancy. These involve, in particular, computing joint rotations based on hand trajectory (the problem of inverse kinematics), computing muscle forces based on joint torque, and computing input signals into a neuronal pool based on its desired output. The last problem is exacerbated by the fact that neurons are threshold elements. This means that the same output can be produced by an infinite number of input signals; it is only required that they are above the activation threshold. This problem is similar to trying to compute the force with which one has to press on a toggle switch to turn on the lights. The huge redundancy of the involved transformations and the presence of threshold elements make solving these problems impossible without imposing some additional constraints, for example, requirements of minimization of certain cost functions (see Chapter 10).

Direct internal models are expected to compensate for the unavoidable time delays involved in the signal transmission. Indeed, if one assumes that a structure in the cerebellum (Wolpert et al., 1998) tries to determine a descending neural signal needed for a planned action, it should be taken into account that this structure receives outdated information on the current state of the system (delayed by afferent

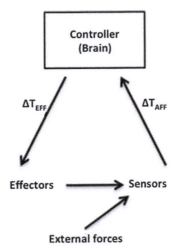

Figure 13.5 The relatively low speed of action potential transmission velocity within the human body leads to nonnegligible time delay associated with conduction of signals from the brain to the effectors (ΔT_{EFF}) and from the peripheral sensors to the brain (ΔT_{AFF}). These delays result in the brain always dealing with outdated sensory information and issuing commands that will be implemented also at a delay. Direct models are used to deal with these problems (see Figure 13.4).

transmission time, T_{AFF} in Figure 13.5), and its signals will be delivered to the muscles after another time delay (efferent transmission time, T_{EFF} in Figure 13.5). During quick actions, such time delays may become comparable to movement time. For the example of moving the cup of tea to one's mouth, to predict changes in the state of the periphery, given its current state and the immediate history, direct models compute expected changes in the activation patterns of alpha-motoneurons, muscle forces, joint moments of force, joint rotations, and hand motion. Some of the current schemes involve cascades of direct and inverse internal models able to predict and implement movement kinematics and kinetics with high accuracy (reviewed in Shadmehr and Wise, 2005).

Three main arguments have been presented against the idea of internal models. One is philosophical, the second one is practical, and the third one is based on experimental observations. The philosophical argument is that accepting that a natural system (e.g., the human body) involves computational processes during its functioning goes against the basics of natural science. Of course, assuming computations can describe any natural phenomenon (e.g., assuming computation of forces by stars and planets, instead of the law of gravity, can account for the regular motion of celestial bodies), but this does not move us closer to understanding how these phenomena come about.

The practical argument deals with the fact that internal models are assumed to solve the inverse process using computational means. As mentioned before, this implies that the numerous problems of redundancy are somehow solved and computation of an input into a threshold system is somehow computed based on the system's output. Such computations are not impossible if one introduces additional constraints on the system and/or applies optimization principles (see Chapter 10). However, no physiological data suggest that such methods are used within the central nervous system.

Computational means have been used traditionally in all the branches of the natural science. In particular, the impressive progress in physics over the past two centuries has been to a large degree facilitated by the development of the corresponding chapters of mathematics. The physical/physiological approach to motor control also relies on a computational apparatus for analysis of models and theories. Within this approach, writing an equation is assumed to reflect natural processes within the object of study, not that this equation is actually solved by the object. Assigning computational processes to structures within the object of study, for example, to the cerebellum (Wolpert et al., 1998), seems to go against the spirit of the natural science. This is particularly striking because many of such computations are based on optimization approaches that involve computation of a cost function over the course of the movement, and, therefore, movement cannot be initiated before the computational process is finished.

In many studies, the notion of an internal model implies generality of a formal account of the system's behavior (Wolpert et al., 1998; Shadmehr and Wise, 2005). In particular, if a set of internal models is elaborated in the course of practice in novel external conditions, these models are assumed to be applicable to other actions performed in the same conditions. In other words, the idea of internal models implies strong transfer effects of practice. Several studies, however, have provided evidence against such strong transfer effects (Malfait et al., 2002; Rochet-Capellan et al., 2012). For example, when a subject practices center-out reaching movements in natural condition, the trajectories are relatively straight (Figure 13.6(A)). If a force field is turned on, which produces forces acting on the hand orthogonal to its trajectory and proportional to the movement speed (Figure 13.6, insert), the trajectories become curved (Figure 13.6(B)). With practice, the trajectories become straight again (Figure 13.6(C)) resembling natural trajectories without the force field. Turning the force field off results in trajectories curved in the opposite direction (Figure 13.6(D)). Since their original publication (Shadmehr and Mussa-Ivaldi, 1994), these results have been reproduced many times.

If a subject practices such a task using hand movement constrained to a narrow range of directions (Figure 13.7(A)), effects of practice are seen for those movements and also for movements in other directions within the same range. However, if the subject tries to perform a movement in a direction outside the range, the movement shows a curved trajectory (Figure 13.7(B)), as if no practice took place (Mattar and Ostry, 2007). Similarly, poor generalization of training has been reported for movements performed over different amplitudes (Mattar and Ostry, 2010). The very limited transfer of effects of practice speaks against using the term *internal model* when one tries to describe neural processes associated with motor learning. This means that internal models, if they exist, are typically highly specific.

Altogether, this brief review suggests that neural processes in the central nervous system are very unlikely to perform computations assumed in the specific definition of internal models. The brain cannot (see Bernstein, 1935, 1967) and does not predict and prescribe peripheral variables, mechanical or electromyographical, produced by the elements involved in natural movements. The nonspecific definition of internal models,

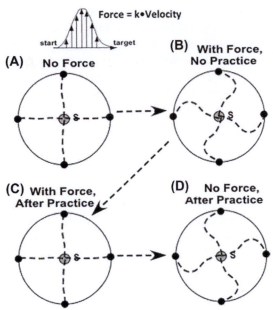

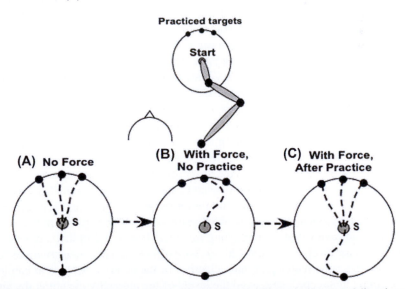

Figure 13.6 The subject performs movements from a center position (S) to four different targets. In natural condition, the trajectories are relatively straight (A). In a force field (external force acting on the hand is proportional to velocity and acting orthogonal to the trajectory; top insert), the trajectories become curved (B). With practice, the trajectories become straight again (C) resembling those in panel A. Turning the force field off results in trajectories curved in the opposite direction (D).

Figure 13.7 The subject practices movements to targets within a small range of directions in an unusual external force field (as in Figure 20.6). With practice, the initially curved movement trajectories (panel B) become straight for movements within the same range (cf. panels A and C). However, if the subject tries to perform a movement in a direction outside the range, the movement shows a curved trajectory as if no learning took place (panel C, bottom trajectory).

on the other hand, is synonymous with the old term *neural representation* (e.g., Perret et al., 1989). Both terms imply that some processes in the central nervous system represent important aspects of actions such as the external force field, the mechanical properties of the effectors, and maybe some others. These representations allow living beings to behave in a predictive way; for example, a predator commonly tries to intercept the prey at its predicted future location, not at the current location. The term *internal model* has additional connotations of neural computation and generality; these connotations make it more loaded and less attractive as compared to *neural representation*.

13.3.3 Do time profiles of referent coordinates qualify as motor programs?

There is little argument that some neural processes within the central nervous system are associated with specific motor actions. According to the idea of control with referent body configurations (see Chapter 12), for a given effector, from the whole body to a single muscle, neural signals define a time profile of a spatial referent coordinate (Feldman and Levin, 1995; Feldman, 2009). At the upper, task-specific, level, a relatively low-dimensional set of such coordinates is defined corresponding to the task at hand. Further, a hierarchical sequence of few-to-many transformations results in referent coordinates for the involved effectors such as limbs, digits, joints, and muscles (see Figure 13.3). At the muscle level, referent coordinates are synonymous with the threshold of the tonic stretch reflex, λ, as in the classical equilibrium-point hypothesis (Feldman, 1966, 1986; see Chapter 12).

Should time profiles of task-specific referent coordinates be called motor programs? Accepting an affirmative answer to this question implies that numerous motor programs are involved in any action, for example, the time profiles of referent coordinates at the highest level of the assumed hierarchy, at each of the intermediate levels, ending up with the muscle level. This would mean that $\lambda(t)$ is a motor program for a muscle, $\{r(t); c(t)\}$ form a motor program for a joint, etc. It is well known, however, that even the very well-practiced movement is associated with variable involvement of elements, such as joints and muscles (Bernstein, 1930; see also Chapter 10). So, repeating an action is expected to lead to different $\lambda(t)$ and $\{r(t); c(t)\}$ time profiles in different attempts. According to the suggested definition, repeating a well-learned movement is associated with different motor programs used in different trials at different levels of the assumed hierarchy. This makes little sense.

May one use the expression motor program to address time profiles of referent coordinates only at the highest, task-specific, level of the hierarchy? It depends on the meaning one wants to associate with motor program. For example, a recent series of studies have shown that a transient force perturbation with a dwell time between the application and removal of the perturbing force can lead to a drift in the referent coordinate at the task level (Zhou et al., 2014). So, even when a person is trying to "do the same" (use the same motor program), changes in the external force field can lead to changes in referent coordinates at all the levels of the hierarchy including the highest, task-specific level. This means that associating motor programs with referent trajectories at the task level leads to a conclusion that motor programs are not under the exclusive control of the person but change with changes in the external force field, even if these changes are transient and end up with the same force field as prior to the perturbation. We end up with a conclusion that using the expression motor program in this meaning is also inadequate and potentially misleading.

At the current level of knowledge, motor program seems to have no association with measurable neural processes. While this expression has an intuitive appeal, so far the notion of motor program remains theoretical, or even metaphorical, similar to the mentioned neural representation. It is possible, of course, that in the future currently unknown neurophysiological processes will be identified that will fit the intuitive understanding of motor program and avoid all the inconsistencies and controversies associated with the current theories on the neural control of movement. In the rest of this chapter, we use the term motor program as metaphorical, related to an unknown neural process associated with performing a well-learned action, which cannot be associated with any of the currently available motor control theories.

13.4   Are there motor programs in the spinal cord?

A number of coordinated motor actions are based on neural structures within the spinal cord, which provide a basic pattern of muscle activation. Signals from the brain can turn this basic pattern on and off and also adjust it to specific external conditions and needs of the animals. The most obvious example is locomotion. As described in the chapter on central pattern generators (CPGs; Chapter 9), mammals with the spinal cord surgically separated from the brain are able to demonstrate patterned neural activity leading to different gaits. Typically, this activity is functionally useless because the animal is unable to support the weight of its body, and, without signals from the brain, the locomotor activity cannot be target directed. Activity of spinal CPGs is modulated by signals from the brain stem as demonstrated in the classical experiments of the Moscow group (Orlovsky et al., 1966; Shik et al., 1967). In those experiments, electrical stimulation of an area in the medulla and the brain stem could induce locomotion, while strength of the stimulation was able to modulate the gait.

Can signals produced by a spinal central pattern generator be called motor programs for locomotion? On the one hand, they are adjustable by descending neural signals with respect to the timing and magnitude of the induced motor pattern as assumed within the idea of generalized motor program. On the other hand, without descending signals, the CPG for locomotion is unable to produce functional movements. Besides, gradual increase in the stimulation of midbrain structures causes phase transitions in the involvement of individual limbs (gait changes, see Chapter 9), which is not expected from a generalized motor program. It seems, therefore, that CPGs for locomotion do not satisfy criteria for generalized motor programs.

Other CPGs, in everyday life, can be triggered by either descending neural signals or peripheral sensory signals. For example, mastication is triggered by presence of food in the mouth, but it can be stopped and resumed by the animal at will. Stimulation of an area of the skin can induce a scratching response, which also represents a cyclical action under the control of a spinal CPG (Stein, 1984). If the neural rhythmical patterns

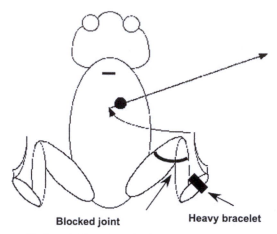

Figure 13.8 The frog with the spinal cord surgically separated from the brain (at the level shown with the black bar) can perform accurate wiping of the stimulus off its back in conditions of loading a distal leg segment with a heavy bracelet and in conditions of blocking motion in one of the major leg joints (knee joint).
Reproduced by permission from Latash (2008).

produced by all those CPGs are called motor programs, then, logically, one has to generalize this notion to all target-directed actions controlled by the spinal cord, whether cyclical or discrete.

If a stimulus (e.g., a small piece of paper soaked in a weak acid solution) is placed on the skin of the back of a sitting spinalized frog, typically the frog performs a coordinated action by the ipsilateral hindlimb and wipes the stimulus off its back (Fukson et al., 1980; Berkinblit et al., 1986, Figure 13.8). This *wiping reflex* can involve a single wiping action or a sequence of rhythmical wiping actions by the hindlimb. These actions are able to wipe the stimulus off the forelimb, even if the forelimb is placed in different positions with respect to the trunk (Fukson et al., 1980). The wiping reflex produces successful wiping actions even in a variety of unexpected hindlimb perturbations such as placing a lead bracelet on a distal hindlimb segment or blocking rotation in one of the major joints (cited in Latash, 2008). In all those conditions, the spinal cord facilitates different joint torques, joint rotations, and muscle activation patterns. Does the frog have motor programs for wiping in its spinal cord? Of course, it depends on the meaning one attaches to the notion of motor program. So far, no neural process or any other measurable process could be associated with such a motor program. This notion remains theoretical and, possibly, metaphorical.

13.5 Do neuronal populations in the brain generate motor programs?

The question "How are functionally important variables represented in the brain?" has been debated literally for centuries (reviewed in Bernstein, 2003). Answers to this question oscillated between two extreme views. The first is that different parts of the brain have specific assigned functions, for example, as reflected by bumps on the skull studied by the currently extinct science of phrenology. The second is that any part of the brain can take part in any function resulting in such well-established phenomena as neuronal plasticity. Since the classical studies of the great Canadian neurosurgeon and scientist Wilder Penfield (1891–1976), motor and sensory brain maps of the body have been documented in various brain structures. Motor maps were drawn based on motor responses induced by electrical stimulation of various brain structures, while sensory maps of different modalities were drawn based on responses of brain neurons to sensory stimuli applied to different body parts. While most early studies represented such brain maps as distorted but recognizable images of the body drawn on specific brain structures, more recent studies emphasize the mosaic structure of such maps and their changes (plasticity) following training or injury (reviewed in Nudo et al., 2001; Wolpaw and Tennison, 2001; Celnik and Cohen, 2004).

Studies of the activity of single brain neurons during voluntary movements performed by awake monkeys were pioneered by Edward Vaughn Evarts (1926–1985) and H. Asanuma (Evarts, 1968; Asanuma, 1973). In those studies, the frequency of firing of cortical neurons in the primary motor area was associated with the force produced by the monkey's hand. Changes in the external movement conditions, however, led to correlations between the neuronal activity and other mechanical variables.

Another major step was made by a Greek-American researcher, Apostolos Georgopoulos, and his team based on the fact that individual cortical neurons may show substantial variability in their firing patterns when the same task is repeated several times. However, the firing patterns of individual neurons may covary in such a way that, as a population, they reflect important performance variable(s) with relatively high consistency. This idea is close to the notion of a synergy as covariation of elemental variables with the purpose to stabilize a salient feature of performance (see Chapter 11). Those remarkable experiments (Georgopoulos et al., 1982, 1986, 1989; Georgopoulos, 1986) involved training monkeys to move an arm in different directions of the workspace. Many electrodes were implanted into the projection of the arm in the primary motor area of the cortex to record the activity of many cortical neurons. Further analysis involved two main steps. First, activity of each neuron was associated with a direction of arm movement in the external space when this neuron showed maximal increase in its baseline activity. This direction was termed *preferred direction* of the neuron. Movements in directions deviating from the preferred direction showed a smaller increase in the activity of that neuron or even a drop in its activity following a cosine-like function (Figure 13.9). After preferred directions for a large number of neurons were identified, the activity of the whole neuronal population was analyzed for a movement in a particular direction. For each neuron, a vector pointing in its preferred direction was drawn multiplied by the number of action potentials that the neuron generated during a brief time interval about the movement initiation. The sum of the vectors for individual neurons represented a vector pointing in the movement direction (as in Figure 13.10).

This method of analysis of the activation patterns in neuronal populations was applied to groups of neurons in various brain structures including different cortical areas, the cerebellum, and nuclei of the basal ganglia. It was also used to explore

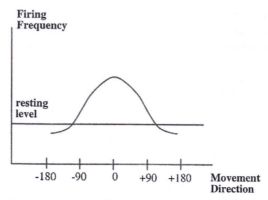

Figure 13.9 The typical cosine tuning of the activity of a cortical neuron with changes in the direction of the limb movement by a monkey. Note that the neuronal activity increases for a certain direction (shown as 0°), this increase drops with deviations of the movement direction from 0°, and it may turn into suppression of the baseline activity for movements in the opposite direction.
Reproduced by permission from Latash (2008).

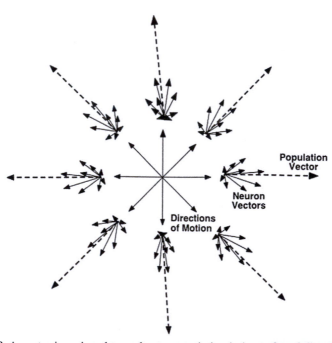

Figure 13.10 A vector is assigned to each neuron pointing in its preferred direction (0° in Figure 13.9). The magnitude of the vector is proportional to the number of action potentials generated by the neuron immediately prior to the action initiation. When neuronal activity vectors are summed up over a pool of cortical neurons, the resultant vector (dashed lines) points close to the movement direction.
Reproduced by permission from Latash (2008).

neuronal population vectors in various tasks including movement and force production tasks. The general finding held—neuronal population vectors pointed in the direction of a mechanical (force or displacement) vector produced by the effector (Georgopoulos et al., 1982; Schwartz, 1993; Coltz et al., 1999; Cisek and Kalaska, 2005). Accuracy of matching the directions of the two vectors, however, differed among brain structures—the highest was for the primary motor area.

The fact that one and the same neuronal population can produce vectors that match different performance variables in different conditions makes such vectors unlikely candidates for being the source of motor programs using the definition from the schema theory. Indeed, the vector itself (its magnitude and direction) does not encode a specific action but rather it can encode different actions in different external conditions.

We would like to end this subsection with a quotation from Nikolai Bernstein (2003): "In the higher motor centers of the brain (very probably in the cortex of the large hemispheres) one can find a localized reflection of a projection of the external space in a form in which the subject perceives the external space with motor means." Note that Bernstein expected neuronal patterns to reflect not aspects of planned actions expressed in mechanical variables but perception by the subject of the external space with motor means. This statement is very close to the idea of direct perception and the notion of affordances as introduced by the highly influential American psychologist James Jerome Gibson (1904—1979) within the framework of ecological psychology and developed by the group of Michael Turvey (Kugler and Turvey, 1987; Turvey, 1998). Within the available experimental material, there seems to be no place for motor program in the patterns of activation of neuronal populations.

13.6   Impaired motor programs

Disorders of motor programs have been commonly invoked in studies of patients with a variety of movement disorders (for reviews see Morris et al., 1994; Abbruzzese and Berardelli, 2003). Such statements have been typically made based on inaccurate performance of well-learned everyday actions, fragmented action trajectories, unusual patterns of involvement of elements (e.g., individual joint rotations or muscle activation patterns), and other atypical features at the level of performance variables. Equating atypical performance in a person with a neurological movement disorder with an impairment of the corresponding motor program assumes that all neural processes contributing to an action are parts of that motor program. This expands the notion of motor program beyond what is implied in its commonly used definitions. For example, changes in spinal reflex mechanisms may by themselves lead to inaccurate, fragmented, and unusual movement patterns even if the descending signals from the brain are unchanged. Similar consequences may also be expected in persons with the large-fiber peripheral sensory neuropathy (so-called "deafferented people," Sainburg et al., 1995) and in persons after a limb amputation due to the changed limb mechanics and absence of some of the sensory receptors.

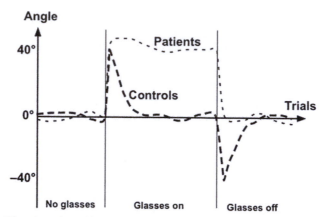

Figure 13.11 Direction of reaching movements to a visual target (located at 0°) prior to wearing prismatic glasses, with the glasses on, and after taking them off. The glasses shifted the perceived directions by 40°. Note the quick adaptation of movement direction to the perceptual distortion induced by the glasses seen in healthy persons (controls, thick dashed line). Note also the aftereffect with movement deviations in the opposite direction immediately after taking the glasses off. The adaptation and aftereffects were absent in patients with cerebellar disorders (thin dashed line). The scheme is drawn based on the data presented in Thach et al. (1992).

Another group of studies that invoked changes in motor programs (or in their cousins, internal models) explored problems with learning new skills and adaptation to unusual mechanical or sensory conditions. In particular, studies of patients with cerebellar disorders have documented poor adaptation of reaching movements to unusual external force fields and to changes in visual feedback produced by prismatic glasses (Thach et al., 1992, Figure 13.11). In those patients, the poor adaptation was accompanied by the lack of aftereffects, which are typically seen in healthy persons after the unusual field is turned off or the prismatic glasses are taken off. Taken together, these observations have been interpreted as pointing at the cerebellum as a likely site for hypothetical internal models (Kawato, 1999; Imamizu et al., 2000).

The notion of impaired internal models has also been used to describe atypical movements, motor learning, and motor adaptation in patients with disorders of other parts of the brain, in particular the basal ganglia (Contreras-Vidal and Buch, 2003), although other studies claimed intact learning of internal models in patients with Huntington's disease (Smith and Shadmehr, 2005). Impaired internal models have also been invoked to address movements by patients with more peripheral disorders, for example, amyotrophic lateral sclerosis (Nowak and Hermsdorfer, 2002) and impaired tactile sensation (Nowak et al., 2003). Such conclusions are commonly drawn based on observations of impaired anticipatory adjustments to planned actions, for example, impaired adjustments of grip force during manipulation of handheld objects or impaired anticipatory postural adjustments, and impaired adaptation to unusual external conditions.

If motor program (or internal model) is understood metaphorically, impaired motor programs may be invoked to describe the mentioned observations of atypical movements. Using this term, however, does not seem to move us closer to understanding physiological mechanisms involved in specific movement disorders.

13.7 Are new motor programs created in the process of motor learning?

The notion of motor program has been traditionally used to describe processes of learning new motor skills (Schmidt and Lee, 2011). Consider, for example, learning how to ride a bicycle. This task requires developing neural processes that ultimately ensure dynamical stability of the body on the very narrow support surface provided by the width of the wheels. This is a nontrivial task, which becomes even harder if one tries to acquire an ability to ensure static stability—it is harder to learn how to keep balance on a bicycle without moving than how to ride it. However, even at that very general and nonspecific level, the notion of motor program is hardly applicable. Indeed, learning how to ride a bicycle may generalize to bicycles of different size and design, but it does not help perform other tasks involving dynamical stability on a narrow surface, such as walking on a tightrope or skating on ice. So, motor program in this context remains a metaphorical notion, with limited generalization and no clear reflection in measurable physiological variables or even in mechanical performance variables.

Along similar lines, processes of learning how to perform a movement in an unusual force field or under distorted sensory information (e.g., while wearing prismatic glasses, see Figure 13.11) have been described using the notions of developing new internal models (Shadmehr and Mussa-Ivaldi, 1994). In those studies, analysis was performed at the level of mechanical variables and the presumed internal models were supposed to produce requisite patterns of muscle forces. First, as mentioned earlier (and emphasized by Bernstein many years ago, Bernstein, 1935), neural processes cannot in principle encode muscle forces. Such encoding would require perfect prediction of movement kinematics and external forces prior to movement initiation. Second, learning how to perform a movement in novel conditions shows poor generalization to unpracticed movements even if the external force field remains the same. For example, practicing movements in a curl force field using a narrow range of movement directions helps perform such movements within the same range but not in directions that deviate significantly from the practiced range (Mattar and Ostry, 2007; see Figure 13.7).

Some of the learned patterns may be hard to change even if external conditions make them functionally inefficient and energetically wasteful. For example, a quick movement in an arm joint is accompanied by properly timed and scaled activation patterns seen in muscles crossing other joints of the arm (Koshland et al., 1991; Latash et al., 1995, Figure 13.12). The purpose of those activations is not to produce movements in those joints but to avoid their motion under the influence of inertial and other joint interaction forces. So, the patterns may be viewed as a reflection of a multijoint (multimuscle) synergy stabilizing the endpoint trajectory. When the nontask joints are fixed with an orthotic device preventing their motion, the mentioned patterns of muscle activation persist even after some practice although their functional role becomes unclear (Koshland et al., 1991). As emphasized earlier in this chapter (Section 13.2), equating muscle activations with a motor program is highly controversial and has obvious limitations.

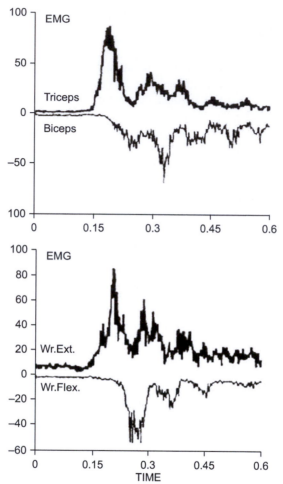

Figure 13.12 A quick movement in one of the joints (wrist or elbow) is accompanied by similar multiphasic patterns of muscle activation in the muscles crossing both joints. The purpose of the muscle activation pattern in the noninstructed joint is to prevent its motion under the action of interaction forces.
Reproduced by permission from Latash et al. (1995), © Elsevier.

Altogether, the available body of literature suggests that the notion of changing motor programs (and internal models) with motor learning is at best metaphorical. This notion means that something changes within the central nervous system when a person learns a new skill—nobody would argue with this statement. It, however, fails to offer a physiologically or physically meaningful interpretation of those unspecified changes. Trying to express them with the language of performance variables has obvious limitations (discussed in Chapters 10–12).

Within the scheme developed based on the idea of control with referent body configurations (Feldman, 2009; Latash, 2010; Chapter 12), learning a new skill may be associated with two potentially independent processes: learning an appropriate shift of the referent body configuration at the task level and learning synergies stabilizing salient performance variables for the new action (see Chapter 11). The former process ensures that the required movement pattern (or required force pattern on the environment) is achieved. The latter process may be viewed as the process of learning adequate few-to-many transformations involved in the scheme producing variables at lower levels of the assumed hierarchy (Chapter 12).

Recent studies have shown that changes in the index of synergy stabilizing an important performance variable (relative amount of intertrial variance compatible with no changes in that variable, V_{UCM}, see Chapter 11) can increase or decrease in the course of practice (reviewed in Wu and Latash, 2014). Early practice is typically associated with an increase in the synergy index, while later it can start to drop, reflecting a floor effect for the variance component that leads to performance variability and the continuing search for more optimal solutions. Reducing these complex time processes to development of a motor program would be a major simplification without a clear gain in understanding what happens within the central nervous system when a person learns a new motor skill.

13.8 The bottom line

The central question, "What do motor programs program?" remains without an answer. It is much easier to say what is not programmed by neural processes at high levels of the central nervous system that are commonly associated with motor programs. The central nervous system definitely does not program in detail displacements of effectors, joint rotations, muscle forces, joint torques, and muscle activation patterns. This is clearly impossible given the dependence of all these variables on the current external conditions, which are never 100% predictable. Associating motor programs with time patterns of referent body configurations is less controversial and more appealing given that there is a theoretical account that links referent configurations with neurophysiological variables such as subthreshold depolarization of neuronal pools (Feldman, 1986, 2009; see Chapter 12). However, as of now, there are no methods to measure referent body configurations during natural movements, and neuronal pools involved in specification of referent body configurations remain unknown. Hence, even in this meaning, motor program remains a theoretical, or even metaphorical, notion.

We come to a conclusion that the current broad usage of the term motor program (and *internal model*) is more misleading than revealing. It implies a level of understanding of mechanisms involved in voluntary motor control, which is currently unavailable. Despite the intuitive appeal of these terms, it is better to avoid using them. The old terms *engram* and *neural representation* are much less controversial. They adequately reflect our current meager knowledge of neurophysiological processes involved in learning, storage, and implementation of motor skills. While mapping engrams (or neural representations) on measurable movement patterns remains elusive, engrams may be associated with referent trajectories for salient performance variables at the hierarchically highest task level.

References

Adams, J.A., 1971. A closed-loop theory of motor learning. Journal of Motor Behavior 3, 111–150.

Abbruzzese, G., Berardelli, A., 2003. Sensorimotor integration in movement disorders. Movement Disorders 18, 231–240.

Adamovich, S.V., Levin, M.F., Feldman, A.G., 1997. Central modifications of reflex parameters may underlie the fastest arm movements. Journal of Neurophysiology 77, 1460–1469.

Asanuma, H., 1973. Cerebral cortical control of movements. Physiologist 16, 143–166.

Berkinblit, M.B., Feldman, A.G., Fukson, O.I., 1986. Adaptability of innate motor patterns and motor control mechanisms. Behavioral and Brain Sciences 9, 585–638.

Bernstein, N.A., 1930. A new method of mirror cyclographie and its application towards the study of labor movements during work on a workbench. Hygiene, Safety and Pathology of Labor, 5, p. 3–9, and 6, p. 3–11. (in Russian).

Bernstein, N.A., 1935. The problem of interrelation between coordination and localization. Archives of Biological Science 38, 1–35 (in Russian).

Bernstein, N.A., 1947. On the Construction of Movements. Medgiz, Moscow (in Russian).

Bernstein, N.A., 1967. The Co-ordination and Regulation of Movements. Pergamon Press, Oxford.

Bernstein, N.A., 1996. On dexterity and its development. In: Latash, M.L., Turvey, M.T. (Eds.), Dexterity and Its Development. Erlbaum Publ, Mahwah, NJ, pp. 1–244.

Bernstein, N.A., 2003. Contemporary Studies on the Physiology of the Neural Process. Smysl, Moscow, Russia.

Braitenberg, V., 1967. Is the cerebellar cortex a biological clock in the millisecond range? Progress in Brain Research 25, 2334–2346.

Celnik, P.A., Cohen, L.G., 2004. Modulation of motor function and cortical plasticity in health and disease. Restorative Neurology and Neuroscience 22, 261–268.

Cisek, P., Kalaska, J.F., 2005. Neural correlates of reaching decisions in dorsal premotor cortex: specification of multiple direction choices and final selection of action. Neuron 45, 801–814.

Coltz, J.D., Johnson, M.T.V., Ebner, T.J., 1999. Cerebellar purkinje cell simple spike discharge encodes movement velocity in primates during visuomotor arm tracking. Journal of Neuroscience 19, 1782–1803.

Contreras-Vidal, J.L., Buch, E.R., 2003. Effects of Parkinson's disease on visuomotor adaptation. Experimental Brain Research 150, 25–32.

Erlhagen, W., Schöner, G., 2002. Dynamic field theory of movement preparation. Psychological Reviews 109, 545–572.

Evarts, E.V., 1968. Relation of pyramidal tract activity to force exerted during voluntary movement. Journal of Neurophysiology 31, 14–27.

Feldman, A.G., 1966. Functional tuning of the nervous system with control of movement or maintenance of a steady posture. II. Controllable parameters of the muscle. Biophysics 11, 565–578.

Feldman, A.G., 1986. Once more on the equilibrium-point hypothesis (λ-model) for motor control. Journal of Motor Behavior 18, 17–54.

Feldman, A.G., 2009. Origin and advances of the equilibrium-point hypothesis. Advances in Experimental Medicine and Biology 629, 637–643.

Feldman, A.G., Levin, M.F., 1995. Positional frames of reference in motor control: their origin and use. Behavioral and Brain Sciences 18, 723–806.

Fukson, O.I., Berkinblit, M.B., Feldman, A.G., 1980. The spinal frog takes into account the scheme of its body during the wiping reflex. Science 209, 1261−1263.

Gelfand, I.M., Tsetlin, M.L., 1966. On mathematical modeling of the mechanisms of the central nervous system. In: Gelfand, I.M., Gurfinkel, V.S., Fomin, S.V., Tsetlin, M.L. (Eds.), Models of the Structural-Functional Organization of Certain Biological Systems. Nauka, Moscow, pp. 9−26 (in Russian, a translation is available in 1971 edition by MIT Press: Cambridge MA).

Georgopoulos, A.P., 1986. On reaching. Annual Review of Neuroscience 9, 147−170.

Georgopoulos, A.P., Kalaska, J.F., Caminiti, R., Massey, J.T., 1982. On the relations between the direction of two-dimensional arm movements and cell discharge in primate motor cortex. Journal of Neuroscience 2, 1527−1537.

Georgopoulos, A.P., Lurito, J.T., Petrides, M., Schwartz, A.B., Massey, J.T., 1989. Mental rotation of the neuronal population vector. Science 243, 234−236.

Georgopoulos, A.P., Schwartz, A.B., Kettner, R.E., 1986. Neural population coding of movement direction. Science 233, 1416−1419.

Gottlieb, G.L., Corcos, D.M., Agarwal, G.C., 1989. Organizing principles for single joint movements. I: a speed-insensitive strategy. Journal of Neurophysiology 62, 342−357.

Gottlieb, G.L., Corcos, D.M., Agarwal, G.C., Latash, M.L., 1990. Organizing principles for single joint movements. III: speed-insensitive strategy as a default. Journal of Neurophysiology 63, 625−636.

Imamizu, H., Miyauchi, S., Tamada, T., Sasaki, Y., Takino, R., Putz, B., Yoshioka, T., Kawato, M., 2000. Human cerebellar activity reflecting an acquired internal model of a new tool. Nature 403, 192−195.

Kawato, M., 1999. Internal models for motor control and trajectory planning. Current Opinions in Neurobiology 9, 718−727.

Keele, S.W., 1968. Movement control in skilled motor performance. Psychological Bulletin 70, 387−403.

Kelso, J.A.S., 1984. Phase transitions and critical behavior in human bimanual coordination. American Journal of Physiology 246, R1000−R1004.

Kelso, J.A.S., 1995. Dynamic Patterns: The Self-organization of Brain and Behavior. MIT Press, Cambridge.

Koshland, G.F., Gerilovsky, L., Hasan, Z., 1991. Activity of wrist muscles elicited during imposed or voluntary movements about the elbow joint. Journal of Motor Behavior 23, 91−100.

Kugler, P.N., Turvey, M.T., 1987. Information, Natural Law, and the Self-assembly of Rhythmic Movement. Erlbaum, Hillsdale, NJ.

Latash, M.L., 2008. Neurophysiological Basis of Movement, second ed. Human Kinetics, Urbana, IL.

Latash, M.L., 2010. Motor synergies and the equilibrium-point hypothesis. Motor Control 14, 294−322.

Latash, M.L., 2012. Fundamentals of Motor Control. Academic Press, New York, NY.

Latash, M.L., Aruin, A.S., Shapiro, M.B., 1995. The relation between posture and movement: a study of a simple synergy in a two-joint task. Human Movement Science 14, 79−107.

Latash, M.L., Scholz, J.P., Schöner, G., 2007. Toward a new theory of motor synergies. Motor Control 11, 276−308.

Malfait, N., Shiller, D.M., Ostry, D.J., 2002. Transfer of motor learning across arm configurations. Journal of Neuroscience 22, 9656−9660.

Mattar, A.A., Ostry, D.J., 2007. Modifiability of generalization in dynamics learning. Journal of Neurophysiology 98, 3321−3329.

Mattar, A.A., Ostry, D.J., 2010. Generalization of dynamics learning across changes in move-ment amplitude. Journal of Neurophysiology 104, 426—438.

Mechsner, F., Kerzel, D., Knoblich, G., Prinz, W., 2001. Perceptual basis of bimanual coor-dination. Nature 414, 69—73.

Meijer, O.G., 2001. Making things happen: an introduction to the history of movement science. In: Latash, M.L., Zatsiorsky, V.M. (Eds.), Classics in Movement Science. Human Kinetics, Urbana, IL, pp. 1—58.

Merton, P.A., 1953. Speculations on the servo-control of movements. In: Malcolm, J.L., Gray, J.A.B., Wolstenholm, G.E.W. (Eds.), The Spinal Cord. Little, Brown, Boston, pp. 183—198.

Morris, M.E., Summers, J.J., Matyas, T.A., Iansek, R., 1994. Current status of the motor program. Physical Therapy 74, 738—748, discussion 748—752.

Nowak, D.A., Hermsdörfer, J., 2002. Impaired coordination between grip force and load force in amyotrophic lateral sclerosis: a case-control study. Amyotrophic Lateral Sclerosis and Other Motor Neuron Disorders 3, 199—207.

Nowak, D.A., Hermsdörfer, J., Marquardt, C., Topka, H., 2003. Moving objects with clumsy fingers: how predictive is grip force control in patients with impaired manual sensibility? Clinical Neurophysiology 114, 472—487.

Nudo, R.J., Plautz, E.J., Frost, S.B., 2001. Role of adaptive plasticity in recovery of function after damage to motor cortex. Muscle and Nerve 24, 1000—1019.

Orlovsky, G.N., Severin, F.V., Shik, M.L., 1966. Locomotion evoked by stimulation of the midbrain. Proceedings of the Academy of Science of the USSR 169, 1223—1226.

Perrett, D.I., Harries, M.H., Bevan, R., Thomas, S., Benson, P.J., Mistlin, A.J., Chitty, A.J., Hietanen, J.K., Ortega, J.E., 1989. Frameworks of analysis for the neural representation of animate objects and actions. Journal of Experimental Biology 146, 87—113.

Rochet-Capellan, A., Richer, L., Ostry, D.J., 2012. Nonhomogeneous transfer reveals specificity in speech motor learning. Journal of Neurophysiology 107, 1711—1717.

Sainburg, R.L., Ghilardi, M.F., Poizner, H., Ghez, C., 1995. Control of limb dynamics in normal subjects and patients without proprioception. Journal of Neurophysiology 73, 820—835.

Schmidt, R.A., 1975. A schema theory of discrete motor skill learning. Psychological Reviews 82, 225—260.

Schmidt, R.A., 2003. Motor schema theory after 27 years: reflections and implications for a new theory. Research Quarterly for Exercise and Sport 74, 366—375.

Schmidt, R.A., Lee, T.D., 2011. Motor Control and Learning: A Behavioral Emphasis, fifth ed. Human Kinetics, Urbana, IL.

Schöner, G., 1995. Recent developments and problems in human movement science and their conceptual implications. Ecological Psychology 8, 291—314.

Schöner, G., 2002. Timing, clocks, and dynamical systems. Brain and Cognition 48, 31—51.

Schöner, G., Thelen, E., 2006. Using dynamic field theory to rethink infant habituation. Psychological Reviews 113, 273—299.

Schwartz, A.B., 1993. Motor cortical activity during drawing movements: population repre-sentation during sinusoid tracing. Journal of Neurophysiology 70, 28—36.

Shadmehr, R., Mussa-Ivaldi, F.A., 1994. Adaptive representation of dynamics during learning of a motor task. Journal of Neuroscience 14, 3208—3224.

Shadmehr, R., Wise, S.P., 2005. The Computational Neurobiology of Reaching and Pointing. MIT Press, Cambridge, MA.

Shik, M.L., Severin, F.V., Orlovskii, G.N., 1967. Structures of the brain stem responsible for evoked locomotion. Sechenov Physiological Journal of the USSR 53, 1125—1132.

Smith, M.A., Shadmehr, R., 2005. Intact ability to learn internal models of arm dynamics in Huntington's disease but not cerebellar degeneration. Journal of Neurophysiology 93, 2809—2821.

Stein, P.S.G., 1984. Central pattern generators in the spinal cord. In: Davidoff, R.A. (Ed.), Handbook of the Spinal Cord, vol. 2-3: Anatomy and Physiology. Marcel Dekker, Inc, New York, Basel, pp. 647—672.

Thach, W.T., Goodkin, H.G., Keating, J.G., 1992. Cerebellum and the adaptive coordination of movement. Annual Reviews in Neuroscience 15, 403—442.

Turvey, M.T., 1998. Dynamics of effortful touch and interlimb coordination. Journal of Biomechanics 31, 873—882.

Vallbo, A., 1970. Discharge patterns in human muscle spindle afferents during isometric voluntary contractions. Acta Physiologica Scandinavica 80, 552—566.

Vallbo, A., 1974. Human muscle spindle discharge during isometric voluntary contractions. Amplitude relations between spindle frequency and torque. Acta Physiological Scandinavica 90, 310—336.

Viviani, P., Terzuolo, C., 1980. Space-time invariance in learned motor skills. In: Stelmach, G.E., Requin, J. (Eds.), Tutorials in Motor Behavior, pp. 525—533 (Amsterdam: N-Holland).

Wolpaw, J.R., Tennissen, A.M., 2001. Activity-dependent spinal cord plasticity in health and disease. Annual Reviews in Neuroscience 24, 807—843.

Wolpert, D.M., Miall, R.C., Kawato, M., 1998. Internal models in the cerebellum. Trends in Cognitive Science 2, 338—347.

Wu, Y.-H., Latash, M.L., 2014. The effects of practice on coordination. Exercise and Sport Sciences Reviews 42, 37—42.

Zhou, T., Solnik, S., Wu, Y.-H., Latash, M.L., 2014. Unintentional movements produced by back-coupling between the actual and referent body configurations: violations of equifinality in multijoint positional tasks. Experimental Brain Research 232, 3847—3859.

Part Four

Examples of Motor Behaviors

Posture

14

Studies of posture are very common. A quick search on PubMed yields over 3000 papers on the topic of *vertical posture* alone. While the notion of *posture* is intuitively appealing, there is no clear definition for this notion in the movement science literature. Postural control may refer to keeping a configuration of a hand (e.g., sign language is commonly viewed as a sequence of hand postures), a limb, or the whole body (e.g., consider the aerial phase of a ski jump). It may also imply keeping a position of a part of the body with respect to either an external reference frame (the environment or an external object moving in the environment) or to the body itself in the process of action or natural, spontaneous changes in the external forces. Consider the following examples of postural control: (1) A figure skater maintaining a beautiful arm configuration while sliding over the skating rink (arm orientation may be maintained with respect to the body but not with respect to the environment); (2) A musician playing the violin while moving the body and the violin (hand posture may be maintained with respect to the violin, not necessarily to the body or the external space); and (3) Holding the handles while riding a bicycle (posture is maintained with respect to a part of the bicycle, not the whole bicycle, e.g., during sharp turns, not the body, and not the environment).

Many postural studies address the problem of keeping vertical orientation of the body in the field of gravity. In those studies, keeping vertical posture means not falling down while the standing person may be performing body movements, for example, swaying, turning, or catching or throwing an object. Imagine a person standing on a tilting platform or slipping while walking on ice. The person may show complex movements of all parts of the body, including the trunk, which help this person not to fall down. In this case, it is hard to identify a salient geometric characteristic corresponding to maintenance of vertical posture. Indeed, during whole-body actions, vertical projection of the center of mass does not have to fall always within the support area (as it happens, for example, during certain phases of walking). Nevertheless, keeping *vertical posture* is commonly viewed as a component of whole-body actions.

The problem of postural control may also be formulated with respect to a system of coordinates moving at varying speed with respect to the Earth; consider, for example, standing on a ship in stormy weather. This example emphasizes an important feature of posture—its stability under changes in the internal body states (including muscle activations) and in external forces. Two aspects of postural control typical of all the mentioned examples will be used in Section 14.2 to create a definition for posture (we do not know how to study a phenomenon without defining it first). These aspects are reflected in the mentioned keywords, *body configuration* (which is commonly equated with *posture* in biomechanics) and *stability*.

Biomechanics and Motor Control. http://dx.doi.org/10.1016/B978-0-12-800384-8.00014-4

14.1   Elements of history

Informal studies of vertical posture (standing posture) probably date back as long as studies of movements in general. Ancient Greek drawings of standing human figures typically depict postures that are stable in the field of gravity. Loss of vertical posture (loss of equilibrium) has always signified a major problem with the central nervous system; e.g., an inability to stand is a typical consequence of an epileptic seizure, and when a person loses consciousness, he/she invariably falls down.

Classical studies of Giovanni Alfonso Borelli (1608—1679), one of the founders of biomechanics, may be viewed as the first attempt to describe maintenance of vertical posture as a mechanical problem. Until the end of the nineteenth century, analysis of vertical posture did not move much beyond the relatively trivial mechanical statement that the projection of the center of mass of the body has to fall within the area of support.

Nikolai Bernstein (1896—1967) viewed posture as a necessary background component for any voluntary motor action. In his multilevel scheme for the construction of movements, postural muscle activity is regulated at the lowest level (Level A or the level of muscle tone, see Chapter 5). Note that Bernstein did not associate postural muscle activity with a geometric body characteristic, such as, joint configuration or body orientation in space. Bernstein also emphasized that the problem of stability in the field of gravity was a cornerstone problem not only with respect to movements but possibly also for the evolution of the central nervous system. In his classic *On Dexterity and Its Development* written in 1947 and published in English in 1996, Bernstein suggested that evolution had discovered two approaches to the problem of postural stability. One of them was realized in insects, animals with an external skeleton that allows maintaining stability in the field of gravity with minimal neural control; e.g., a dead bug is rather stable while standing on its six legs. The other approach used an internal skeleton that required continuous muscle activity to keep the body stable in the field of gravity. This approach placed significant burden on the central nervous system and encouraged its development in the process of evolution. According to Bernstein's guess, this challenging approach to the problem of postural stability was a major factor leading to the emergence of the complex central nervous system typical of vertebrates.

A major, qualitative step in addressing the problem of postural control was made in the middle of the twentieth century by von Holst and Mittelstaedt (1950) who introduced the famous *posture-movement paradox* applicable across postural tasks. By that time, a number of posture-stabilizing mechanisms had been known—in particular, those leading to the generation of forces against external perturbations applied during steady-state motor tasks at very short time delays (under 100 ms), shorter than the shortest simple reaction time (see Section 14.6.2). Von Holst and Mittelstaedt paid attention to the fact that posture-stabilizing mechanisms acted in response to external perturbations (resulting in quick postural corrections) but not to voluntary movements, even if the latter led to very similar deviations of the body or an effector from the pre-existing posture. They asked a seemingly naïve question: How can voluntary movements be performed without triggering resistive action of the posture-stabilizing mechanisms? For instance, why don't muscle stretch reflexes prevent muscle lengthening during voluntary movements? Von Holst and Mittelstaedt offered a solution for the posture-movement paradox based on the idea of *efferent copy* (or *efference copy*), as described in Chapter 8; this solution is unfortunately intrinsically contradictory and cannot solve the paradox. The posture-movement paradox remains a litmus test for motor control hypotheses; it has been reviewed in more detail in Chapter 12.

In the middle of the 1960s, the equilibrium-point hypothesis was introduced (Feldman, 1966; see Chapter 12). This was and still is the only hypothesis that considers posture and movement as different peripheral consequences of a single neurophysiological process. The hypothesis assumes that equilibrium states of the body in the environment are controlled using neurophysiological signals that define parameters of muscle reflexes. Changing these parameters readdresses posture-stabilizing mechanisms to different body positions and configurations. This method of control turns posture-stabilizing mechanisms into movement-producing ones and avoids the posture-movement paradox: The neural process associated with a voluntary action does not turn off or fight posture-stabilizing mechanisms but uses them to produce the movement and stabilize the new posture.

By the end of the twentieth century, two major approaches started to dominate studies of posture. One of them was inspired by the success of classical mechanics and control theory. It led to the analysis of vertical posture as the problem of stabilizing a single-axis or a multiaxis inverted pendulum with the help of feedback loops originating from sensory receptors of different modalities (Winter et al., 1996; Maurer et al., 2006). These schemes considered the force and velocity dependence of muscle forces as a major posture-stabilizing factor helped by adjustable gains in the feedback loops. The other approach was inspired by the equilibrium-point hypothesis and considered postural tasks, including those of standing in the field of gravity, as a particular subgroup of motor tasks controlled by specifying referent body configurations.

14.2   Creating a definition for posture

In different areas of movement science, the word *posture* is used with different meanings, sometimes explicitly specified and sometimes implied. An explicit definition of posture is offered in biomechanics: posture is equated with *joint configuration* (Zatsiorsky, 1998). According to this definition, posture is a component of *body position* (in addition to body *location* and *orientation*). This definition is very much appealing, in particular with respect to such sport activities as diving, when a particular posture (joint configuration) can be maintained during the large-amplitude whole-body motion in space.

In other fields of human movement studies, however, the word *posture* is commonly used in a much broader meaning, going beyond joint configuration. It may include also body orientation (e.g., as in *vertical posture*), maintenance over a certain period of time, and resistance to external perturbations (*local stability*).

For example, the frequently used expression *body posture* means orientation of the trunk (and, sometimes, head) with respect to a specific direction in the external space, e.g., the direction of gravity. Expressions such as *vertical posture* and *standing posture* are used frequently to describe a component of an action performed by a person in the field of gravity and typically related to an ability of that person not to fall down. In the area of movement disorders, patients with Parkinson's disease and some other neurological disorders are commonly described as having *postural problems* (or problems with postural stability). These terms imply an impaired ability to keep vertical posture (*balance, body equilibrium*) during actions such as standing, stepping, turning, etc. They can also imply, however, an inability to keep a steady position (steady orientation) in the space of an effector or an object held by the hand. Such problems can lead to spilling tea from a cup, losing orientation of handheld tools and utensils, etc. Clearly, these examples suggest a much broader understanding of posture as compared to the aforementioned biomechanical definition.

In the field of motor control, any action has been frequently viewed as a combination of postural and movement components. The former refers to keeping a particular geometric body characteristic (e.g., a joint angle or trunk orientation in space) or even a characteristic of an external object steady over the time of action, while the latter refers to time-varying body position changes. For example, the task of hammering a nail into a board is described as having a postural component (keeping the nail position and orientation in space constant) and a movement component (moving the hammer). Typically, humans prefer to use the nondominant arm to perform postural components of such actions and the dominant arm to perform the movement components (cf. the dynamic dominance hypothesis, Sainburg, 2002).

Within the aforementioned classical posture-movement paradox (von Holst and Mittelstaedt, 1950), the word *posture* implies a spatial (geometric) body characteristic (e.g., muscle length, joint angle, coordinates of the endpoint of a multijoint limb, etc.) stabilized by certain physiological mechanisms. This word has been applied to a range of objects, from a single muscle to the whole-body configuration. The word *posture* has been used in a similar meaning within the equilibrium-point hypothesis (Feldman, 1966, 1986): A postural state has been equated with an equilibrium state.

All these examples suggest that there is no single consistent meaning of posture in the movement science literature. We will start with a very general definition of posture as a state characterized by a certain position (configuration) of body parts—from a single muscle to multimuscle, multijoint systems, and to the whole body—with respect to a reference frame. Within this broad definition, movement is a time sequence of postures, while posture is an inherent component of any movement.

Typically, a body state is addressed as posture if it is maintained within a certain error margin over a certain observation time. Note that small perturbations are happening all the time due to the spontaneous variability of internal states of the body, including muscle activations and forces, and variations in external forces. Therefore, we will supplement the definition in the previous paragraph with the requirement of local stability. Two concepts discussed earlier in the book (Chapters 11 and 12) are directly related to this definition. These are the concept of *referent body configuration* and the concept of *synergy* as a neural organization providing for *stability* of particular characteristics of motor actions. Further, in this chapter we will explore relations between these concepts and issues relevant to postural control.

14.3   Posture as a steady-state process: postural sway

Keeping steady vertical posture in the field of gravity is not a trivial task from the point of view of mechanics. It can be compared to balancing an inverted pendulum—a naturally unstable system—with a few joints along its axis on a relatively small supporting surface (on the order of 0.3×0.3 m; Figure 14.1). One has to avoid moving the center of mass (COM) projection outside the support area and, at the same time, prevent individual joints from collapsing. This is achieved using a variety of posture-stabilizing mechanisms, from those inherent to the peripheral muscle properties (such as the length-, and velocity-dependence of muscle force, see Chapters 2 and 3) to those mediated by feedback loops originating from sensory receptors of different modalities (Chapters 6 and 7).

The importance of vision, vestibular system, and somatosensation for the control of posture is well established, in particular, based on observations of patients with a variety of sensory and motor disorders. Several disorders lead to disruption of feedback loops from somatosensory receptors in the lower body. Earlier in the twentieth century, untreated syphilis could lead to a state called *tabes dorsalis*, a disruption of

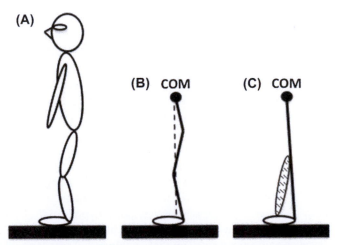

Figure 14.1 (A) The body of a standing person may be viewed in a sagittal plane as a chain of body segments connected by joints. (B) The body is sometimes modeled as a series of rigid links connecting the center of mass (COM) to the feet. To keep equilibrium during standing, the COM projection has to fall within the support area. Such a model neglects that the COM can be displaced by movements of the body parts that are located above the COM. (C) The single-joint inverted pendulum model reduces the system to the center of mass moving about a single joint served by spring-like muscles.

transmission of sensory signals via the dorsal columns of the spinal cord. Those patients could only keep vertical posture with their eyes open. Similar observations have been made in patients with advanced stages of diabetes and large-fiber peripheral neuropathy (Sanes et al., 1985; van Deursen and Simoneau, 1999; Bonnet et al., 2009).

The importance of vision is exemplified by the fact that, even in a healthy person, closing the eyes makes prolonged standing challenging, particularly if the person is standing on a compliant surface (Riemann et al., 2003; Patel et al., 2008). Vertical posture can also show major deviations from the vertical under controlled motion of the visual field. On the other hand, people who were born blind can stand and perform the whole variety of actions while standing without showing major differences from people who can see (Schieppati et al., 2014). These observations suggest that the role of vision may be compensated for by signals of other sensory modalities.

The role of the vestibular system in postural control is less obvious. On the one hand, signals from the vestibular system inform the central nervous system on the position and motion of the head in space, which may or may not be relevant for the task of quiet standing. A healthy person can easily perform rather fast head movements without losing balance. On the other hand, vestibular disorders commonly lead to postural problems and, sometimes, to an inability to maintain vertical posture. One of the hypotheses is that the vestibular system participates in the creation of a reference frame within which sensory signals of other modalities, such as vision and somatosensation, are evaluated with respect to the task of keeping vertical posture (Mergner, 2007).

When a healthy person stands quietly, the body posture shows spontaneous deviations addressed as *postural sway*. Typically, postural sway is studied using deviations of the COM of the body in a horizontal plane or of the coordinate of the application of the resultant vertical force acting on the body from the supporting surface addressed as *center of pressure* (COP). Figure 14.2 shows typical time profiles of the COM and COP deviations in the anterior-posterior direction for a healthy, young person standing with eyes open. Both time series show seemingly irregular deviations of relatively small amplitude, on the order of a few millimeters.

There are two main views on the nature and importance of postural sway. The first considers postural sway as a consequence of noise within the neuromotor system, a sign of imperfection in the design of the human body, which requires sophisticated neural control to keep the sway low and avoid losing balance (e.g., Peterka et al., 2011). The second views sway as a reflection of a purposeful design of the neurophysiological system for postural control (Riccio, 1993; Riley et al., 1997).

There seem to be several problems with the "noise hypothesis" of postural sway. The first is philosophical. Given our current measly understanding of the central nervous system, it takes some nerve to claim that evolution led to a faulty design of the system for postural control, and contemporary engineers are able to identify the flaw. The second is the observation that low sway does not necessarily mean better postural stability. Arguably, the best-known example is the markedly reduced sway in patients with advanced Parkinson's disease, a disorder characterized by dramatic problems with postural stability. These patients may lose balance in response to even a very small external force perturbation. The third is the counterintuitive increase in postural

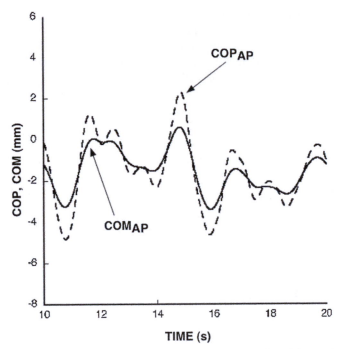

Figure 14.2 During quiet standing on a horizontal surface, the point of the vertical ground reaction force application (center of pressure, COP) shows spontaneous migration in the horizontal plane (thin trace). This process is accompanied by migration of the center of mass (COM) projection on the horizontal plane.
Modified by permission from Latash (2008).

sway when a healthy person is required to stand on a board with a narrow supporting surface (Mochizuki et al., 2006). This is observed even when the narrow dimension of the support is much larger than the typical sway magnitude. These (and some other) observations suggest that postural sway is a purposeful process, possibly related to exploration of the stability conditions (Mochizuki et al., 2006; Murnaghan et al., 2014).

There is no one-to-one correspondence between postural sway and postural stability. On the one hand, sway is increased in some populations with postural problems, from healthy older adults to persons with atypical development, and to patients with different movement disorders (e.g., Hadders-Algra and Carlberg, 2008; Stylianou et al., 2011; Sarabon et al., 2013). On the other hand, the mentioned reduced sway in advanced Parkinson's disease presents an impressive counterexample. Note that stability is a property of a system to return to an equilibrium state after a transient perturbation, while sway is a spontaneous phenomenon, a behavioral characteristic of the system, observed in the absence of any external perturbation.

Analysis of postural sway uses a large number of metrics quantified in the time series of the COM or COP displacements. Commonly, sway is analyzed separately in the anterior-posterior (AP) and mediolateral (ML) directions. Among the most commonly used parameters of sway in the AP and ML directions are its amplitude, average velocity, and frequency characteristics. The overall sway area is also commonly used with the sway trajectory in a horizontal plane reduced to an ellipse containing a certain, reasonably defined, percentage of the data points (Oliveira et al., 1996). There are also more sophisticated methods of sway analysis that use methods from the field of nonlinear time series analysis and explore such features of sway as its regularity and predictability (Collins and De Luca, 1993; Riley et al., 1997; Stergiou, 2004). The large number of outcome characteristics of sway do not necessarily contribute to better understanding of its mechanisms and changes in these mechanisms with age, pathology, training, etc. The problem is in associating changes in sway parameters with possible changes in the involved anatomical structures and physiological processes.

14.3.1 Inverted pendulum model

Postural sway has been viewed as oscillation of an inverted pendulum about the ankle joints produced by spontaneous variations in the muscle activation levels (Winter et al., 1996) leading to time variations of parameters describing the length and velocity dependence of the muscle forces (Chapter 3; see Figure 14.1(C)). According to this view, the resulting time changes of the ankle joint apparent stiffness (and apparent damping) modify the natural frequency of the inverted pendulum and produce the typical irregular sway patterns.

This very simple model has several problems. First, the notion of joint stiffness is not well defined (see Chapter 2). Second, experimental estimations of the apparent ankle joint stiffness have produced conflicting results, typically much lower than those necessary to stabilize the inverted pendulum representing the body (Morasso and Sanguineti, 2002; Casadio et al., 2005). Third, it has been shown that the length of muscle fibers may change in the opposite direction to what could be concluded by observation of joint motion (Loram et al., 2005). In particular, body sway forward (stretching the plantarflexors) is, at certain phases, associated with shortening of the muscle fibers in the triceps surae group. This means that muscle fiber length changes may be in the opposite direction to the changes in the muscle-tendon length. These observations question whether reflexes originating from sensory signals in the muscle spindles (which produce signals reflecting muscle fiber length and its changes, not changes in the muscle-tendon complex) can help modulate muscle activation to provide body stability.

The idea of a single-joint inverted pendulum has been supplemented by an assumption of a neurally controlled torque generator at the ankle joint acting in parallel to the "ankle joint spring," with torque magnitudes implemented with the help of feedback loops typical of control theory approaches. Such control schemes are sometimes rather complex involving several loops acting in parallel driven by sensory signals of different modalities, with different characteristic time delays and adjustable gains (Maurer et al., 2006; Goodworth and Peterka, 2012). These models have been used to analyze changes in the effects of signals of different sensory modalities on posture in different populations and with modifications of the postural task. Relative changes in the gains of particular feedback loops within such a model are commonly addressed as *sensory reweighting* (Oie et al., 2002; Maurer et al., 2006; Assländer and Peterka, 2014). Note, however, that these changes are computed within an assumed model and may be reflective of the structure of the model, not only of actual processes within the human body.

14.3.2 Rambling and trembling

A major assumption within the aforementioned computational approaches is that posture is stabilized with respect to a fixed coordinate corresponding to the unstable equilibrium of the inverted pendulum. A number of studies have provided strong evidence that the body sways not about a fixed point but about a migrating point (Duarte and Zatsiorsky, 1999). One such approach to postural sway has been developed as *rambling-trembling* (Rm-Tr) decomposition of the COP trajectory (Zatsiorsky and Duarte, 1999, 2000). This method analyzes the time series of the COP trajectory in the AP and ML directions separately. For each direction, points in time are identified when the horizontal force acting on the body in the selected direction is zero. By definition, the body is in an instantaneous equilibrium state in each of those points. An interpolation of those points is Rm, while the difference between the original COP trajectory and Rm is defined as Tr (Figure 14.3).

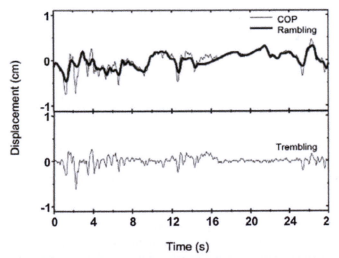

Figure 14.3 Top panel: A typical trajectory of the point of the vertical ground reaction force application (center of pressure, COP) in the anterior-posterior direction during quiet standing (thin trace). The points when the resultant force acting on the body in the anterior-posterior direction is zero were identified on the COP curve and interpolated to produce the rambling trajectory (Rm, thick trace). Bottom panel: The difference between the COP and Rm trajectories is trembling (Tr).
Reproduced by permission from Latash (2008).

The Rm-Tr decomposition assumes that COP trajectory represents a superposition of two processes, one of which (Rm) reflects migration of the equilibrium set point while the other (Tr) reflects body oscillations about that moving set point. The migration of the equilibrium set point does not induce restoring forces (as can be judged from the horizontal component of the ground reaction force) while the body oscillation about the moving set point does this (Figure 14.4).

The above oscillations may be functions of many variables including the mechanical body properties (e.g., inertia and activation-dependent muscle apparent stiffness)

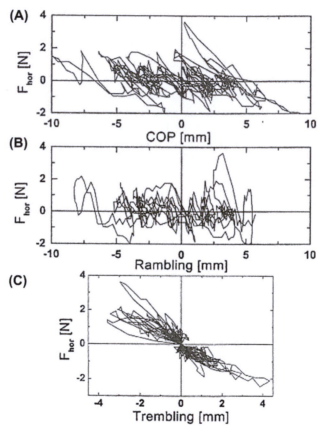

Figure 14.4 Relations between the horizontal component of the ground reaction force, F_{HOR}, and (A) the center of pressure (COP) migration, (B) rambling (Rm), and (C) trembling (Tr). A representative example is shown. For the COP and Rm, the coefficients of correlation are low ($r < -0.24$). In contrast, a large negative correlation ($r = -0.89$) was observed for the Tr. The negative correlation signifies existence of the restoring force that acts against the displacement. The large deviations of the COP and Rm do not give rise to restoring forces while the small deviations in Tr do. This fact confirms the idea that Rm represents migration of the equilibrium reference, which does not require correction, while Tr represents the deviations from the moving equilibrium trajectory that induce corrections.
Reproduced by permission from Zatsiorsky and Duarte (2000), © Human Kinetics.

and reflex feedback effects. This view is readily compatible with the referent config-uration hypothesis (Chapter 12): Migration of the equilibrium set point reflects changes in the referent body configuration, while body mechanics in combination with reflex loops lead to deviations of the actually observed COP trajectory from the equilibrium trajectory.

Several studies have shown contrasting changes in characteristics of rambling and trembling with age, manipulations of visual information, and other manipulations of the task of quiet standing (Danna-Dos-Santos et al., 2008b; Sarabon et al., 2013). Re-lations of Rm and Tr to physiological processes remain speculative, however.

14.4   Posture and movement: two outcomes of control with referent configurations

Human muscles are length (and velocity) dependent force generators. These properties are inherent to both peripheral tissues (muscles and tendons, Chapters 2 and 3) and the action of reflex loops (Chapter 6). As a result, equilibrium in one spatial reference frame may be violated in another spatial reference frame. Indeed, if muscle force bal-ances the external force, and the system is in equilibrium, a shift in the referent value of muscle length is expected to lead to a change in the muscle force (since it depends on the difference between actual muscle length and its referent length), violation of the equilibrium conditions, and movement of the system. In other words, a shift of the reference frame can by itself produce movement of an object that was in equilibrium in another reference frame. This basic fact allows considering posture (equilibrium) and movement as two peripheral consequences of control with spatial referent coordinates.

As far as we know, only one theory in the field of motor control handles this basic property of the neuromuscular system in a noncontradictory way that is compatible with the laws of physics and the known physiology. This is the equilibrium-point hy-pothesis (Feldman, 1966, 1986) that has recently been developed in the form of the referent configuration hypothesis (Feldman and Levin, 1995; Chapter 12). Within this hypothesis, neurophysiological signals produce shifts in the referent body config-uration. These shifts produce changes in the equilibrium states of the body and/or its segments given the external force field. As a result, earlier equilibrium states are discarded and new ones are established leading to movements from the former to the latter.

Consider, for simplicity, the problem of keeping posture in a joint with one degree of freedom controlled by two muscles. Figure 14.5 illustrates dependences of the active muscle force on muscle length transferred into the torque-angle units for two muscles, agonist and antagonist, crossing the joint. Note that one of the muscles, the antagonist, generates negative torque values. Referent configuration for this system may be viewed as a combination of two variables, one per muscle, corresponding to the thresholds (λ) of the tonic stretch reflex (Liddell and Sherrington, 1924; Feldman, 1986). Another, equivalent, pair of variables can be used corresponding to the

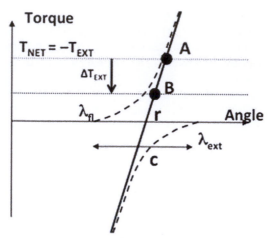

Figure 14.5 Muscle force—length characteristics for two muscle crossing a joint, agonist (positive torque values) and antagonist. Referent configuration for this system is a combination of two variables, λ_{ag} and λ_{ant}, corresponding to the thresholds of the tonic stretch reflex. An equivalent pair may be used, $\{r; c\}$—the reciprocal and coactivation commands. Given external torque (T_{EXT}), the system will be in equilibrium in point A where the net torque $T_{NET} = -T_{EXT}$. A change in T_{EXT} (ΔT_{EXT}) leads to a shift in the equilibrium to point B.

midpoint between the two λs and to the distance between their values (the reciprocal command and the co-activation command, $\{r, c\}$ (Feldman, 1980). This system will be in a state of stable equilibrium against an external torque (T_{EXT}), if the algebraic sum of the active torques produced by the two muscles (T_{NET}) is equal in magnitude and opposite in sign to T_{EXT}. A change in T_{EXT} is expected to move the system to a new equilibrium position where its net muscle torque balances the new value of T_{EXT} (Figure 14.5); the system is expected to move back to the initial state as soon as T_{EXT} returns to its preperturbation value assuming that the subject is not reacting to this perturbation (in other words, that the $\{r, c\}$ values remain unchanged). We will refer to all the involuntary mechanisms that contribute to stability of the equilibrium state as *posture-stabilizing* ones.

Performing a movement within this framework is associated with changing one of the variables $\{r, c\}$ or both. Figure 14.6 illustrates consequences of changing r. Note that a change in r leads to the emergence of a new equilibrium state (B), and the old state (A) becomes a deviation from the new one. The same posture-stabilizing mechanisms that returned the system back to state A in the earlier example (Figure 14.6) now move the system to its new equilibrium at point B. These examples show that posture and movement may be viewed as consequences of the same neural control process—specification of RC, such as values of $\{r, c\}$ for a joint or respective values for multijoint systems (for more detail, see Chapter 12). Studies over the past 50 years have shown that this description is applicable to a variety of motor tasks, including whole-body tasks such as standing and walking (reviewed in Latash, 1993; Feldman, 2009).

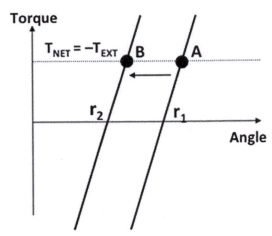

Figure 14.6 Voluntary movement in a two-muscle, single-joint system is associated with a change in r-command and/or in c-command (see Figure 18.5). A shift in r leads to disappearance of the old equilibrium state (point A) and emergence of a new equilibrium state (point B). Both stabilization of a postural state and transition to a new state are due to the same posture-stabilizing mechanisms. The solid lines show the torque-angle characteristics of the joint.

14.5   Postural synergies

As described in Chapter 11, the word *synergy* has at least three different meanings in motor control. One of them is directly related to stability of salient performance variables during motor actions. This makes this meaning most relevant to the problem of postural stability. Therefore, in this chapter, we are going to use the word *synergy* to imply a neural organization of a redundant set of elemental variables with the purpose of stabilizing a particular characteristic of an action.

The framework of the uncontrolled manifold (UCM) hypothesis (Scholz and Schöner, 1999; Chapter 11) has been used to identify and quantify synergies in a variety of tasks including those performed by standing persons (Krishnamoorthy et al., 2003b) or persons involved in another postural task (Krishnamoorthy et al., 2007). Within this framework, analysis is performed in a multidimensional space of elemental variables chosen based on the selected level of analysis. For example, elemental variables may represent activation levels of muscle groups (M-modes, Krishnamoorthy et al., 2003a), joint rotations, or other kinematic or kinetics variables. Variance across repetitive trials in the space of elemental variables is used as a reflection of stability of the action in different directions within that space. Assuming natural variability in the initial conditions, directions of low stability are expected to show large intertrial variance magnitudes while directions of high stability are expected to show low intertrial variance. Within the UCM-based method, variance is estimated in two subspaces. One of them (the UCM) corresponds to no changes in a potentially important performance variable, while the other, orthogonal to the UCM, subspace (ORT) corresponds to changes in that variable.

If variance within the UCM (V_{UCM}) is larger than variance orthogonal to the UCM (V_{ORT}), both quantified per degree of freedom, a conclusion can be drawn that a multielement synergy stabilizes the variable with respect to which the analysis was performed.

Postural tasks, as all natural motor tasks, are based on redundant sets of elements. For example, if one considers the kinematics of keeping the projection of the COM of the body within the support area, there are several joints that link the COM with the support area. As a result, an infinite number of joint configurations can lead to the same COM horizontal coordinates. Figure 14.7 illustrates three joint configurations in a sagittal plane that all correspond to the same horizontal coordinate of COM. The same is true if muscle forces or activation levels are used as elemental variables. A number of studies have documented multijoint kinematic synergies stabilizing COM coordinates and trunk orientation with respect to the vertical (Freitas et al., 2006), and multi-M-mode synergies stabilizing forces and moments of force acting on the body from the supporting surface as well as COP coordinates (Krishnamoorthy et al., 2003b; Klous et al., 2010).

As discussed in more detail in Chapter 11, the notion of elemental variables is nontrivial. In particular, these variables may be viewed as reflections of basic, elemental, shifts of body referent configurations. This view is linked to the notion of *primitives* that has been developed in the field of motor control. This notion assumes that the motor repertoire is based on a limited set of elementary actions (primitives) that can be scaled and combined to match specific tasks. Primitives have been associated with force fields, combinations of muscle activation patterns, and combinations of kinematic trajectories (Tresch et al., 1999; Hart and Giszter, 2010; Hogan and Sternad, 2012). Let us emphasize one more time (see Chapter 12) that neural signals within the central nervous system (CNS) cannot encode peripheral mechanical variables because the latter are functions of the external force field,

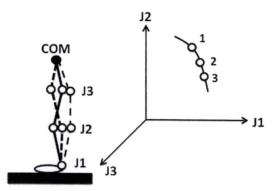

Figure 14.7 Three joint configurations in a sagittal plane corresponding to the same horizontal coordinate of the center of mass (COM). Note that this schematic does not show effects of the body segments above COM on the COM coordinate. These configurations belong to the uncontrolled manifold in the three-dimensional joint space, {J1; J2; J3}, corresponding to no COM shifts. This manifold is shown schematically as a curve in the right drawing.

which typically changes in an unpredictable fashion. Hence, primitives can be viewed as small sets of elementary changes in body RCs used to produce desired actions. This definition makes primitives a specific example of *modes* as defined in analysis of movement synergies (see Chapter 11). Note that this set may still be redundant, and synergies in the space of elemental RC shifts may be organized reflected in synergic relations among kinetic, kinematic, and/or muscle activation variables.

A specific example of possible elemental RCs has been described based on the idea of decomposing the equation of motion in the three main leg joints (ankle, knee, and hip) into independent equations along new coordinates referred to as eigenmovements (Alexandrov et al., 2001). Each eigenmovement produces motion involving a linear combination of rotations in all three joints. In the equations of motion written for eigenmovements, motion along each eigenmovement depends only on torques directed along the same eigenmovement. The resulting whole-body elementary motions (kinematic primitives) correspond roughly to the *ankle strategy*, *hip strategy* (Horak and Nashner, 1986), and a third motion involving flexion (or extension) in all three joints. Note that the ankle strategy is not limited to ankle movement while the hip strategy in not limited to hip movement; the names refer to joints showing the largest excursion.

Recent studies of activation patters in multimuscle systems can be interpreted as resulting from superposition of several basic elemental RC shifts. A number of those studies used various matrix factorization techniques to demonstrate that changes in the activation patterns within a large set of muscles during whole-body tasks (including postural and locomotor tasks) can be described using a smaller number of *factors*, *synergies*, or *modes* (Krihsnamoorthy et al., 2003b; Ivanenko et al., 2004; Ting and Macpherson, 2005). We will use the word *M-mode* to avoid confusion with the earlier introduced notion of synergy. Each M-mode may reflect a primitive at the level of control, i.e., an elemental shift in the body RC (including RC changes leading to muscle co-contraction). Since the number of modes may still be larger than the number of task constraints, synergies in the mode space may be created to stabilize important performance variables produced by all the muscles together.

The composition of M-modes shows consistency across a variety of tasks (Ivanenko et al., 2005; Torres-Oviedo and Ting, 2010). In certain conditions, M-mode composition can change, however. In particular, during postural tasks performed while standing, M-modes commonly show reciprocal patterns. This means that a mode is likely to involve muscles on the frontal or dorsal surface of the body but not agonist-antagonist muscle pairs acting at a joint. In challenging conditions and in older persons, so-called co-contraction M-modes become more and more common; these involve parallel changes in the activation levels of opposing (agonist and antagonist) muscles (Danna-Dos-Santos et al., 2008a). Such patterns of unusual M-modes observed in challenging tasks have been linked directly to different sets of elemental body RC shifts (Robert et al., 2008).

RCs are not easy to record in an experiment. Even for a simple single-joint action, reconstruction of RC time profiles requires assuming a mechanical model of the moving effector and performing multiple trials under changing external conditions

(Latash and Gottlieb, 1991). To the best of our knowledge, only two studies so far reconstructed RCs for an action with an explicit postural component (Latash et al., 1999; Domen et al., 1999). In the former experiment, the subjects performed a quick action in one of the joints of the elbow-wrist system without a special instruction about the behavior of the other joint (Figure 14.8(A)). In such conditions, humans naturally keep the noninstructed joint nearly motionless. This is not a trivial observation. Indeed, the mechanical joint coupling leads to the generation of time-varying interaction joint torque changes in the apparently postural joint during motion of the instructed joint (see Section 1.3.3 in Chapter 1).

Several studies reported triphasic muscle activation patterns in the agonist—antagonist pair acting at the postural joint resembling the muscle activation patterns at the instructed joint (Koshland et al., 1991; Latash et al., 1995; Figure 14.8(B)). The purpose of the muscle action at the postural joint is not to move that joint but to avoid its motion expected from the action of motion-dependent torques. RCs were reconstructed in this simple two-joint system (Figure 14.9). They showed similar time patterns at both instructed and noninstructed (postural) joints. These observations suggest that there was indeed a synergy between the $\{r; c\}$ pair changes at the two joints stabilizing the equilibrium of the postural joint at its initial position, which required keeping the net torque at that joint at a value close to zero at all times.

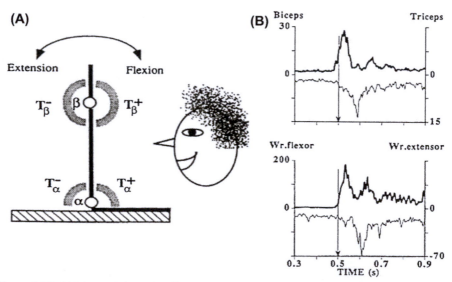

Figure 14.8 (A) A two-joint wrist-elbow system. Motion in one joint of this system generates torques in the other joint due to the mechanical joint coupling. (B) A quick action in one joint is characterized by a triphasic pattern of muscle activation seen in muscles crossing both the focal joint (wrist, bottom traces) and the apparently postural joint (elbow, top traces). Note the similar triphasic patterns in both joints. The triceps and wrist extensor signals are inverted for better visualization.

Modified by permission from Latash et al. (1995), © Elsevier.

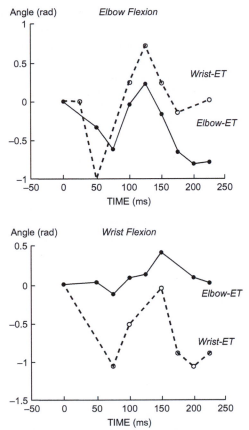

Figure 14.9 Equilibrium trajectories for the two-joint system illustrated in Figure 18.7 associated with motion of the elbow joint (top) and of the wrist joint (bottom). Note the similarly timed equilibrium trajectories (Wrist-ET, equilibrium trajectory for the wrist joint; Elbow-ET, equilibrium trajectory for the wrist joint) in both focal and postural joints.
Reproduced by permission from Latash et al. (1999), © Elsevier.

14.6   Postural preparation to action

Animals commonly behave in a predictive fashion. The fox does not run toward the spot where the rabbit is but rather to a spot where the fox plans to intercept the rabbit. Such predictive control is common in a variety of motor tasks performed by humans, including postural tasks. Any action by a person involved in a postural task is the source of a potential perturbation for the postural task. This is due, in particular, to the mechanical coupling of body segments. Besides, an action by a standing person commonly leads to a shift in the location of the COM, for example, when the person accepts or releases a heavy load, or moves one of the effectors. This could also require a corrective action to avoid losing equilibrium.

The first line of defense of the relatively fragile postural equilibrium involves pre-dictive changes at the level of neural control in anticipation of a predictable pertur-bation. When the timing of a perturbation is predictable (for example, when the perturbation is triggered by an action performed by the same person), there are feed-forward adjustments in action of elements (muscles, joints, digits, etc.) involved in the postural task observed before the perturbation actually happens. One of the components of anticipatory postural control—anticipatory postural adjustments (APAs)—has been known for about 50 years (Belenkiy et al., 1967) while the other one—anticipatory synergy adjustments (ASAs)—has been described only recently (Olafsdottir et al., 2005; Klous et al., 2011; Krishnan et al., 2011).

14.6.1 Anticipatory postural adjustments

Anticipatory postural adjustments were originally described as changes in the activa-tion of postural muscles in preparation to a self-generated action that produced a postural perturbation (reviewed in Massion, 1992). These changes are seen about 100 ± 50 ms prior to the perturbation if the perturbation time is perfectly predictable (e.g., if it is triggered by the action of the standing person). Many later studies explored APAs under such actions as quick arm movements and load manipulations. From the very first studies, the primary function of APAs has been assumed to produce forces and moments of force that would counteract the expected perturbations. This hypoth-esis received support in both experimental and modeling studies (Cordo and Nashner, 1982; Ramos and Stark, 1990). There were, however, exceptions to this general rule. For example, when the expected perturbation moved the body from a precarious posture to a more stable posture, the APAs could reverse and act in the direction of the perturbation (Forssberg and Hirschfeld, 1988; Krishnamoorthy and Latash, 2005).

APAs have been shown to scale with factors other than the site, magnitude, and di-rection of the postural perturbation. In particular, they scale with the magnitude of the action that triggered a standard perturbation, with changes in postural stability, and with time pressure (when the person had little time to prepare for a perturbation). APAs are shifted toward the action initiation time when a person is required to perform a quick action as quickly as possible after a signal (Lee et al., 1987); they are also reduced when the same action is performed in unstable conditions as well as in very stable conditions (Nardonne and Schieppati, 1988; Nouillot et al., 1992; Aruin et al., 1998). These observations show that APAs are a luxury, and that similar actions can be produced with dramatically attenuated APAs or even without APAs. This conclusion is supported by the very small changes in mechanical variables associated with APAs; for example, COP shifts by about 1 mm during a typical APA seen prior to a very quick bilateral shoulder movement, which is smaller than its spontaneous changes during postural sway (see Section 14.4).

As any other action, APA may be viewed as produced by a shift in the referent body configuration. Such a shift may be based on a set of basic (elemental) changes in RC and associated with multimuscle (multi-M-mode) synergies stabilizing the time pro-files of important performance variables, such as COP shifts and changes in the shear forces acting on the body (Krishnamoorthy et al., 2003a). Typical patterns of changes in muscle activations during APAs include reciprocal patterns, i.e., an increase in the activation level of a muscle is accompanied by a drop in the activation level of its antagonist (Aruin and Latash, 1995). Such patterns are compatible with changes in the r-command to a postural joint (Chapter 12). However, adjustments in those basic patterns to changes in conditions of postural stability are more consistent with adjustments in the other, c-command (Slijper and Latash, 2000). APAs in challenging conditions and in populations with impaired postural stability commonly show co-contraction patterns in agonist-antagonist muscle groups acting at postural joints (e.g., Woollacott et al., 1988).

Sometimes, APAs become the source of postural perturbations. This may happen if COP shifts produced by an APA move beyond the reduced support area or produce shear forces that are too large for the friction coefficient, for example, when standing on a slippery surface. Another example is APAs that are preparing the posture for a perturbation that does not come. A typical example is the so-called "broken escalator" phenomenon (Bronstein et al., 2009). A typical person with experience of entering a moving escalator can do this task without visible changes in the vertical body posture. Note that stepping on the moving escalator produces a major postural perturbation, which may lead to balance problems (seen in persons who enter the moving escalator for the first time). The lack of such balance problems is due to appropriate APAs. If the escalator is broken, however, an experienced person is suddenly at a disadvantage—it is hard to step on a broken escalator without major postural problems induced by the APAs.

By their nature, APAs are based on predictions of upcoming perturbations. As a result, they are always suboptimal and compensate only for a fraction of the actual perturbation (or overcompensate it). The residual perturbation is handled by reactive posture-stabilizing mechanisms described in Section 14.7.

14.6.2 Early postural adjustments

There are postural adjustments prior to an action that have also sometimes been addressed as APAs. For example, when a person prepares to make a step from a standing posture, changes in the muscle activation patterns (and in mechanical variables) can be seen 500−1000 ms prior to the toe-off of the leading foot (Brenier and Do, 1986; Crenna and Frigo, 1991). These adjustments lead to a nonmonotonic shift of the COP toward the stepping foot followed by its shift toward the supporting foot with a simultaneous COP shift backward. The described COP shift is necessary to initiate stepping—it allows unloading the stepping leg and generating a moment of the ground reaction force, rotating the body forward. Postural adjustments with a similar timing are seen prior to a voluntary body sway (Klous et al., 2012). To avoid confusion, it is better to address these postural adjustments using a different term, for example, early postural adjustments (EPAs; Krishnan et al., 2011; Klous et al., 2012).

APAs and EPAs differ in several important aspects. The most striking one is their timing—EPAs start much earlier than APAs. The second is the function—APAs counteract an expected perturbation, while EPAs adjust posture to make the planned action possible in the absence of any identifiable perturbation. As mentioned earlier, APAs may be viewed as a luxury; in most situations, balance can be kept in the absence

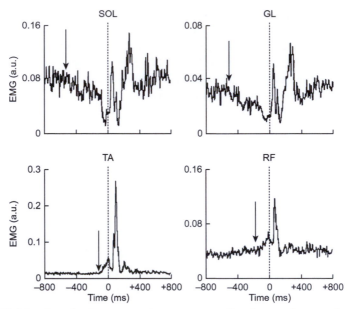

Figure 14.10 Muscle activation patterns in a person standing in a semi-squatted posture and watching a passive object approaching the person and hitting him at the shoulders. EPAs are seen as an early drop in the background activation levels of the dorsal muscles (SOL, soleus; GL, lateral gastrocnemius) starting about 500 ms prior to the impact (top panels), while APAs are seen in both dorsal (an abrupt drop in the activation level) and ventral muscles (a burst of activation, bottom panels, TA, tibialis anterior; RF, rectus femoris) seen about 100 ms prior to the impact.
Modified by permission from Krishnan et al. (2011), © Springer.

of APAs (or when APAs are significantly delayed and/or reduced in magnitude). EPAs are mechanically necessary to make the planned action possible. As a result, EPAs show relatively minor timing changes under the reaction time conditions (Klous et al., 2012), in contrast to the mentioned major changes in APAs.

Figure 14.10 illustrates muscle activation patterns seen in a person standing in a semi-squatted posture and watching a passive object approaching the person and hitting him at the shoulders (Krishnan et al., 2011). In these conditions, both APAs and EPAs can be observed as a sequence of events. In Figure 14.10, EPAs are seen as an early drop in the background activation levels of the dorsal muscles starting about 500 ms prior to the impact (top panels), while APAs are seen in both dorsal (an abrupt drop in the activation level) and ventral muscles (a burst of activation, bottom panels) seen about 100 ms prior to the impact.

14.6.3 Anticipatory synergy adjustments

Anticipatory synergy adjustments are defined as changes in the index of synergy stabilizing a performance variable in preparation to a quick change of that variable

(Olafsdottir et al., 2005). Their purpose is related to the fact that high stability of a variable naturally resists its quick change. For example, if a person plans to shift the COP quickly, having a strong synergy stabilizing the current COP coordinate becomes counterproductive. ASAs start before APAs, 200–300 ms prior to the planned change in the performance variable; they have been demonstrated in postural tasks prior to both APAs and EPAs (Klous et al., 2011; Krishnan et al., 2011). Figure 14.11 illustrates typical APAs (seen as changes in muscle activations), and ASAs observed prior to the ASAs. In that experiment, the subjects stood leaning backward and then performed a quick bilateral arm movement forward; the initiation of this action is shown as time zero. The left panel shows typical APAs in biceps femoris under self-paced (solid lines) and reaction-time (dashed lines) movements. The right panel shows changes in the synergy index stabilizing the center of pressure coordinate in the anterior-posterior direction (ΔV); the three lines show the changes in ΔV computed with three different methods. Note that the drop in ΔV (ASA) starts about 200 ms prior to time zero, while the APA starts much later, about 100 ms prior to time zero.

There are several similarities between characteristics of ASAs and APAs. Both APAs and ASAs are delayed when a person has to perform an action as quickly as possible to a signal with unpredictable timing, as in typical simple reaction time experiments (Lee et al., 1987; Olafsdottir et al., 2005). Observations of such actions show that APAs and ASAs are both not absolutely necessary to initiate an action; they represent elements of luxury, possibly optimizing aspects of actions such as their smoothness and possibly energy expenditure.

Both APAs and ASAs are delayed and reduced in magnitude in a variety of subpopulations characterized by impaired postural stability. In particular, smaller and delayed APAs and ASAs have been reported in the healthy elderly (Woollacott et al., 1988; Olafsdottir et al., 2007) and in patients with neurological disorders such as Parkinson's

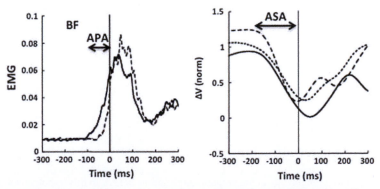

Figure 14.11 The left panel shows typical APAs in biceps femoris (BF) under self-paced (solid lines) and reaction time (dashed lines) movements. The standing subject performed a quick bilateral arm movement forward. The right panel shows changes in the synergy index stabilizing the center of pressure coordinate in the anterior-posterior direction (ΔV) computed with three different methods. Note the early drop in ΔV (ASA) followed by the APA.
Modified by permission from Klous et al. (2011), © American Physiological Society.

disease and olivo-ponto-cerebellar atrophy (Park et al., 2012, 2013). These changes may contribute to postural impairment during everyday actions characterized by frequent self-triggered postural perturbations, such as those during quick arm movements, picking up loads, releasing loads, etc.

On the other hand, there is an important qualitative difference between APAs and ASAs. ASAs facilitate a quick change in a performance variable independently of the direction of its change. As a result, ASAs are a universal mechanism of preparing for a quick action, even in conditions when the direction of the action is unknown in advance (Zhou et al., 2013). This may be seen in some sports as an increased body sway (destabilization of the vertical posture) in preparation to a quick action, e.g., getting ready for a fast tennis serve or for a penalty kick in football (soccer). In contrast, APAs make sense only if they generate forces and moments in a direction counteracting an expected perturbation. An APA in a wrong direction may exacerbate destabilizing effects of a perturbation on posture.

The existence of both ASAs and APAs fits well the recent scheme of the control of redundant systems (reviewed in Latash, 2010). Within that scheme, neural control of a multielement action may be associated with setting two groups of variables (Figure 14.12). One group (NV1) is related to characteristics of the planned action. Variables within this group define a common input into the redundant set of elements. The other group (NV2) defines stability properties of that action. These variables define gains in back-coupling projections in a hypothetical network that produces synergic adjustments in the elemental variables (for example, magnitudes of muscle modes). With respect to postural control, preparation to action (PTA) may involve two components: anticipatory changes in NV1 lead to APAs with nonzero changes in the produced forces and moments of force while changes in NV2 lead to ASAs without direct effects on the magnitude of mechanical variables.

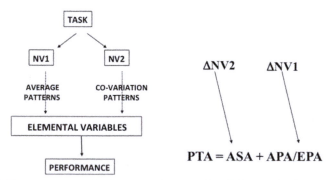

Figure 14.12 Two components in the preparation to action (PTA). In this hypothetical scheme, two kinds of neural variables, NV1 and NV2, result in average (across trials) patterns of performance variables and in their intertrial covariation, respectively. Preparation to action (PTA) involves changes in both NVs. A change in NV1 leads to changes in muscle activations and net mechanical effects resulting in APAs and EPAs. Changes in NV2 result in indices of covariation (synergy indices) without affecting average net performance.

14.7   Posture-stabilizing mechanisms

Anticipatory mechanisms of postural stabilization described in the previous section are almost always suboptimal because they are based on predictions of expected perturbations. These predictions are almost never 100% correct because of the variability in the state of the neural and muscle elements of the body and in external forces. As a result, a residual perturbation always takes place, and its effects on posture have to be compensated by reactive actions. There is a sequence of such actions that differ in their characteristic time delays and efficacy. Typically, the short-delay responses are relatively crude, while the longer-latency responses are more task specific and context specific. It is possible to state that the shorter-latency responses produce approximate corrections of the effects of perturbations, which gives the central nervous system time to generate more accurate and appropriate longer-latency corrections. Ultimately, voluntary corrections at relatively long time delays, on the order of 150–200 ms, have to deal with residual effects of the perturbation.

14.7.1   Zero-delay mechanisms of postural stability: the preflexes

It is well known that muscle forces are length and velocity dependent (Chapters 2 and 3). These dependences produce changes in muscle force that act against the imposed position changes at zero time delay. As a result, these properties of muscle force are posture-stabilizing factors. The coefficients in the length and velocity dependences are functions of the muscle activation level—higher activation levels lead to higher slopes of these dependences (larger "apparent stiffness" and damping, see Chapters 2 and 3). This property gives the CNS the ability to modulate resistance of muscles to mechanical perturbations by changing their activation levels. For example, this can be done by keeping the same posture and changing the level of co-contraction in the agonist—antagonist groups crossing individual joints. This method of feedforward modulation of muscle reactions to perturbations has been addressed as "preflexes" (Loeb, 1999).

Preflexes have the advantage of always acting against external perturbations at zero time delay, thus avoiding potential problems with instability when loops with substantial time delays are involved in reactions to perturbations. Another advantage of preflexes is that they stabilize posture against any perturbation independently of its direction. For example, if a person prepares for a perturbation in a certain direction, but the actual perturbation acts in the opposite direction, APAs would exacerbate the effects of the perturbation, while preflexes would reduce its effects on posture.

On the other hand, changing preflexes involves increased opposing muscle contractions and, therefore, is associated with higher energy expenditure and, potentially, with higher fatigue. As a result, modulation of the co-contraction levels in postural muscles is seen in young healthy subjects only in challenging conditions; such subjects more commonly show the mentioned reciprocal changes in the muscle activation patterns in preparation to a perturbation. Co-contraction patterns are more commonly seen in populations with impaired postural control such as the healthy elderly, persons with Down syndrome, and patients with various neurological and peripheral disorders (Woollacott et al., 1988; Aruin and Almeida, 1996).

14.7.2 Reflexes and preprogrammed reactions

The next line of defense of posture from effects of perturbations is associated with transmission of sensory signals to the CNS and the generation of responses at various time delays. Some of these responses come at a relatively short time delay (under 50 ms) and show relatively stereotypical patterns; they are commonly addressed as reflexes (for a discussion on the dichotomy "reflexes vs voluntary actions" see Prochazka et al., 2000; also see Chapter 6). Other responses come at delays between 50 and 100 ms; these have been addressed using different names including long-loop reflexes, M_{2-3}, triggered reactions, and preprogrammed reactions (Chapter 7). Reflexes typically produce increased muscle contractions in stretched muscles and, as a result, oppose external perturbations experienced by muscles. This makes reflex action relatively local. In contrast, preprogrammed reactions can be seen in muscles whose length is not affected, or even shortened, by the perturbation. They can also be seen in remote muscles as long as these muscles contribute to stabilization of an important posture-related variable, for example, they can be seen in arm muscles of a person standing in a bus and holding on to a railing when the bus starts to move unexpectedly. Preprogrammed reactions are suboptimal due to their nature: They are prepared by the central nervous system prior to a perturbation. As a result, there are typically residual effects of a perturbation on posture that have to be compensated by voluntary corrections.

Preprogrammed reactions are instruction and context dependent. For example, if a person stands in a bus, and the bus suddenly begins to move, preprogrammed reactions will be seen in postural muscles of the legs and trunk. If the same person holds onto a railing, preprogrammed reactions will be seen in arm muscles at a comparable time delay. Note that in the last example, the length of the arm muscles may be increased, decreased, or left unaffected by the perturbation.

The two components of preprogrammed reactions, M_2 and M_3, come at delays of about 50–60 and 70–90 ms, respectively. They are likely to be of different neurophysiological origins, reflected, in particular, by their contrasting changes under fatigue (Balestra et al., 1992). M_2 is likely subcortical, while M_3 has been assumed to involve transcortical loops (Dietz et al., 1984).

Within the scheme of control with referent body configurations, preprogrammed reactions are similar in nature to voluntary postural corrections. They both may be viewed as corrective actions resulting from RC shifts for appropriate effectors or for the whole body. The difference is that preprogrammed reactions are prepared in advance, in anticipation of a likely perturbation, and triggered by a salient sensory stimulus, while voluntary actions have to be generated in response to actual perturbations without the benefit of preprogramming.

14.8 The bottom line

Posture can be characterized by two key words, *configuration* and *stability*. The ideas of control with referent body configurations and synergic control are perfectly suited to address these two key issues. The RC-hypothesis remains the only hypothesis in the field of motor control that solves the posture-movement paradox in a way compatible with physics and physiology of the human body. The ideas of synergic control and the computational approach to analysis of tasks performed by redundant systems developed within the uncontrolled manifold hypothesis offer a way to quantify stability of postural states with respect to task-specific variables.

Studies of postural control have been focused on three issues: (1) How do people stand without falling down? (2) How do people prepare to actions or predictable perturbations? and (3) How do people deal with actual perturbations to postural tasks? Within the first issue, studies of postural sway have been recently supplemented with the rambling-trembling decomposition of the sway that allows to link aspects of sway to shifts of the equilibrium state of the body and to oscillations about that state. Recent studies of the second issue have focused on anticipatory adjustments of posture to expected perturbations (APAs and EPAs) and also on anticipatory adjustments of synergies (ASAs) stabilizing posture-related variables in preparation to an action or reaction by effectors involved in a steady-state, postural task. A number of mechanisms relevant to the third issue have been studied involving zero-delay preflexes, reflex, and preprogrammed responses to perturbations, and voluntary postural corrections. While there has been substantial progress in conceptual understanding on postural control and its relations to the control of movements, large gaps remain in the knowledge of physiological mechanisms involved in postural tasks.

References

Alexandrov, A.V., Frolov, A.A., Massion, J., 2001. Biomechanical analysis of movement strategies in human forward trunk bending. I. Modeling. Biological Cybernetics 84, 425–434.

Aruin, A.S., Almeida, G.L., 1996. A coactivation strategy in anticipatory postural adjustments in persons with Down syndrome. Motor Control 1, 178–191.

Aruin, A.S., Forrest, W.R., Latash, M.L., 1998. Anticipatory postural adjustments in conditions of postural instability. Electroencephalography and Clinical Neurophysiology 109, 350–359.

Aruin, A.S., Latash, M.L., 1995. Directional specificity of postural muscles in feed-forward postural reactions during fast voluntary arm movements. Experimental Brain Research 103, 323–332.

Assländer, L., Peterka, R.J., 2014. Sensory reweighting dynamics in human postural control. Journal of Neurophysiology 111, 1852–1864.

Balestra, C., Duchateau, J., Hainaut, K., 1992. Effects of fatigue on the stretch reflex in a human muscle. Electroencephalography and Clinical Neurophysiology 85, 46–52.

Belenkiy, V.E., Gurfinkel, V.S., Paltsev, E.I., 1967. On the elements of control of voluntary movements (in Russian). Biofizika 12, 135–141.

Bonnet, C., Carello, C., Turvey, M.T., 2009. Diabetes and postural stability: review and hypotheses. Journal of Motor Behavior 41, 172–190.

Brenier, Y., Do, M.C., 1986. When and how does steady state gait movement induced from upright posture begin? Journal of Biomechanics 19, 1035–1040.

Bronstein, A.M., Bunday, K.L., Reynolds, R., 2009. What the "broken escalator" phenomenon teaches us about balance. Annals of New York Academy of Science 1164, 82–88.

Casadio, M., Morasso, P.G., Sanguineti, V., 2005. Direct measurement of ankle stiffness during quiet standing: implications for control modeling and clinical application. Gait and Posture 21, 410–424.

Collins, J.J., De Luca, C.J., 1993. Open-loop and closed-loop control of posture: a random-walk analysis of center-of-pressure trajectories. Experimental Brain Research 95, 308–318.

Cordo, P.J., Nashner, L.M., 1982. Properties of postural adjustments associated with rapid arm movements. Journal of Neurophysiology 47, 287–302.

Crenna, P., Frigo, C., 1991. A motor programme for the initiation of forward-oriented movements in humans. Journal of Physiology 437, 635–653.

Danna-Dos-Santos, A., Degani, A.M., Latash, M.L., 2008a. Flexible muscle modes and synergies in challenging whole-body tasks. Experimental Brain Research 189, 171–187.

Danna-Dos-Santos, A., Degani, A.M., Zatsiorsky, V.M., Latash, M.L., 2008b. Is voluntary control of natural postural sway possible? Journal of Motor Behavior 40, 179–185.

Dietz, V., Quintern, J., Berger, W., 1984. Corrective reactions to stumbling in man: functional significance of spinal and transcortical reflexes. Neuroscience Letters 44, 131–135.

Domen, K., Zatsiorsky, V.M., Latash, M.L., 1999. Reconstruction of equilibrium trajectories during whole-body movements. Biological Cybernetics 80, 195–204.

Duarte, M., Zatsiorsky, V.M., 1999. Patterns of center of pressure migration during prolonged unconstrained standing. Motor Control 3, 12–27.

Feldman, A.G., 1966. Functional tuning of the nervous system with control of movement or maintenance of a steady posture. II. Controllable parameters of the muscle. Biophysics 11, 565–578.

Feldman, A.G., 1980. Superposition of motor programs. I. Rhythmic forearm movements in man. Neuroscience 5, 81–90.

Feldman, A.G., 1986. Once more on the equilibrium-point hypothesis (λ-model) for motor control. Journal of Motor Behavior 18, 17–54.

Feldman, A.G., 2009. Origin and advances of the equilibrium-point hypothesis. Advances in Experimental Medicine and Biology 629, 637–643.

Feldman, A.G., Levin, M.F., 1995. Positional frames of reference in motor control: their origin and use. Behavioral and Brain Sciences 18, 723–806.

Forssberg, H., Hirschfeld, H., 1988. Phasic modulation of postural activation patterns during human walking. Progress in Brain Research 76, 221–227.

Freitas, S.M.S.F., Duarte, M., Latash, M.L., 2006. Two kinematic synergies in voluntary whole-body movements during standing. Journal of Neurophysiology 95, 636–645.

Goodworth, A.D., Peterka, R.J., 2012. Sensorimotor integration for multisegmental frontal plane balance control in humans. Journal of Neurophysiology 107, 12–28.

Hadders-Algra, M., Carlberg, E.B. (Eds.), 2008. Posture: A Key Issue in Developmental Disorders. MacKeith Press, London, UK.

Hart, C.B., Giszter, S.F., 2010. A neural basis for motor primitives in the spinal cord. Journal of Neuroscience 30, 1322–1336.

Hogan, N., Sternad, D., 2012. Dynamic primitives of motor behavior. Biological Cybernetics 106, 727–739.

Horak, F.B., Nashner, L.M., 1986. Central programming of postural movements: adaptation to altered support-surface configurations. Journal of Neurophysiology 55, 1369–1381.

Ivanenko, Y.P., Cappellini, G., Dominici, N., Poppele, R.E., Lacquaniti, F., 2005. Coordination of locomotion with voluntary movements in humans. Journal of Neuroscience 25, 7238−7253.

Ivanenko, Y.P., Poppele, R.E., Lacquaniti, F., 2004. Five basic muscle activation patterns account for muscle activity during human locomotion. Journal of Physiology 556, 267−282.

Klous, M., Danna-dos-Santos, A., Latash, M.L., 2010. Multi-muscle synergies in a dual postural task: evidence for the principle of superposition. Experimental Brain Research 202, 457−471.

Klous, M., Mikulic, P., Latash, M.L., 2011. Two aspects of feed-forward postural control: anticipatory postural adjustments and anticipatory synergy adjustments. Journal of Neurophysiology 105, 2275−2288.

Klous, M., Mikulic, P., Latash, M.L., 2012. Early postural adjustments in preparation to whole-body voluntary sway. Journal of Electromyography and Kinesiology 22, 110−116.

Krishnamoorthy, V., Goodman, S.R., Latash, M.L., Zatsiorsky, V.M., 2003a. Muscle synergies during shifts of the center of pressure by standing persons: identification of muscle modes. Biological Cybernetics 89, 152−161.

Krishnamoorthy, V., Latash, M.L., Scholz, J.P., Zatsiorsky, V.M., 2003b. Muscle synergies during shifts of the center of pressure by standing persons. Experimental Brain Research 152, 281−292.

Krishnamoorthy, V., Latash, M.L., 2005. Reversals of anticipatory postural adjustments during voluntary sway. Journal of Physiology 565, 675−684.

Krishnamoorthy, V., Scholz, J.P., Latash, M.L., 2007. The use of flexible arm muscle synergies to perform an isometric stabilization task. Clinical Neurophysiology 118, 525−537.

Krishnan, V., Aruin, A.S., Latash, M.L., 2011. Two stages and three components of postural preparation to action. Experimental Brain Research 212, 47−63.

Koshland, G.F., Gerilovsky, L., Hasan, Z., 1991. Activity of wrist muscles elicited during imposed or voluntary movements about the elbow joint. Journal of Motor Behavior 23, 91−100.

Latash, M.L., 1993. Control of Human Movement. Human Kinetics, Urbana, IL.

Latash, M.L., 2008. Neurophysiological Basis of Movement, second ed. Human Kinetics, Urbana, IL.

Latash, M.L., 2010. Motor synergies and the equilibrium-point hypothesis. Motor Control 14, 294−322.

Latash, M.L., Aruin, A.S., Shapiro, M.B., 1995. The relation between posture and movement: a study of a simple synergy in a two-joint task. Human Movement Science 14, 79−107.

Latash, M.L., Aruin, A.S., Zatsiorsky, V.M., 1999. The basis of a simple synergy: reconstruction of joint equilibrium trajectories during unrestrained arm movements. Human Movement Science 18, 3−30.

Latash, M.L., Gottlieb, G.L., 1991. Reconstruction of elbow joint compliant characteristics during fast and slow voluntary movements. Neuroscience 43, 697−712.

Lee, W.A., Buchanan, T.S., Rogers, M.W., 1987. Effects of arm acceleration and behavioral conditions on the organization of postural adjustments during arm flexion. Experimental Brain Research 66, 257−270.

Liddell, E.G.T., Sherrington, C.S., 1924. Reflexes in response to stretch (myotatic reflexes). Proceedings of the Royal Society of London, Series B 96, 212−242.

Loeb, G.E., 1999. What might the brain know about muscles, limbs and spinal circuits? Progress in Brain Research 123, 405−409.

Loram, I.D., Maganaris, C.N., Lakie, M., 2005. Active, non-spring-like muscle movements in human postural sway: how might paradoxical changes in muscle length be produced? Journal of Physiology 564, 281−293.

Massion, J., 1992. Movement, posture and equilibrium — interaction and coordination. Progress in Neurobiology 38, 35—56.

Maurer, C., Mergner, T., Peterka, R.J., 2006. Multisensory control of human upright stance. Experimental Brain Research 171, 231—250.

Mergner, T., 2007. Modeling sensorimotor control of human upright stance. Progress in Brain Research 165, 283—297.

Mochizuki, L., Duarte, M., Amadio, A.C., Zatsiorsky, V.M., Latash, M.L., 2006. Changes in postural sway and its fractions in conditions of postural instability. Journal of Applied Biomechanics 22, 51—66.

Morasso, P.G., Sanguineti, V., 2002. Ankle muscle stiffness alone cannot stabilize balance during quiet standing. Journal of Neurophysiology 88, 2157—2162.

Murnaghan, C.D., Squair, J.W., Chua, R., Inglis, J.T., Carpenter, M.G., 2014. Cortical contributions to control of posture during unrestricted and restricted stance. Journal of Neurophysiology 111, 1920—1926.

Nardonne, A., Schiepatti, M., 1988. Postural adjustments associated with voluntary contractions of leg muscles in standing man. Experimental Brain Research 69, 469—480.

Nouillot, P., Bouisset, S., Do, M.C., 1992. Do fast voluntary movements necessitate anticipatory postural adjustments even when equilibrium is unstable? Neuroscience Letters 147, 1—4.

Oie, K.S., Kiemel, T., Jeka, J.J., 2002. Multisensory fusion: simultaneous re-weighting of vision and touch for the control of human posture. Brain Research: Cognitive Brain Research 14, 164—176.

Oliveira, L.F., Simpson, D.M., Nadal, J., 1996. Calculation of area of stabilometric signals using principal component analysis. Physiological Measures 17, 305—312.

Olafsdottir, H., Yoshida, N., Zatsiorsky, V.M., Latash, M.L., 2005. Anticipatory covariation of finger forces during self-paced and reaction time force production. Neuroscience Letters 381, 92—96.

Olafsdottir, H., Yoshida, N., Zatsiorsky, V.M., Latash, M.L., 2007. Elderly show decreased adjustments of motor synergies in preparation to action. Clinical Biomechanics 22, 44—51.

Park, J., Lewis, M.M., Huang, X., Latash, M.L., 2013. Effects of olivo-ponto-cerebellar atrophy (OPCA) on finger interaction and coordination. Clinical Neurophysiology 124, 991—998.

Park, J., Wu, Y.-H., Lewis, M.M., Huang, X., Latash, M.L., 2012. Changes in multi-finger interaction and coordination in Parkinson's disease. Journal of Neurophysiology 108, 915—924.

Patel, M., Fransson, P.A., Lush, D., Gomez, S., 2008. The effect of foam surface properties on postural stability assessment while standing. Gait and Posture 28, 649—656.

Peterka, R.J., Statler, K.D., Wrisley, D.M., Horak, F.B., 2011. Postural compensation for unilateral vestibular loss. Frontiers in Neurology 2, 57.

Prochazka, A., Clarac, F., Loeb, G.E., Rothwell, J.C., Wolpaw, J.R., 2000. What do reflex and voluntary mean? Modern views on an ancient debate. Experimental Brain Research 130, 417—432.

Ramos, C.F., Stark, L.V., 1990. Postural maintenance during movement: simulations of a two-joint model. Biological Cybernetics 63, 363—375.

Riccio, G.E., 1993. Information in movement variability about the qualitative dynamics of posture and orientation. In: Newell, K.M., Corcos, D.M. (Eds.), Variability and Motor Control. Human Kinetics, Champaign, IL, pp. 317—358.

Riemann, B.L., Myers, J.B., Lephart, S.M., 2003. Comparison of the ankle, knee, hip, and trunk corrective action shown during single-leg stance on firm, foam, and multiaxial surfaces. Archives of Physical Medicine and Rehabilitation 84, 90—95.

Riley, M.A., Wong, S., Mitra, S., Turvey, M.T., 1997. Common effects of touch and vision on postural parameters. Experimental Brain Research 117, 165—170.

Robert, T., Zatsiorsky, V.M., Latash, M.L., 2008. Multi-muscle synergies in an unusual postural task: quick shear force production. Experimental Brain Research 187, 237−253.

Sainburg, R.L., 2002. Evidence for a dynamic-dominance hypothesis of handedness. Experimental Brain Research 142, 241−258.

Sanes, J.N., Mauritz, K.H., Dalakas, M.C., Evarts, E.V., 1985. Motor control in humans with large-fiber sensory neuropathy. Human Neurobiology 4, 101−114.

Sarabon, N., Panjan, A., Latash, M.L., 2013. The effects of aging on the rambling and trembling components of postural sway: effects of motor and sensory challenges. Gait and Posture 38, 637−642.

Schieppati, M., Schmid, M., Sozzi, S., 2014. Rapid processing of haptic cues for postural control in blind subjects. Clinical Neurophysiology 125, 1427−1439.

Scholz, J.P., Schöner, G., 1999. The uncontrolled manifold concept: identifying control variables for a functional task. Experimental Brain Research 126, 289−306.

Slijper, H., Latash, M.L., 2000. The effects of instability and additional hand support on anticipatory postural adjustments in leg, trunk, and arm muscles during standing. Experimental Brain Research 135, 81−93.

Stergiou, N. (Ed.), 2004. Innovative Analyses of Human Movement: Analytical Tools for Human Movement Research. Human Kinetics, Urbana, IL.

Stylianou, A.P., McVey, M.A., Lyons, K.E., Pahwa, R., Luchies, C.W., 2011. Postural sway in patients with mild to moderate Parkinson's disease. International Journal of Neuroscience 121, 614−621.

Ting, L.H., Macpherson, J.M., 2005. A limited set of muscle synergies for force control during a postural task. Journal of Neurophysiology 93, 609−613.

Torres-Oviedo, G., Ting, L.H., 2010. Subject-specific muscle synergies in human balance control are consistent across different biomechanical contexts. Journal of Neurophysiology 103, 3084−3098.

Tresch, M.C., Saltiel, P., Bizzi, E., 1999. The construction of movement by the spinal cord. Nature Neuroscience 2, 162−167.

Van Deursen, R.W., Simoneau, G.G., 1999. Foot and ankle sensory neuropathy, proprioception, and postural stability. Journal of Orthopedics, Sports, and Physical Therapy 29, 718−726.

Von Holst, E., Mittelstaedt, H., 1950/1973. Daz reafferezprincip. Wechselwirkungen zwischen Zentralnerven-system und Peripherie, Naturwiss 37, 467−476. The reaference principle. In: The Behavioral Physiology of Animals and Man. The Collected Papers of Erich von Holst. Martin, R., (translator) University of Miami Press, Coral Gables, Florida, 1, pp. 139−173.

Winter, D.A., Prince, F., Frank, J.S., Powell, C., Zabjek, K.F., 1996. Unified theory regarding A/P and M/L balance in quiet stance. Journal of Neurophysiology 75, 2334−2343.

Woollacott, M., Inglin, B., Manchester, D., 1988. Response preparation and posture control. Neuromuscular changes in the older adult. Annals of the New York Academy of Sciences 515, 42−53.

Zatsiorsky, V.M., 1998. Kinematics of Human Motion. Human Kinetics, Champaign, IL.

Zatsiorsky, V.M., Duarte, M., 1999. Instant equilibrium point and its migration in standing tasks: rambling and trembling components of the stabilogram. Motor Control 3, 28−38.

Zatsiorsky, V.M., Duarte, M., 2000. Rambling and trembling in quiet standing. Motor Control 4, 185−200.

Zhou, T., Wu, Y.-H., Bartsch, A., Cuadra, C., Zatsiorsky, V.M., Latash, M.L., 2013. Anticipatory synergy adjustments: preparing a quick action in an unknown direction. Experimental Brain Research 226, 565−573.

Grasping

15

People interact daily with various objects: workings tools, utensils, glasses with liquids, door handles, a computer mouse, and many others. Hand actions include three sequential parts: (1) selection of a specific grasp pattern—depending on the performance goal and the object shape, people grasp objects differently; (2) a reach-to-grasp movement (Jeannerod, 1984; reviewed in Bennett and Castiello, 1994); and (3) object manipulation—exerting forces on the object and changing its position, that is, orientation and location, in space. This chapter deals only with the latter part of typical hand actions.

The grasps completely restraining the grasped objects are called *closures*. The closures are classified as *form closures* in which the object cannot move without changing the finger(s) position and *force closures*, the grasps that can be broken under external force without finger movement. With rare exceptions, for example, keeping a small pebble in the hand, the human grasps are force closures. The simplest classification of the human grasps (or *grips*) includes two varieties (Napier, 1956): *power grips* when the object is in contact with the palm of the hand (e.g., holding a tennis racket) and *precision grips* when only the digit tips are in contact with the object.

15.1 Elements of history

Arm function has intrigued scientists for centuries (Figure 15.1). Until the 1980s, studies of grasp were mainly limited to verbal narratives not substantiated by experimental evidence, recordings of the maximal grasping force (commonly called the *hand force*), and various grasp classifications (Napier, 1956).

The contemporary period of studies started almost simultaneously in robotics and motor control/biomechanics. In robotics, a mathematical theory of stable multifinger grasps was formulated (Salisbury and Craig, 1982; Mason and Salisbury, 1985;

Figure 15.1 A seventeenth-century tribute to hand dexterity.
From Mario Bettini (1645), Apiaria universae philisophiae mathematicae, Bologna, Italy.

Biomechanics and Motor Control. http://dx.doi.org/10.1016/B978-0-12-800384-8.00015-6

Kerr and Roth, 1986). In motor control, seminal papers by researchers from Umea University, Sweden (Johansson and Westling, 1984; Westling and Johansson, 1984) made a strong impact on the field; as of April 2015, the paper by Johansson and Westling (1984) was cited over 1150 times. These and subsequent numerous studies concentrated mainly on the pinch grasps. The tasks involved: (1) two digits only, usually the thumb and the index finger; (2) parallel contact surfaces; and (3) vertically oriented objects that are either at rest or are moved in a vertical direction. In such grasps (*prismatic pinch grasps*), the normal forces are exerted horizontally while the load force is directed vertically and hence is manifested as the shear (tangential) force acting on the contact surface. The prismatic pinch grasps have been the object of most intensive research. An advantage of studying these tasks is the mechanical independence of the normal and tangential forces, in a sense that any change of the normal force is due to motor control and is not a pure mechanical consequence of changes in tangential forces. In other grasps, for instance, in grasps of nonvertically oriented objects, the normal forces can be due to both gravity and neural control.

Gradually the interest of researchers shifted to more intricate cases—multifinger grasps and objects of diverse shapes that are differently oriented with respect to gravity and being moved in various directions. To date, the experimental data are mainly obtained for the so-called *prismatic grasps* in which the thumb opposes the fingers and the contact surfaces are parallel to each other (Figure 15.2). The contact forces and moments are typically recorded with six-component force and moment sensors.

For multifinger prismatic grasps (Figure 15.2), in a number of studies, the forces of the fingers opposing the thumb were reduced to a resultant force and a resultant moment of force. This is equivalent to replacing a set of fingers with a *virtual finger* (VF), an imagined digit (Arbib et al., 1985; Iberall, 1987). A VF generates the same mechanical effect as the set of actual fingers.

Since 1998, several dozens of papers on multifinger grasps have been published (reviewed in Zatsiorsky and Latash, 2008a,b, 2009; Latash and Zatsiorsky, 2009).

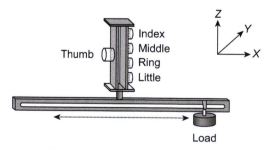

Figure 15.2 The "inverted-T" handle/beam apparatus commonly used to study the prismatic precision grips. Five six-component force sensors (black rectangles) are used to register individual digit forces. During testing, the suspended load and its location along the horizontal bar could vary across trials. When the handle is oriented vertically, the force components in the X direction are the *normal* forces and the forces in the Z and Y directions are the *shear* (or *tangential*) forces, respectively. The handle allows for the independent change of the external load and external torque exerted on the handle. The figure is not drawn to scale.

Many recent studies have used grasping tasks to explore general issues of motor control such as hierarchical control and synergies (see Chapter 11).

15.2 Basic mechanics of grasps

15.2.1 Single-finger contacts

15.2.1.1 Contact modeling

Finger contacts are modeled in one of three ways (Mason and Salisbury, 1985), as a *point contact*, a *hard-finger contact*, or as a *soft-finger contact*. In the first case, the digit tip force is assumed to act at a point. In the point model with friction, the digit can exert a force but not a torque (moment of force) on the sensor. In the hard-finger model, the contact takes place over an area but the fingertip deformation and rolling are neglected. In the soft-finger contact model (recommended as more realistic), the contact takes place over a certain area, the finger deforms, and the point of force application can change during performance. It is assumed that the digits do not stick to the object and hence they can only push but not pull the object. As a result, the digits cannot exert force couples (free moments) on the sensor in planes other than the plane of contact. An attempt to generate such a moment will result in a digit-tip rolling over the contact surface.

Within a soft-finger model, the digit–object interaction is characterized by six variables (a 6×1 vector): three orthogonal force components (the normal force component is unidirectional and the two tangential force components are bidirectional), free moment in the plane of contact, and two coordinates of the point of force application on the sensor. To obtain these data, six-component force sensors are necessary; the sensors yield three orthogonal force component and three moment component values. The moments are reported with respect to the sensor center (not with respect to the point of force application that can be different from the center). If f_X is normal force (along the X-axis that is perpendicular to the sensor surface) and m_Y is moment about an axis Y that goes through the sensor center in the plane of contact (Figure 15.2), the coordinate of force application along axis Z can be found as $z = m_Y / f_X$.

Digit tips deform under force during grasping. Normal deformation of 2 mm has been reported for loads as low as 1.0 N (Serina et al., 1997), and tangential deformation of 3.5 mm has been reported for loads of approximately 2 N (Nakazawa et al., 2000). The tangential deformation depends not only on the tangential force but also on the normal force—the "apparent tangential stiffness" (see Chapter 2) increases linearly with the normal force (Pataky et al., 2005). The fingertip deformation: (1) increases the area of contact and consequently the magnitude of frictional moments that can be produced by the digits; (2) decreases the distances from the distal phalanx bones to the objects and hence the moment arms of the forces (as muscles exert forces on the bones and not at the finger tips); and (3) allows for large displacements of the points of force application, up to 12 mm for the thumb and 5–6 mm for the fingers (Zatsiorsky et al., 2003a).

In the human hand, tangential (ab/adduction) forces at the fingertips are exerted by the active torque production at the MCP joints (Pataky et al., 2004c). In this regard, human hands are different from the present-day robotic hands where the MCP joints are simple hinges and tangential loads are supported passively by the joint structures, without active control.

15.2.1.2 Slip prevention

Commonly, both the normal and tangential forces act at the digit contacts. If tangential force is too large (or the friction is too low), the object may slip. Friction resists lateral motion of a digit tip in contact. The coefficient of friction μ is usually determined as the tangential force/normal force ratio ($\mu = F_t/F_n$) at the instant of slip. In most studies, it is assumed that the skin obeys Coulomb's law, that is, the coefficient μ magnitude does not depend on the normal force. Some researchers (e.g., Comaish and Bottoms, 1971), however, have reported that skin obeys Coulomb's law over a limited range of loads only while others (Savescu et al., 2008) found that the law is not valid for some materials, such as sandpaper. Assuming that Coulomb's law is valid (such an assumption greatly simplifies the analysis), the admissible range of the tangential force magnitude that is below the slipping threshold is determined by $|F_t| \leq \mu F_n$, where $\mu > 0$ is the static coefficient of friction, F_t is tangential force and F_n is normal force. In three dimensions, this relation can be represented geometrically as a *friction cone* (Figure 15.3). To avoid slipping, the contact forces must lie within the friction cone (*FC*).

To quantify the risk of slipping, two measures have been suggested (Westling and Johansson, 1984; Johansson and Westling, 1984): (1) the tangential force/normal force ratio; to avoid slipping the ratio should be smaller than the coefficient of friction; and (2) the *safety margin (SM)*, which can be estimated as another ratio (Burstedt et al., 1999):

$SM = (F_i^n - |F_i^t|/\mu)/F_i^n$, where i is a digit, F^n is the digit normal force, and F^t is the digit tangential force. For grasping of a vertically oriented object, the reported safety margin values are typically between 0.3 and 0.5 (Burstedt et al., 1999) and can be even as high as 0.6–0.7 (Pataky et al., 2004a). Hence, when people manipulate objects they use a "better safe than sorry" strategy. To prevent slipping, performers adjust digit normal forces both to the load (tangential forces) and friction.

SM is a useful measure for vertically oriented pinch grasps wherein the normal forces of the thumb and the finger are directed horizontally; however, in nonvertical

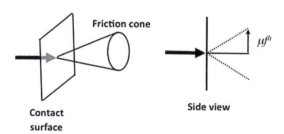

Figure 15.3 Geometric representation of the Coulomb friction model.

grasps these forces also have nonhorizontal components either assisting or resisting the gravity force and the notion of SM has to be reformulated (see Pataky et al., 2004b). In multifinger grasps, the SM is usually computed as the average SM of all involved digits (e.g., Solnik et al., 2014); however, it may happen that forces exerted by individual fingers can be either above or below the slipping threshold.

15.2.2 Multifinger precision grasps

15.2.2.1 Grasp equations

Consider maintaining in the air at rest an "inverted-T" handle shown in Figure 15.2. The contact surfaces are parallel, the handle is oriented vertically, and an external moment acts in the *plane of the grasp* (a vertical plane that contains all the points of digit contact with the object). The planar case is only considered. To hold the handle at rest: (1) the normal force of the thumb (F_{th}^n) should be equal and opposite to the total F^n of the opposing fingers; (2) the sum of the tangential forces (F^t) should be equal and opposite the gravity load L; (3) the sum of the moments exerted by all digit forces should be equal and opposite to the external moment M acting on the handle; and (4) the F^n of the thumb and opposing fingers should be sufficiently large to prevent slip. In the planar case, the equations of equilibrium are:

$$0 = F_{th}^n + \left(F_i^n + F_m^n + F_r^n + F_l^n \right) \tag{15.1a}$$

$$0 = \left(F_{th}^t + F_i^t + F_m^t + F_r^t + F_l^t \right) + L \tag{15.1b}$$

$$0 = M + \left(\underbrace{F_{th}^n d_{th} + F_i^n d_i + F_m^n d_m + F_r^n d_r + F_l^n d_l}_{\text{Moment of the normal forces} \equiv M^n} \right.$$

$$\left. + \underbrace{F_{th}^t r_{th} + F_i^t r_i + F_m^t r_m + F_r^t r_r + F_l^t r_l}_{\text{Moment of the tangential forces} \equiv M^t} \right) \tag{15.1c}$$

where the subscripts th, i, m, r, and l refer to the thumb, index, middle, ring, and little fingers, respectively; the superscripts n and t stand for the normal and tangential force components, respectively; L is load (weight of the object); and coefficients d and r stand for the moment arms of the normal and tangential force with respect to a preselected center, respectively. In Eqn (15.1), the forces are considered vectors and can assume both positive and negative values. Mechanically, the digits work as parallel manipulators.

Because the moment constraint (Eqn (15.1c)) includes all the normal and tangential digit forces, whereas other constraints deal with the normal and tangential force components separately, the torque constraint (Eqn (15.1c)) is the core constraint—any change of the digit forces must satisfy the condition of rotational equilibrium.

Equation (15.1) are linear equations that can be presented in a matrix form. In five-digit grasps, vector of individual digit forces and moments $\mathbf{F}_{digits}$ is a 30×1 vector (5 digits $\times$ 6 force and moment components). Its relation with a 6×1 vector $\mathbf{F}$ of the resultant forces and moments acting on the object is described by

$$- \mathbf{F} = [\mathbf{G}]\mathbf{F}_{digits}, \tag{15.2}$$

where $[\mathbf{G}]$ is a 6×30 *grasp matrix*, also known as the *matrix of moment arms* (Salisbury and Craig, 1982; Kerr and Roth, 1986). The rows of matrix $[\mathbf{G}]$ correspond to the resultant forces and moments acting on the object, and the columns correspond to the digit forces and moments. The elements in the first three rows of $[\mathbf{G}]$ relate the orthogonal components of the digit forces to the components of the resultant force; they are cosines of the angles formed by the lines of force action and the reference axes (see Zatsiorsky, 2002; Chapter 1). In the prismatic grasp, the force components of the same name, for example, x components of the digit forces and the X component of the resultant force, are parallel and hence the direction cosines equal either 1 or -1; the forces either add to or subtract from one another. The elements in the last three rows are the moment arms of the digit forces with respect to a preselected pivot, for example, the center of mass of the handheld object. The moment arms depend on (1) the digit sensor location with respect to the pivot and (2) the displacement of the point of digit force application on the sensor during the object manipulation. The digits do not stick to the sensors and hence they press but not pull on the object (the normal forces are unidirectional) and do not generate force couples (free moments) in the planes perpendicular to the grasp plane. Hence, the latter moments can be disregarded and vector $\mathbf{F}$ can be reduced to a 20×1 vector just as $[\mathbf{G}]$ can be reduced to a 6×20 matrix.

Equation (15.2) is based on a simplifying assumption that the elements of the grasp matrix (the coefficients in the equations) are constant, that is, the points of digit force applications do not displace during the period of observation. If they migrate, the elements of $[\mathbf{G}]$ are not constant anymore and the equations become nonlinear: variable values of digit forces are multiplied by the variable values of moment arms. This obstacle can be avoided if the (10×1) $\mathbf{F}_{digits}$ vector is expanded to a (15×1) vector where the added elements are the moments exerted by the individual digits with respect to the corresponding sensor centers. Within such an approach, matrix $[\mathbf{G}]$ in planar case is 3×15.

In nonvertical grasps, the gravity force is resisted not only by the tangential forces but also by a normal force exerted on the object. Equation (15.1) can be adjusted to such a task; the equation can also include the nonslip requirements (Pataky et al., 2004b).

15.3 Basics of grasp control

15.3.1 *Virtual finger and hierarchical control*

In the prismatic grasps, the forces of the four fingers are commonly reduced to a resultant force and a moment of force. This is equivalent to replacing a set of fingers with a

VF (Arbib et al., 1985; Iberall, 1987; Baud-Bovy and Soechting, 2001). VF generates the same mechanical effect (the same *wrench*, see Zatsiorsky, 2002; Chapter 1) as a set of actual fingers. There are substantial differences between VF and individual finger (IF) forces: (1) force directions are, as a rule, dissimilar (for a review, see Zatsiorsky and Latash, 2008a). The IF forces can be exerted in disparate directions such that only their resultant (i.e., VF) force is in the desired direction, Figure 15.4 (Zatsiorsky et al., 2003a; Gao et al., 2005a); (2) VF and IF forces adjust differently to modified task conditions (Zatsiorsky et al., 2002a,b); and (3) IF forces are much more variable than VF forces (Shim et al., 2003). The desired performance at the VF level is achieved by a synergic covariation (see Chapter 11) among individual finger forces at the IF level (this phenomenon is also observed in multifinger pressing tasks, Kapur et al., 2010). The control of prehension has regularly been analyzed at the two levels of a hypothetical control hierarchy, at the upper TH−VF level, also called the *task level*, and the lower IF level.

15.3.2 Internal forces

An *internal force* is a set of contact forces that can be applied to an object without disturbing its equilibrium (Mason and Salisbury, 1985; Murray et al., 1994). The

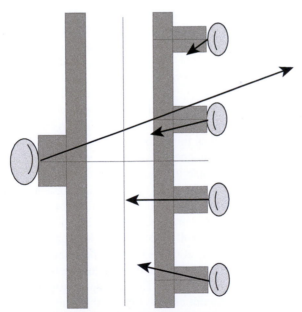

Figure 15.4 Forces at the digit tips, group average. A representative example. The subjects maintained at rest a handle exerting a supination (clockwise) torque 1.5 Nm. The total mass of the apparatus with the load suspended was 1.6 kg (weight 15.68 N). Note: The index and middle fingers exert forces in the downward direction and hence do not support the load. They also exert moments of force about the thumb in counterclockwise direction. These fingers act opposite to the required force direction (upward) and moment direction (clockwise).
Adapted by permission from Zatsiorsky et al. (2003), © Elsevier.

elements of an internal force vector act in opposite directions. They cancel each other and, hence, do not contribute to the *manipulation force*, that is, the resultant force and moment acting on the grasped object. An internal force is not a single force; it is a set of forces and moments that, when acting together, generate a zero resultant force and a zero resultant moment.

The internal forces lie in the so-called *null space* of the grasp matrix [**G**]; for a simple introduction to the concept of null space, see Zatsiorsky (2002), Section 2.4. This is another way of saying that the elements of the internal force vector cancel each other and do not affect the manipulation force. For readers who prefer mathematical notation, it can be said that the vector of internal forces $\mathbf{F_i}$ lies in the null space of [**G**]; the null space of an m by n matrix [**G**] is a set of all vectors **F** in R^n such that [**G**] $\mathbf{F} = \mathbf{0}$. (The symbol R^n designates an n-dimensional space of real numbers.) Any linear combination of the internal force vectors is also an internal force. In particular, multiplication of an internal force vector by a constant results in an internal force, and the sum of two or several internal force vectors also lies in the null space of [**G**]. Hence, internal force vectors are innumerable and looking for all of them does not make sense. However, it is possible to find independent vectors of a prescribed length, for example, of unit length, spanning the null space of [**G**], the *orthonormal basis vectors*. Analysis of the grasp matrix [**G**] provides a convenient tool for discovering orthonormal basis vectors. Because the rank of a 6×30 matrix is at most 6, the dimensionality of the null space of the grasp matrix (its nullity) is at least 24. A vector-by-vector analysis of all of them is a daunting task. Because of that, the internal force analysis is done mainly at the VF level. At this level, for the prismatic grasps, mathematical analysis yields three internal forces (Gao et al., 2005b): (1) the grasp force; (2) the internal moment, that is, the *moment of normal force—moment of tangential force* combination; and (3) the twisting moments about the axes normal to the surfaces of the contacts. The latter combination cannot, however, be realized in single-hand grasping because people cannot twist the thumb and finger(s) in opposite directions (in two-hand grasping this option can be realized). When individual finger forces are considered, the internal tangential force component—when some fingers exert force downward while other fingers produce upward force, see an example in Figure 15.4—can be rather large. All three internal force components—the grasp force, internal moment, and the tangential component—are larger in older persons (Solnik et al., 2014).

In multidigit grasping, the resultant force vector (manipulation force) and the vector of the internal force are mathematically independent (Kerr and Roth, 1986; Yoshikawa and Nagai, 1991). The mathematical independence of the above forces allows for their independent (decoupled) control. Such a decoupled control is realized in robotic manipulators (e.g., Zuo and Qian, 2000). The decoupled control saves computational resources. People, however, do not use this option; they modulate the internal forces, in particular the grasp force, with the manipulation force (Gao et al., 2005c).

15.3.3   Grasp force and its control

The concept of grasp force looks intuitively simple—everybody is familiar with the hand dynamometer and knows how to measure the maximal grasp force—the dynamometer should be gripped with the highest effort. Note, however, that the grip force measured in such a way represents two forces acting on the dynamometer from two sides. The forces act along the same line; they are equal and opposite. When the measurements are performed at equilibria, the forces cancel each other, and because of that the dynamometer does not accelerate. These two forces are collectively called the *grip force*. If instead of a spring dynamometer that yields a single force value, the efforts are applied to a prismatic handle similar to shown in Figure 15.2 and the forces are reduced to the thumb force and the VF force, two forces rather than one will be recorded from the force sensors. At rest or during vertical handle movement these forces are equal and can be called collectively a grip force. When the handle is inclined or when the handle moves in a horizontal direction, the forces are not equal. The grasp force is then defined as an internal force; it equals the smallest force of the two opposing forces. For instance, if the two forces equal 15 N and −10 N, the grasp force is 10 N and the manipulation force equals the resultant 15 N + (−10 N) = 5 N.

The above examples deal with collinear forces. In power grasps, the contact occurs over large curved surfaces, and the forces are not collinear. Although in statics they still cancel each other, a single internal force cannot be determined. Hence, strictly speaking, the grasp force in grasping a tennis racket or a golf club cannot be specified. Those interested in measuring the grasp strength should somehow redefine it. For instance, in experiments of Pataky et al. (2013) subjects grasped a circular handle; a flexible high-resolution pressure mat was used for the measurements. The following two-step procedure was employed for computations. First, from the original two-dimensional (2D) pressure data one-dimensional (1D) radial force distributions (units: N/rad) were computed. Then, the values obtained at the first step were summed up over a 360° arc (2π radians). The found "grasp force" is different from the grasp force discussed above; it is a scalar quantity (it has no direction) and it is not an internal force (manipulation force, if it exists, is added to the computed values).

Return back to the prismatic grasps. When performers move a vertically oriented object in the vertical direction, they vary the grip force in parallel with the load force (Johansson and Westling, 1984; reviewed in Flanagan and Johansson, 2002). The load force comprises (1) the static weight of the lifted object and (2) the inertial load due to the object acceleration (ma). People adjust differently to these two components of the load force. Also, at instances of zero acceleration during the handle up-and-down movements they exert larger force than at rest. This inspired decomposing of the grasp force into the *static*, *dynamic*, and *statodynamic fractions* (Figure 15.5).

To prevent object slipping, the performers adjust the grasp force to the friction at the object-digit contact. Grasp force is increased with a decrease in friction resulting in higher *grip force* to *load force* ratios at low friction while the SM is relatively constant (Johansson and Westling, 1984; Jaric et al., 2005). When the friction at the two sides of the object is different, for example, it is high under the thumb and low under the fingers, the grip force falls between the forces seen for the high friction and low friction conditions applied for all digits (Aoki et al., 2006).

During horizontal movement of a vertically oriented object, the maximal grasping force is observed at the instances of minimal acceleration and maximal velocity. This is

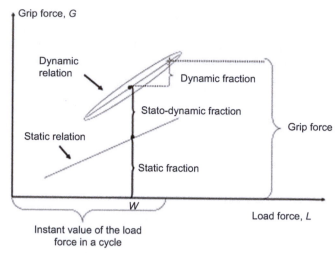

Figure 15.5 Decomposing grip force into three fractions: static, statodynamic, and dynamic. *W* is the object weight. The static relation is represented by a straight line. The static relation (*static fraction*) is obtained be registering grip force *G* at different weights of the load. During oscillation of the object in the vertical direction, the grip force varies with the object acceleration (the *dynamic relation*). At the instant of zero acceleration, the load force equals the weight of the object. At this instant, the grip force is, however, larger than in statics. The difference represents the *statodynamic fraction* of the grip force. The *dynamic fraction* represents the changes in the grip force that are solely due to the forces of inertia.
Adapted by permission from Zatsiorsky et al. (2005), © Springer.

valid both for the three-digit grasps from above (Smith and Soechting, 2005) and prismatic grasps (Gao et al., 2005b; Figure 15.6).

Due to the different dependencies of the grasp force on movement kinematics during object motion in the vertical and horizontal directions, these relations may

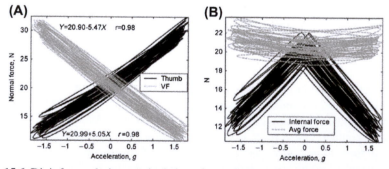

Figure 15.6 Digit forces during manipulation of a vertically oriented object in a horizontal direction. (A) Normal forces of the thumb and VF versus the handle acceleration in the horizontal direction. A representative trial, the load was 11.3 N, the frequency was 3 Hz. (B) Internal force (grip force) and average normal force versus the handle acceleration. Note that the average normal force is almost constant and is not very informative.
Adapted by permission from Gao et al. (2005b), © Springer.

become quite complicated during composite, for example, circular, movements (Figure 15.7, upper panel). However, the dependencies are so strong that, with an appropriate mathematical model, it is possible to predict the grasp force changes from the known movement kinematics (Figure 15.7, bottom panel).

15.3.4   Grasp moments and tilt prevention

In this section, the grasp moment is understood as the moment in the $X–Z$ plane around the Y-axis (see Figure 15.2). The moment is associated with the

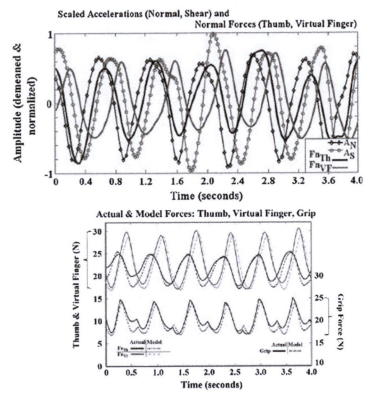

Figure 15.7 Accelerations and forces acting on the instrumented handle during circular arm movements. Upper panel: Normalized accelerations (a_N: normal, Z-axis, a_S: shear, XY plane) and normal forces (F^n_{Th}, F^n_{Vf}) during a counterclockwise circular movement in the vertical lateromedial plane (YZ plane). A representative example, frequency 1.5 Hz, circle diameter 20 cm. Note: a_S lags a_N, F^n_{Vf} lags F^n_{Th}, and no two are in phase. Bottom panel: Comparison between actual normal forces (Th, Vf, Grip) and modeled values for a circular arm movement. Shown example data set: 10 cm, 1.5 Hz counterclockwise movement. The grip force is plotted on the same relative scale but shifted down for viewing convenience as otherwise it would overlap on the normal force plots. Note the good correspondence between the actual grip force changes and the force from the mathematical model.
Adapted by permission from Slota et al. (2011), © Springer.

supination-pronation efforts. Maintaining object orientation is achieved by an appropriate combination of: (1) moments of the tangential forces M^t that result from unequal thumb and VF forces, and (2) moments of the normal forces. The moment of the normal forces M^n is due to the normal forces of the thumb and the VF. These forces are equal in magnitude (otherwise the handle would accelerate along the line of force action) but they may not act along the same line. Therefore, they form a force couple that generates a free moment that remains unchanged under parallel translation (see, for instance, Zatsiorsky, 2002, p. 20). Hence the moments of the tangential and normal forces can be added to obtain the total moment exerted by the performer (total moment, $M_{TOT} = M^t + M^n$).

In multidigit grasps: (1) both M^n and M^t are always present; (2) in zero-torque tasks, M^n and M^t act in opposite directions and cancel each other (Shim et al., 2003; Gao et al., 2005b,c; Solnik et al., 2014); and (3) in nonzero torque tasks, M^n and M^t act in the same direction assisting each other (Zatsiorsky et al., 2002a,b). When M^n and M^t cancel each other, they constitute a component of an internal force vector, or more specifically an *internal moment*.

The individual fingers are located either above or below the thumb and hence their normal forces generate moments (with respect to the thumb as a pivot) in opposite directions, either clockwise or counterclockwise (see Figure 15.4). The fingers resisting external torque and hence helping to keep rotational equilibrium are called *torque agonists* while the fingers assisting external torque are *torque antagonists*. The moments generated by these fingers are the *agonist* and *antagonist moments*, respectively (Figure 15.8). Hence, some fingers "work in the wrong direction." In order to counterbalance the antagonist moments, the fingers producing moments in the

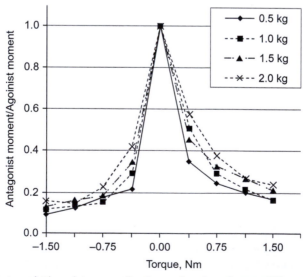

Figure 15.8 "Antagonist/agonist moment" ratio as a function of torque. When small torques are produced, the magnitudes of the antagonist moments are close to 60% of the agonist moment. Adapted by permission from Zatsiorsky et al. (2002a), © Springer.

intended direction should generate larger forces. Such performance does not appear to be optimal.

The moment of normal finger forces is generated by two mechanisms: (1) changing the sharing pattern and, consequently, the coordinate of the point of application of the resultant force; and (2) increasing the magnitude of the resultant force. When the external torque is small, for example, <0.5 Nm, the first mechanism dominates; while for larger torque, for example, >1.0 Nm, the point of force application does not change substantially and the increase in torque production is achieved by a rise in the VF force magnitude.

Individual fingers are activated according to their mechanical advantage. During torque efforts, the "peripheral" fingers with relatively large moment arms (index and little fingers) produce larger torques than the "central" fingers with smaller moment arms. When both the load and the external torque change, the force of the peripheral fingers depends mainly on the torque while the force exerted by the "central" (middle and ring) fingers depends both on the torque and load.

In nonzero torque tasks, when the hand with the grasped object is being accelerated vertically the external moment M is changing. To maintain the object orientation, the grasp moment should counterbalance these changes. When the handle acceleration increases, both the load (L) and M rise. An increase in L should induce an increase in the grasp force and hence an increase in all digit normal forces. In contrast, an increase in M requires increasing the force of only the agonist fingers. When both L and M increase, the normal forces of the agonist fingers get higher while the forces generated by the antagonist fingers do not change substantially (Figure 15.9). Hence, for the antagonist fingers, the classical grasp force—load relations can be overridden by torque production requirements.

15.4 Motor control constraints in hand and digit actions

Human movements are evidently constrained by anatomical restrictions (e.g., people cannot extend the knee joint beyond approximately 180°), laws of mechanics, and physiological limitations. They also can be constrained by motor control phenomena. One of the examples is the *finger interdependence* (reviewed in Schieber and Santello, 2004), which is manifested as *force deficit* and *finger enslaving*.

15.4.1 Force deficit

In multifinger tasks, the fingers generate smaller maximal forces than in the single-finger tasks (Ohtsuki, 1981). This phenomenon is called the *force deficit*. The force deficit is similar across tasks with the same number of explicitly involved fingers (reviewed in Danion et al., 2003). The deficit is usually parameterized as the ratio of the actual total force in N-finger pressing tasks ($1 \leq N \leq 4$) to the predicted force computed as the sum of forces exerted in single-finger tasks. For two-finger tasks (IM, IR, IL, MR, ML, and RL) the ratio ranges from 0.61 to 0.64, for three-finger tasks

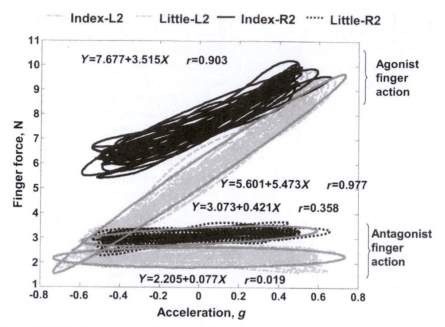

Figure 15.9 The dependence of individual finger forces on object acceleration during vertical arm oscillation, a representative example. Frequency 2 Hz.
Adapted by permission from Gao et al. (2005a), © Springer.

(IMR, IML, IRL, and MRL) it ranges from 0.43 to 0.45, and for the four-finger task (IMRL) it is about 0.38. The dependence of the above ratio on N, the number of fingers explicitly involved in the task, is described by an empirical equation: *Force deficit ratio* $= 1/N^{0.712}$ (Danion et al., 2003).

A hypothesis has been offered (Li et al., 1998) that explains the deficit by a "ceiling effect." According to the hypothesis, the control is realized by a two-level control system. A central neural drive (CND) arrives at the level of synergies and is distributed among muscles. An additional assumption is that CND has a certain limit, a *ceiling* that cannot be exceeded. The more effectors are involved, the smaller the neural drive that is available for each individual effector. If this hypothesis holds: (1) the force magnitude produced by each finger in a multifinger task should be less than 100% of its force in the one-finger task; (2) each finger force should decrease when the number of explicitly involved fingers increases; and (3) a saturation effect may be seen when many effectors are involved, that is, an addition of a new effector to the task does not substantially change the total output. All of these predictions were confirmed in experiments.

15.4.2 Finger enslaving

When one of the fingers moves or exerts force other fingers also move or generate force even when they are not required to do that by instruction. This phenomenon has been called *enslaving* (Li et al., 1998). To demonstrate the enslaving turn your palm up and wiggle the ring finger. You will see that other fingers also move. At least three mechanisms contribute to the finger enslaving: (1) peripheral mechanical connections between the muscles and tendons; (2) multidigit motor units in the extrinsic flexor and extensor muscles; and (3) diverging central commands. The enslaving was successfully modeled by a three-layer neural network (Figure 15.10).

The network modeling is based on the idea that the central commands are sent to both multidigit (extrinsic) hand muscles and the single-finger (intrinsic) hand muscles. It is assumed that the enslaving is mainly due to the activity of the extrinsic hand muscles. The neural network yields a relation between the central commands and the individual finger forces. The relation is expressed as a matrix equation:

$$\mathbf{F} = {}^{1}\!/_{n}\,[w]\mathbf{c} + [v]\mathbf{c} \tag{15.3}$$

where $\mathbf{F}$ is a (4×1) vector of the finger forces, $[w]$ is a (4×4) matrix of weight coefficients (the matrix models the multidigit muscles); $\mathbf{c}$ is a (4×1) vector of the dimensionless central commands; $[v]$ is a (4×4) diagonal matrix with the gain coefficients that models the input–output relations for the single-digit muscles; and n is the number of fingers that are intended to produce force. For $n = 4$, Eqn (15.3) can be reduced to

$$\mathbf{F} = [\mathbf{W}]\mathbf{c} \tag{15.4}$$

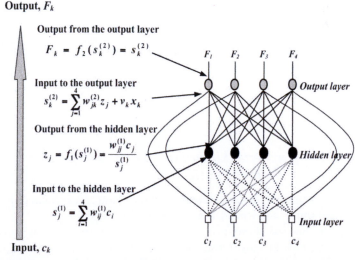

Output, F_k

Output from the output layer

$$F_k = f_2(s_k^{(2)}) = s_k^{(2)}$$

Input to the output layer

$$s_k^{(2)} = \sum_{j=1}^{4} w_{jk}^{(2)} z_j + v_k x_k$$

Output from the hidden layer

$$z_j = f_1(s_j^{(1)}) = \frac{w_{ij}^{(1)} c_j}{s_j^{(1)}}$$

Input to the hidden layer

$$s_j^{(1)} = \sum_{i=1}^{4} w_{ij}^{(1)} c_i$$

Input, c_k

$F_1 \quad F_2 \quad F_3 \quad F_4$

Output layer

Hidden layer

Input layer

$c_1 \quad c_2 \quad c_3 \quad c_4$

Figure 15.10 Neural network and the associated mathematical formulations. The index, middle, ring, and little fingers correspond to 1, 2, 3, and 4, respectively. The mathematical background of the network is explained in Zatsiorsky et al. (1998) and Li et al. (2002).

where [W] is a (4 × 4) matrix of weight coefficients (*interfinger connection matrix*) that relates hypothetical central commands to individual fingers with actual finger forces; **F** is a (4 × 1) vector of the normal finger forces; and **c** is a (4 × 1) vector of the *central (neural) commands (modes)*. The elements of vector **c** equal 1.0 if the finger is intended to produce maximal force (maximal voluntary activation) or 0.0 if the finger is not intended to produce force (no voluntary activation). The elements of [W]—when multiplied by commands c_i—represent: (1) elements on the main diagonal—the forces exerted by finger i in response to the command sent to this finger (*direct*, or *master, forces*); (2) elements in rows—force of finger i due to the commands sent to all the fingers (sum of the direct and enslaved forces); and (3) elements in the columns: forces exerted by all four fingers caused by a single command sent to one of the fingers (a *mode*). On the whole, the off-diagonal elements of the matrix portray the finger enslaving. Danion et al. (2003) suggested a simplified method of estimating the elements of [W] not requiring neural network modeling.

If matrix [W] is known and actual finger forces in a prehension task are recorded, the vector of neural commands **c** can be reconstructed by inverting Eqn (15.4): $\mathbf{c} = [\mathbf{W}]^{-1}\mathbf{F}$ (Zatsiorsky et al., 2002b). When the vector **c** is reconstructed, forces generated by individual fingers can be decomposed into components that are due to (1) direct commands to the targeted fingers and (2) the enslaving effects, that is, the commands sent to other fingers, Figure 15.11.

Due to enslaving, fingers that generate moment in the direction opposite to the direction required by the task (*torque antagonists*) are often activated even though such activation is mechanically not efficient (Zatsiorsky et al., 2002b). When mathematical optimization methods were employed to determine whether the force patterns observed in human prehension agree with certain optimization criteria—a broad range of such criteria was used—the cost functions based on finger forces were not able to predict the activation of the torque antagonists. In contrast, minimization of the Euclidian distance in the space of finger modes predicted nonzero torque-antagonist finger forces. The criterion based on neural commands exhibited better performance because it accounts for enslaving effects, whereas other criteria do not (Zatsiorsky et al., 2002b). However, Martin et al. (2013) have found that in multifinger *pressing* tasks the analytical inverse optimization (ANIO; Terekhov et al., 2010; Terekhov and Zatsiorsky, 2011; see also Chapter 10) worked equally well on both the finger forces and modes.

15.4.3 Enslaving in serial kinematic chains: grasp force and grasp moment enslaving

The tendons of extrinsic finger muscles cross the wrist joint. As a result, grip force and wrist torque may show interdependence (as shown by Paclet et al., 2014): An intentional change in one of these two variables may be expected to produce an unintentional change in the other (enslaving). The patterns of this enslaving are, however, not simple. The effects have been shown to be asymmetrical—a change in wrist torque (quantified using hand force exerted on a stop) produced consistent changes in the grip force in the same direction; the effects of the changes in the grasp force on the

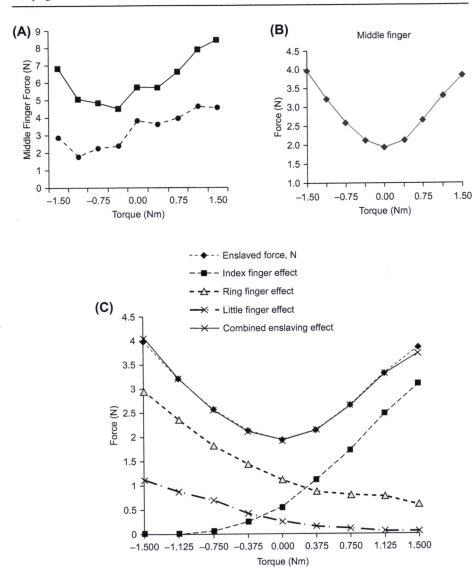

Figure 15.11 Decomposition of the normal forces of the middle finger during holding a 2.0 kg load at different external torques. The data are from a representative subject. (A) Actual and "direct" finger forces. The direct forces (dashed line) were computed as the products of the diagonal elements of the interfinger connection matrix times the corresponding finger commands. (B) Enslaved forces, that is, the difference between the actual and "direct" forces. (C) Decomposition of the enslaving effects. Effects of the commands to other fingers on the middle finger force are presented.
Adapted by permission from Zatsiorsky et al. (2002b), © Springer.

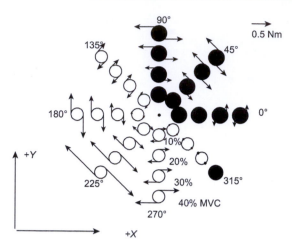

Figure 15.12 The enslaved moments of force exerted on the cylindrical handle (grasp moments) during arm force production. The moments are with respect to the handle longitudinal axis. Note that these moments are different from the grip moments discussed previously; they are exerted about dissimilar axes. The arrows represent the moment magnitude (10−40% of MVC) and direction. Subjects sat in a chair, grasped a circular handle with a power grip, and generated forces of different magnitudes in different directions. No instruction was given on grasp moment production. The right arm was restricted to a horizontal plane at the shoulder height, with the upper arm flexed at 45° from the frontal plane and the elbow flexed at 65°. Visual feedback was provided for the arm force only. Black circles—moments in the counterclockwise direction; empty circles—clockwise moments (as seen from the top). The locations of the circles correspond to the target force direction and magnitude. Counterclockwise grasp moments are in the direction of the flexion joint torques.
Adapted by permission from Xu et al. (2012), © Springer.

wrist flexion/extension action were much smaller. The indices of enslaving also showed significant effects of such factors as the direction of intentional force change (increase or decrease) and the baseline force level of the enslaved variable.

Another example of the enslaving in the arm concerns production of nonzero *grasp moments* during the arm force production. Such moments are generated even when the moment production is not required by an instruction. The moments are strongly affected by the direction and magnitude of the arm force (Xu et al., 2012; Figure 15.12).

15.5   Prehension synergies

The term "prehension synergies" describes conjoint changes in digit forces and moments during multifinger grasping tasks (Santello and Soechting, 2000; Zatsiorsky et al., 2003b) as well as the neural organization responsible for such coordinated adjustments. To study prehension synergies, researchers have employed several experimental techniques, they: (1) applied external perturbations (Cole and Abbs, 1987, 1988; Zatsiorsky et al., 2006); (2) inflicted self-induced perturbations—the subjects varied the number of grasping fingers during a static prehension task (Budgeon et al., 2008); (3) computed correlations among output variables in single trials of long duration (Santello and Soechting, 2000; Vaillancourt et al., 2002); (4) varied the task parameters, in particular the object geometry, resisted torque, and/or load (Zatsiorsky et al., 2002a,b; Zatsiorsky et al., 2003); and (5) studied trial-to-trial variability (Shim et al., 2003, 2005; Shim and Park, 2007). In all those cases, the individual digit force changes have always been interrelated.

When local perturbations are applied, for instance, the friction under one digit is changed, not only *local effects* but also *synergic effects* in unperturbed digits are observed (Figure 15.13).

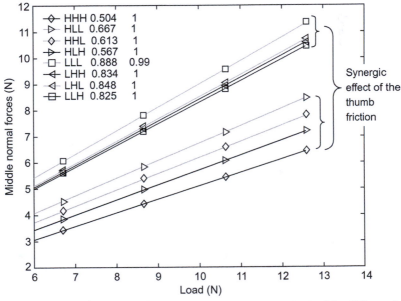

Figure 15.13 Dependence of the middle finger normal force on the load for different friction sets in three-digit grasps. The experimental design included 32 combinations of 8 friction conditions and 4 loads. The eight friction conditions were HHH, HLL, HHL, HLH, LLL, LHH, LHL, and LLH, where the letters correspond to the high (H) and low (L) friction condition for the thumb, index, and middle fingers, respectively. Group averages are shown. The numbers in the figure are the dimensionless coefficients representing a relation between a normal force of a digit and the load and the coefficients of correlation squared (all $r^2 > 0.98$). The dotted lines designate the low friction contact at the middle finger. The large figure bracket indicates the *synergic effect*, the effect of thumb friction on the middle finger normal force. Two small figure brackets show the *local friction effects*, that is, the spreading induced by the high or low friction contact at the middle finger. At a given thumb friction, the forces were larger at the low friction contact at the middle finger.
Adapted by permission from Niu et al. (2007), © Springer.

In static tasks, the variations in elemental variables—for example, individual digit forces and points of their application—to a large extent intercompensate for each other's effects on the resultant variables applied to the handheld object. As a result, the variation of performance variables specified at a higher level of the assumed control hierarchy is decreased. For instance, the deviations in the resultant forces and moments acting on the grasped object are much smaller than the variations of the thumb and VF forces, and the variations of forces at the VF level are smaller than at the IF level. This is valid both for the force magnitude and force direction. The stabilizing effects of the synergies can be assessed within the framework of the *uncontrolled manifold hypothesis* (UCM hypothesis; Scholz and Schöner, 1999; see Chapter 11).

Some of the adjustments are mechanically necessary (without them the task cannot be performed), whereas others are results of choice by the central nervous system. People adjust the digit forces to changing conditions, for example, variable external moments, and they also do not perform the same prehensile task in one and the same way; in each trial, the digit forces are different (Figure 15.14). In the example in Figure 15.14 the VF and thumb tangential forces change but their sum stays constant—as required by the task mechanics—whilst other variables, for example, relative contribution of the moments of the normal and tangential forces into the total moment, fluctuate representing a choice made by the central controller.

Two important features of prehension synergies have to be mentioned: (1) *chain effects* and (2) the *principle of superposition*.

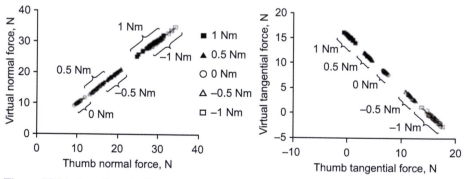

Figure 15.14 Coordinated adjustments of the VF and thumb normal forces (left panel) and tangential forces (right panel). The subjects were asked to hold the handle statically. They performed 25 trials at each of the five external torques: -1.0, -0.5, 0, 0.5, and 1.0 Nm. The data are from a representative subject. Left panel: The thumb and VF force magnitudes were equal in every trial. Right panel: The sum of the recorded forces was constant and equal to the object weight, 14.9 N. All the coefficients of correlation were close to 1.0 and -1.00, respectively. The trial-to-trial variations of the moment of tangential forces were compensated by the matching changes of the moments of normal forces (not shown in the figure).
Adapted by permission from Shim et al. (2003), © Springer.

1. The *chain effects* are sequences of local cause—effect adjustments necessitated by the task mechanics. The word "sequence" in the present context does not imply a chronological order; it only refers to the cause—effect relations. For instance, when handles with the expanding/contracting width were used, the normal digit forces changed with the width of the handle although the load L and external moment M stayed put (Zatsiorsky et al., 2006). The finding was explained by the following chain effect: increasing/decreasing the handle width changed the moment arms of the tangential forces $\rightarrow$ even though the tangential forces did not alter, the M^t changed $\rightarrow$ to maintain the total moment constant, M^n had to show opposite changes $\rightarrow$ the individual normal finger forces were adjusted. The chain effects arise from the necessity to find a solution that simultaneously satisfies a variety of constraints. A similar situation occurs in the control of the muscles serving individual fingers (Valero-Cuevas et al., 1998).

Chain effects are regularly present in prehension tasks. For instance, in three dimensions, an increase of the external torque in one direction changes the forces exerted in all three directions (see Figures 15.15 and 15.16). A set of chain effects explains this finding (Shim et al., 2005). The forces change because the line of action of the VF normal force acting in Z direction is not collinear with the line of action of the thumb normal force (this major experimental finding could not be predicted from the grasp mechanics). The VF and the thumb normal forces are equal in magnitude; they form a force couple that generates rotational effects about the X- and Y-axes.

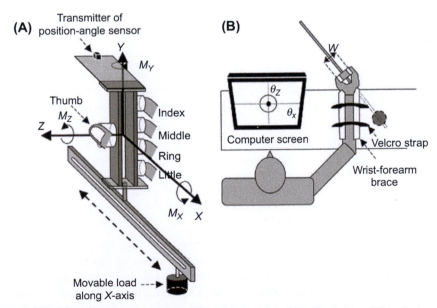

Figure 15.15 Experimental setup for studying prehension in three dimensions. When the handle is grasped with five digits, the external torque acts in a plane perpendicular to the plane of grasp. The task is similar to holding a book vertically in the air where the center of mass of the book is located farther from the hand than the points of the digit contacts.
Adapted by permission from Shim et al. (2005), © Springer.

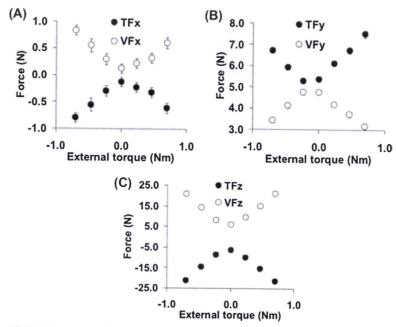

Figure 15.16 VF and thumb forces in three directions at different external torques. The torque about the Z-axis was changed by varying the load location along the X-axis. Note that forces in three directions were changed.
Adapted by permission from Shim et al. (2005), © Springer.

To preserve the rotational equilibrium of the handle, the moments of the couple must be counterbalanced. It is achieved by moments generated by the X- and Y-force components. Thus, the systematic relations in three dimensions are mechanically necessary.

2. The *principle of superposition* (suggested in robotics by Arimoto et al. (2001)) refers to decomposing complex actions into elemental actions that are controlled independently by different controllers. Such a control decreases the computation time. It has been shown that dexterous manipulation of an object by two soft-tip robot fingers can be realized by a linear superposition of two commands, one command for the stable grasping and the second one for regulating the orientation of the object.

When applied to human performers and multifinger grasps, the principle of superposition claims that forces and moments during prehension are defined by two independent commands: "Grasp the object stronger/weaker to prevent slipping" and "Maintain the rotational equilibrium of the object." The effects of the two commands are summed up. The commands correspond to the two internal forces discussed previously, the grip force and the internal moment.

Several experiments confirmed validity of the principle. In one of the experiments, the subjects exerted clockwise (negative) and counterclockwise torques of −1.0, −0.5, 0, 0.5, and 1.0 Nm. The load was always 14.8 N. At each torque, the subjects

performed 25 trials. The forces and moments exerted by the digits on the object maintained statically in the air were recorded and analyzed. All the performance variables
belonged to one of the two subsets (Figure 15.17). The variables within each subset
highly correlated with each other over repetitions of a task while the variables from
different subsets did not correlate. The first subset included normal forces of the thumb
and VF. The second subset included tangential forces of the thumb and VF, the
moments produced by the tangential and normal forces, and the moment arm of the
VF normal force D_{vf}^n. In particular, trial-to-trial changes of the VF normal force F_{vf}^n
did not correlate with the variations of the moment of the normal force M_{vf}^n
(Figure 15.17(A-2)). Because the moment of the normal force is simply the product
of the VF normal force and its moment arm, this lack of correlation is counterintuitive.

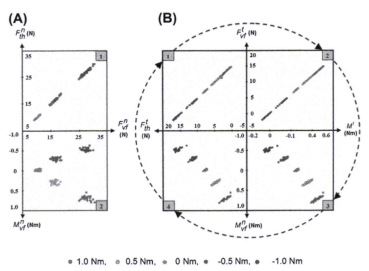

● 1.0 Nm, ○ 0.5 Nm, ● 0 Nm, ● -0.5 Nm, ● -1.0 Nm

Figure 15.17 Demonstration of the principle of superposition and the chain effects.
Interrelations among the experimental variables. A representative example. F and M designate
the force and moment; superscripts n and t refer to the normal and tangential force components;
subscripts th and vf refer to the thumb and virtual finger, respectively. Left panels: (A-1) F_{th}^n
correlated closely with F_{vf}^n. These two forces represent the first subset of variables mentioned in
the text. (A-2) F_{vf}^n versus M_{vf}^n. The correlation coefficients are close to zero. Right panels: (B-1)
F_{th}^t versus F_{vf}^t. The values of F_{th}^t and F_{vf}^t are on a straight line. This correlation was expected
because $F_{th}^t + F_{vf}^t = $ Constant (weight of handle). The different location of F_{th}^t and F_{vf}^t values
along the straight line signifies the different magnitude of M^t. (B-2) F_{vf}^t versus $M^t [M^t = $
$0.5(F_{vf}^t - F_{th}^v)d$, where $d = 68$ mm]. As the sum F_{th}^t and F_{vf}^t is constant, a change in one of
these forces determines the difference between their values and, hence, the moment that these
force produce. (B-3) M^t versus M_{vf}^n. (B-4) M_{vf}^n versus F_{th}^n. The variables in the panel B
(F_{th}^t, F_{vf}^t, M^t, M_{vf}^n) plus D_{vf}^n constitute the second subset of variables mentioned in the text. The
arrows signify the sequence of events resulting in the high correlation between F_{th}^t and M_{vf}^n
("chain effects"). Such a correlation does not exist between F_{vf}^n and M_{vf}^n; see panel (A-2).
Adapted by permission from Zatsiorsky et al. (2004), © Cambridge University Press.

In contrast, high correlation between M_{vf}^n and the tangential force F_{th}^t was discovered (Figure 15.17(B-4)) as well as high correlation between F_{th}^t and D_{vf}^n (not shown in the figure).

Functionally, the fine-tuning of the variables of the first subset prevents the object from slipping out of the hand and from moving in the horizontal direction. Conjoint adjustments of the variables of the second subset maintain the torque and vertical orientation of the handle constant (they also prevent the object from moving in the vertical direction). Hence, the data conform to the principle of superposition—preventing the object from slipping out of the hand and maintaining the object orientation are controlled by two separate commands whose effects do not correlate with each other.

The principle allows explaining the digit force adjustments to different factors such as (1) the load force and its modulation associated with the handle acceleration; (2) the external torque and its modulation; (3) the object orientation in the gravity field; (4) friction at the digit tips; and (5) some other variables. The variations of the above factors may require similar or opposite adjustments. For instance, an increase of the object weight and a decrease in friction both require a larger gripping force while a decrease of the load and a decrease in friction require opposite grasp force changes, a force decrease and increase, respectively. It has been suggested that the responses of the central nervous system to a mixture of similar or opposite requirements follow the rule: *Adjustment to the sum equals the sum of the adjustments* (reviewed in Zatsiorsky and Latash, 2008).

Prehension synergies change with old age and some neurological disorders, for example, Parkinson's disease. In both cases indices of prehension synergies (described in Chapter 11) are lower than in young healthy subjects (Park et al., 2011; Skm et al., 2012; Wu et al., 2013).

15.6   The bottom line

Multidigit coordination during prehension follows all the attributes of motor synergies described in Chapter 11. The synergies are observed at the two levels of control hierarchies: (1) on the higher VF level—in this analysis mechanical actions of individual fingers are reduced to their resultant force and moment, and (2) on the lower individual finger level when actions of the individual fingers are analyzed. A specific feature of the prehension synergies is the task redundancy. In five-digit grasps, a vector of individual digit forces and moments $\mathbf{F}_{digits}$ is a 30×1 vector (5 digits $\times$ 6 force and moment components). Its relation with a 6×1 vector $\mathbf{F}$ of the resultant forces and moments acting on the object is described by the *grasp equation* $\mathbf{F} = -[\mathbf{G}]\mathbf{F}_{digits}$ where $[\mathbf{G}]$ is a 6×30 *grasp matrix*. The equation neglects possible displacements of the points of digit force applications.

The main components of prehensile tasks include *slip prevention* and *tilt prevention*. The slip prevention is influenced by the contact mechanics. The object-digit contacts are *soft contacts*. Such contacts take place over a certain area, are accompanied by finger deformation, and allow displacement of the points of force application during performance. To avoid slipping, the tangential force/normal force ratio should be smaller than the coefficient of friction. The risk of slipping is usually quantified by the *safety margin*, which equals the ratio $(f^n_i - |f^t_i|/\mu)/f^n_i$, where i is an arbitrary digit, f^n is the digit normal force, and f^t is the digit tangential force. The notion of SM can be applied only to the vertical grasps. To prevent slipping, performers adjust digit normal forces (*grasp force*) both to the load (tangential forces) and friction.

Grasp force is an *internal force*, a set of contact forces that cancel each other. The grasp force does not affect the object equilibrium. During vertical movement of vertically oriented objects, performers increase the grasp force with object acceleration. In contrast, during the horizontal object transport, the grasp force is maximal at the instant of maximal velocity and hence at the instant of zero acceleration.

Maintaining object orientation (*tilt prevention*) is achieved by an appropriate combination of: (1) moments of the tangential forces that result from unequal thumb and VF forces, and (2) moments of the normal forces. Frequently, the individual fingers generate the moments (with respect to the thumb as a pivot) in opposite directions, either clockwise or counterclockwise. The fingers resisting external torque and hence helping to keep rotational equilibrium are called *torque agonists* while the fingers assisting external torque are *torque antagonists*. Hence, some fingers "work in the wrong direction."

Fingers actions are constrained by *finger interdependence*, specifically by force deficit and finger enslaving. The term *force deficit* refers to the decreased levels of maximal force in multifinger tasks as compared with the single-finger tasks. *Enslaving* describes involuntary activity of fingers when other fingers exert force or move. Hence, central commands sent to one of fingers affect other fingers' activity. The relations can be described by matrix equations. For five-digit grasps the equation is $\mathbf{F} = [\mathbf{W}]\mathbf{c}$, where $[\mathbf{W}]$ is a (4×4) matrix of weight coefficients (*interfinger connection matrix*) that relates the central commands to individual fingers with finger forces; $\mathbf{F}$ is a (4×1) vector of the normal finger forces, and $\mathbf{c}$ is a (4×1) vector of the *central (neural) commands*. Inverting this equation allows decomposing the finger forces into the *direct forces* (force of a finger due to the central command to this finger) and *enslaved forces*, force of finger i due to the activation of finger j.

Prehension synergies are manifested as conjoint changes of digit forces and moments in gripping tasks. They are just a specific case of motor synergies described in Chapter 11. Specific traits of the prehension synergies include *chain effects* and the *principle of superposition*. The chain effects are sequences of local cause-effect adjustments necessitated by the task mechanics. The word "sequence" in the present context does not imply a chronological order; it only refers to the cause-effect relations. The principle of superposition (first explored in robotic grasping) describes decomposing complex actions into elemental actions that are controlled independently by different controllers. In particular in human grasping preventing the object from slipping out of the hand and maintaining the object orientation are controlled by two separate commands whose effects do not correlate with each other.

References

Aoki, T., Niu, X., Latash, M.L., Zatsiorsky, V.M., 2006. Effects of friction at the digit-object interface on the digit forces in multi-finger prehension. Experimental Brain Research 172, 425—438.

Arbib, M.A., Iberall, T., Lyons, D., 1985. Coordinated control programs for movements of the hand. In: Goodwin, A.W., Darian-Smith, I. (Eds.), Hand Function and the Neocortex. Springer Verlag, Berlin, pp. 111—129.

Arimoto, S., Tahara, K., Yamaguchi, M., Nguyen, P.T.A., Han, H.Y., 2001. Principles of superposition for controlling pinch motions by means of robot fingers with soft tips. Robotica 19, 21—28.

Baud-Bovy, G., Soechting, J.F., 2001. Two virtual fingers in the control of the tripod grasp. Journal of Neurophysiology 86, 604—615.

Bennett, K.M.B., Castiello, U. (Eds.), 1994. Insights into the Reach to Grasp Movement. Elsevier Science, Amsterdam.

Budgeon, M.K., Latash, M.L., Zatsiorsky, V.M., 2008. Digit force adjustments during finger addition/removal in multi-digit prehension. Experimental Brain Research 189, 345—359.

Burstedt, M.K., Flanagan, J.R., Johansson, R.S., 1999. Control of grasp stability in humans under different frictional conditions during multidigit manipulation. Journal of Neurophysiology 82, 2393—2405.

Cole, K.J., Abbs, J.H., 1987. Kinematic and electromyographic responses to perturbation of a rapid grasp. Journal of Neurophysiology 57, 1498—1510.

Cole, K.J., Abbs, J.H., 1988. Grasp force adjustments evoked by load force perturbations of a grasped object. Journal of Neurophysiology 60, 1513—1522.

Comaish, S., Bottoms, E., 1971. The skin and friction: deviations from Amonton's laws, and the effects of hydration and lubrication. British Journal of Dermatology 84, 37—43.

Danion, F., Schoner, G., Latash, M.L., Li, S., Scholz, J.P., Zatsiorsky, V.M., 2003. A mode hypothesis for finger interaction during multi-finger force-production tasks. Biological Cybernetics 88, 91—98.

Flanagan, J.R., Johansson, R.S., 2002. Hand movements. In: Ramshandran, V.S. (Ed.), Encyclopaedia of the Human Brain. Academic Press, San Diego, pp. 399—414.

Gao, F., Latash, M.L., Zatsiorsky, V.M., 2005a. Control of finger force direction in the flexion-extension plane. Experimental Brain Research 161, 307—315.

Gao, F., Latash, M.L., Zatsiorsky, V.M., 2005b. Internal forces during object manipulation. Experimental Brain Research 165, 69—83.

Gao, F., Latash, M.L., Zatsiorsky, V.M., 2005c. In contrast to robots, in humans internal and manipulation forces are coupled (ThP01-18). In: Proceedings of the 2005 IEEE 9th International Conference on Rehabilitation Robotics, Chicago, IL, USA, pp. 404—407.

Iberall, T., 1987. The nature of human prehension: three dexterous hands in one. In: Proceedings of 1987 IEEE International Conference on Robotics and Automation, Raleigh, NC, pp. 396—401.

Jaric, S., Russell, E.M., Collins, J.J., Marwaha, R., 2005. Coordination of hand grip and load forces in uni- and bidirectional static force production tasks. Neuroscience Letters 381, 51—56.

Jeannerod, M., 1984. The timing of natural prehension movements. Journal of Motor Behavior 16, 235—254.

Johansson, R.S., Westling, G., 1984. Roles of glabrous skin receptors and sensorimotor memory in automatic control of precision grip when lifting rougher or more slippery objects. Experimental Brain Research 56, 550—564.

Kapur, S., Friedman, J., Zatsiorsky, V.M., Latash, M.L., 2010. Finger interaction in a three-dimensional pressing task. Experimental Brain Research 203, 101−118.

Kerr, J.R., Roth, B., 1986. Analysis of multifingered hands. Journal of Robotic Research 4, 3−17.

Latash, M.L., Zatsiorsky, V.M., 2009. Multi-finger prehension: control of redundant mechanical system. In: Sternad, D. (Ed.), Progress in Motor Control: A Multidisciplinary Perspective. Advances in Experimental Biology and Medicine, vol. 629. Springer, pp. 597−618.

Li, Z.-M., Latash, M.L., Zatsiorsky, V.M., 1998. Force sharing among fingers as a model of the redundancy problem. Experimental Brain Research 119, 276−286.

Li, Z.-M., Zatsiorsky, V.M., Latash, M.L., Bose, N.K., 2002. Anatomically and experimentally based neural networks modelling force coordination in static multi-finger tasks. Neurocomputing 47, 259−275.

Martin, J.R., Terekhov, A.V., Latash, M.L., Zatsiorsky, V.M., 2013. Optimization and variability of motor behavior in multifinger tasks: what variables does the brain use? Journal of Motor Behavior 45, 289−305.

Mason, M.T., Salisbury, J.K., 1985. Robot Hands and the Mechanics of Manipulation. The MIT Press, Cambridge, MS.

Murray, R.M., Li, Z., Sastry, S.S., 1994. A Mathematical Introduction to Robotic Manipulation. CRC Press, Boca Raton.

Nakazawa, N., Ikeura, R., Inooka, H., 2000. Characteristics of human fingertips in the shearing direction. Biological Cybernetics 8, 207−214.

Napier, J.R., 1956. The prehensile movements of the human hand. Journal of Bone and Joint Surgery 38-B (4), 902−913.

Niu, X., Latash, M.L., Zatsiorsky, V.M., 2007. Prehension synergies in the grasps with complex friction patterns: local versus synergic effects and the template control. Journal of Neurophysiology 98, 16−28.

Ohtsuki, T., 1981. Inhibition of individual fingers during grip strength exertion. Ergonomics 24, 21−36.

Paclet, F., Ambike, S., Zatsiorsky, V.M., Latash, M.L., 2014. Enslaving in a serial chain: interactions between grip force and hand force in isometric tasks. Experimental Brain Research 232, 775−787.

Park, J., Sun, Y., Zatsiorsky, V.M., Latash, M.L., 2011. Age-related changes in optimality and motor variability: an example of multifinger redundant tasks. Experimental Brain Research 212, 1−18.

Pataky, T.C., Latash, M.L., Zatsiorsky, V.M., 2004a. Prehension synergies during non-vertical grasping, I: experimental observations. Biological Cybernetics 91, 148−158.

Pataky, T.C., Latash, M.L., Zatsiorsky, V.M., 2004b. Prehension synergies during non-vertical grasping, II: modeling and optimization. Biological Cybernetics 91, 231−242.

Pataky, T.C., Latash, M.L., Zatsiorsky, V.M., 2004c. Tangential load sharing among fingers during prehension. Ergonomics 47, 876−889.

Pataky, T.C., Latash, M.L., Zatsiorsky, V.M., 2005. Viscoelastic response of the finger pad to incremental tangential displacements. Journal of Biomechanics 38, 1441−1449.

Pataky, T.C., Slota, G.P., Zatsiorsky, V.M., Latash, M.L., 2013. Is power grasping contact continuous or discrete? Journal of Applied Biomechanics 29, 554−562.

Salisbury, J.K., Craig, J.J., 1982. Articulated hands: force control and kinematic issues. International Journal of Robotic Research 1, 4−17.

Santello, M., Soechting, J.F., 2000. Force synergies for multifingered grasping. Experimental Brain Research 133, 457−467.

Savescu, A.V., Latash, M.L., Zatsiorsky, V.M., 2008. A technique to determine friction at the fingertips. Journal of Applied Biomechanics 24, 43—50.

Schieber, M.H., Santello, M., 2004. Hand function: peripheral and central constraints on performance. Journal of Applied Physiology 96, 2293—2300.

Scholz, J.P., Schöner, G., 1999. The uncontrolled manifold concept: identifying control variables for a functional task. Experimental Brain Research 126, 289—306.

Skm, V., Zhang, W., Zatsiorsky, V.M., Latash, M.L., 2012. Age effects on rotational hand action. Human Movement Science 31, 502—518.

Serina, E.R., Mote Jr., C.D., Rempel, D., 1997. Force response of the fingertip pulp to repeated compression—effects of loading rate, loading angle and anthropometry. Journal of Biomechanics 30, 1035—1040.

Shim, J.K., Latash, M.L., Zatsiorsky, V.M., 2003. Prehension synergies: trial-to-trial variability and hierarchical organization of stable performance. Experimental Brain Research 152, 173—184.

Shim, J.K., Latash, M.L., Zatsiorsky, V.M., 2005. Prehension synergies: trial-to-trial variability and principle of superposition during static prehension in three dimensions. Journal of Neurophysiology 93, 3649—3658.

Shim, J.K., Park, J., 2007. Prehension synergies: principle of superposition and hierarchical organization in circular object prehension. Experimental Brain Research 180, 541—556.

Slota, G.P., Latash, M.L., Zatsiorsky, V.M., 2011. Grip forces during object manipulation: experiment, mathematical model, and validation. Experimental Brain Research 213, 125—139.

Smith, M.A., Soechting, J.F., 2005. Modulation of grasping forces during object transport. Journal of Neurophysiology 93, 137—145.

Solnik, S., Zatsiorsky, V.M., Latash, M.L., 2014. Internal forces during static prehension: effects of age and grasp configuration. Journal of Motor Behavior 46, 211—222.

Terekhov, A.V., Pesin, Y.B., Niu, X., Latash, M.L., Zatsiorsky, V.M., 2010. An analytical approach to the problem of inverse optimization with additive objective functions: an application to human prehension. Journal of Mathematical Biology 61, 423—453.

Terekhov, A.V., Zatsiorsky, V.M., 2011. Analytical and numerical analysis of inverse optimization problems: conditions of uniqueness and computational methods. Biological Cybernetics 104, 75—93.

Vaillancourt, D.E., Slifkin, A.B., Newell, K.M., 2002. Inter-digit individuation and force variability in the precision prip of young, elderly, and Parkinson's disease participants. Motor Control 6, 113—128.

Valero-Cuevas, F.J., Zajac, F.E., Burgar, C.G., 1998. Large index-fingertip forces are produced by subject-independent patterns of muscle excitation. Journal of Biomechanics 31, 693—703.

Westling, G., Johansson, R.S., 1984. Factors influencing the force control during precision grip. Experimental Brain Research 53, 277—284.

Wu, Y.H., Pazin, N., Zatsiorsky, V.M., Latash, M.L., 2013. Improving finger coordination in young and elderly persons. Experimental Brain Research 226, 273—283.

Xu, Y., Terekhov, A.V., Latash, M.L., Zatsiorsky, V.M., 2012. Forces and moments generated by the human arm: variability and control. Experimental Brain Research 223, 159—175.

Yoshikawa, T., Nagai, K., 1991. Manipulating and grasping forces in manipulation by multi-fingered robot hands. IEEE Transactions on Robotics and Automation 7, 67—77.

Zatsiorsky, V.M., 2002. Kinetics of Human Motion. Human Kinetics, Champaign, IL.

Zatsiorsky, V.M., Gao, F., Latash, M.L., 2003a. Finger force vectors in multi-finger prehension. Journal of Biomechanics 36, 1745—1749.

Zatsiorsky, V.M., Gao, F., Latash, M.L., 2003b. Prehension synergies: effects of object geometry and prescribed torques. Experimental Brain Research 148, 77—87.

Zatsiorsky, V.M., Latash, M.L., Gao, F., Shim, J.K., 2004. The principle of superposition in human prehension. Robotica 22, 231—234.

Zatsiorsky, V.M., Gao, F., Latash, M.L., 2006. Prehension stability: experiments with expanding and contracting handle. Journal of Neurophysiology 95, 2513—2529.

Zatsiorsky, V.M., Gregory, R.W., Latash, M.L., 2002a. Force and torque production in static multifinger prehension: biomechanics and control. I. Biomechanics. Biological Cybernetics 87, 50—57.

Zatsiorsky, V.M., Gregory, R.W., Latash, M.L., 2002b. Force and torque production in static multifinger prehension: biomechanics and control. II. Control. Biological Cybernetics 87, 40—49.

Zatsiorsky, V.M., Gao, F., Latash, M.L., 2005. Motor control goes beyond physics: differential effects of gravity and inertia on finger forces during manipulation of hand-held objects. Experimental Brain Research 162, 300—308.

Zatsiorsky, V.M., Li, Z.-M., Latash, M.L., 1998. Coordinated force production in multi-finger tasks. Finger interaction and neural network modeling. Biological Cybernetics 79, 139—150.

Zatsiorsky, V.M., Latash, M.L., 2008a. Multifinger prehension: an overview. Journal of Motor Behavior 40, 446—476.

Zatsiorsky, V.M., Latash, M.L., 2008b. Human hand as a parallel manipulator. In: Wu, H. (Ed.), Parallel Manipulators. Towards New Applications. I-Tech: Vienna, Austria, pp. 449—468.

Zatsiorsky, V.M., Latash, M.L., 2009. Digit forces in multi-digit grasps. In: Nowak, D.A., Hermsdörfer, J. (Eds.), Sensorimotor Control of Grasping: Physiology and Pathophysiology. Cambridge University Press, Cambridge, UK, pp. 33—54.

Zuo, B.R., Qian, W.H., 2000. A general dynamic force distribution algorithm for multi-fingered grasping. IEEE Transactions on Systems, Man, and Cybernetics, Part B-Cybernetics 30, 185—192.

Glossary

A

Abduction Movement of a body part away from the midsagittal plane.

Abundance Availability of extra degrees of freedom in a multielement system, which afford the controller a rich repertoire of solutions.

> *Principle of abundance* Systems with more degrees of freedom than task constraints do not try to eliminate redundant degrees of freedom but organize them to produce flexible and stable task-specific behaviors.

Acceleration The rate of change of velocity, a vector.

> *Centripetal acceleration* The acceleration pointing toward the center of curvature in a curvilinear motion.

> *Coriolis acceleration* The acceleration acting on a body moving with respect to a rotating reference frame.

> *Normal acceleration* See *Centripetal acceleration*.

> *Tangential acceleration* The rate of change of the velocity magnitude; t.a. is directed along the movement trajectory.

Accelerometer A device that produces an electrical signal proportional to the acceleration of the device.

Achilles tendon The tendon connecting the triceps surae and plantaris muscles to the calcaneus.

Actin One of two contractile proteins, the second is *myosin*.

Action potential A relatively standard, short-lasting time profile of the transmembrane potential changes in muscle cells and neurons, which is generated when the membrane potential is brought to its threshold value.

> *Complex* An unusual action potential with multiple peaks; can be produced by Purkinje cells.

> *Simple* A typical action potential consisting of a single peak reaching a positive magnitude of the transmembrane potential followed by a period of hyperpolarization (lower potential as compared to the equilibrium potential).

Active elements (of the musculoskeletal system) Muscles and joints that can change their length (or angle) without external forces.

Active state State of a muscle during the latent period, from an arriving stimulus to a recordable force rise at the muscle–tendon junction.

Acton Muscles or muscle parts with point-to-point attachments that generate moments about a single joint axis.

Actuator A device that generates forces or converts one form of energy into another.

> *Force actuator* A device that is able to generate a force. Muscles are force actuators.

> *Linear actuator* See *Force actuator*.

> *Torque actuator* A real or imaginary generator that produces a moment of force with respect to a joint axis.

Adaptation (of movement) Changes in performance following practice of a novel task and/or in unusual external conditions and/or with unusual sensory input.

Aftereffects Changes in performance seen after adaptation to a novel external force field or unusual sensory feedback seen after the novel conditions have been replaced by typical ones.

To force field Changes in trajectories, torques, forces, patterns of muscle activation, and pattern of neuronal discharge after practice in a novel force field.

To visual feedback Changes in trajectories, torques, forces, patterns of muscle activation, and pattern of neuronal discharge after practice with distorted visual feedback, for example, while wearing glasses that rotate or flip the visual field.

Adduction Movement of a body part toward the midsagittal plane.

Adequate language A small set of basic notions, specific for a class of problems, that allow research of those problems using the scientific method (formulating and testing hypotheses).

Afferent *adj.* Conveying or transmitting toward the central nervous system. *n.* A neural fiber carrying sensory information toward the central nervous system.

Afferent fibers (afferents) Axons of sensory neurons that carry sensory information from receptors to the central nervous system. *More generally*: Fibers that carry information from a peripheral part of a system of interest into its more central part.

Ia Afferents that deliver action potentials from the primary endings of muscle spindles; the fastest conducting afferents.

Ib Afferents that deliver action potentials from the Golgi tendon organs.

II Afferents that deliver action potentials from the secondary endings of muscle spindles.

Agonist A muscle that generates a moment at a joint in the same direction as: (a) another muscle (see *Anatomic agonists*); or (b) resultant joint moment (see *Joint agonists*); or (c) actively shortens during propulsive phase of movement or lengthens being active during the yield phase (see *Task agonists*).

Anatomical agonists Muscles that lie on the same side of a bone, act on the same joint, and move it in the same direction (cf. *Joint agonists* and *Task agonists).*

Biomechanical agonists See *Joint agonists.*

Complete agonist A two-joint muscle that is a joint agonist at both joints that it serves.

Joint agonists Muscles that generate the moment in the same direction as the resultant joint moment (cf. *Anatomical agonists*).

Task agonists Muscles that assist the ongoing task.

All-or-none law The law according to which an excitable structure either generates a standard output (if brought to the threshold) or does not generate any output at all.

Alpha motoneurons Large neural cells innervating extrafusal muscle fibers.

Analytic inverse optimization (ANIO) A recently developed method of solving the inverse optimization problem.

Anisometric Non-isometric.

Antagonists Muscles that generate moments at a joint in opposite directions (*Anatomic antagonists*); see also, *Joint antagonists* and *Task antagonists*.

Anatomical antagonists Muscles that act on the same joint but move it in opposite directions (cf. *Joint antagonists* and *Task antagonists*).

Biomechanical antagonists See *Joint antagonists.*

Joint antagonists Muscles that generate the moment of force opposite to the resultant joint moment (cf. *Anatomical antagonists*).

Task antagonists Muscles that resist the ongoing task.

Anatomic position (of a body) An erect body stance with the arms at the sides and the palms facing forward.

Angiography A method of visualizing blood vessels by injecting a contrast agent with an ability to absorb X-rays into the blood stream.

Angular momentum (of a rigid body) The product of the body *moment of inertia* and *angular velocity*.

Anticipatory postural adjustments (APAs) Changes in baseline activation of postural muscles in anticipation of an expected perturbation or an action triggering a postural perturbation.

Anticipatory synergy adjustment (ASAs) Changes in a synergy index in anticipation of an event that requires a quick change in the variable stabilized by the synergy.

Ascending limb (of a force-length curve) A section of the force-length relation where the muscle generates larger active force at a larger length.

Asynergia A clinical term synonymous with discoordination; originally used to describe a feature of motor disorders after an injury to (dysfunction of) the cerebellum.

Ataxia Loss of joint coordination during whole-body tasks typical of movement disorders after an injury to (dysfunction of) the cerebellum.

ATP Adenosine triphosphate, the immediate source of energy for cross-bridge formation.

Attractor A point or trajectory toward which a dynamical system evolves over time.

    Fixed point attractor A final state that a dynamical system evolves toward.

    Limit cycle attractor A periodic orbit that a dynamical system evolves toward.

Axes

    Pointing axis The line along the distal segment of a kinematic chain.

    Radial axis The line intersecting the center of the proximal joint and the endpoint.

B

Basal ganglia A set of paired (left and right) subcortical nuclei involving the globus pallidus, caudate nucleus, putamen, substantia nigra, and the subthalamic nucleus.

    Direct loop A loop from the cortex through the striatum, external part of the globus pallidus, thalamus, and back to cortex.

    Disorders Sensorimotor and cognitive disorders associated with impaired function of the basal ganglia; motor disorders include hypokinetic (e.g., Parkinson's disease) and hyperkinetic (e.g., dystonia, hemiballismus, and chorea) ones.

        Dystonia A motor disorder characterized by clumsy postures and twisting component of involuntary movements.

        Hemiballismus A disorder produced by an injury to a subthalamic nucleus. Leads to large-amplitude, poorly controlled movements on the side contralateral to the injury.

        Huntington's disease A genetic disorder associated with degeneration of projections from the striatum to the external part of the globus pallidus. It leads to poorly coordinated excessive movements, in particular a dance-like gait (chorea).

        Parkinson's disease A progressive disorder associated with death of dopamine-producing neurons in substantia nigra.

            Bradykinesia Slowness in movement initiation and execution.

            Gait freezing Episodes of inability to initiate a step.

            Postural instability Poor balance associated with delayed and reduced anticipatory postural adjustments and poorly modulated, high-amplitude long-latency responses to perturbations.

            Rigidity Increased resistance to externally imposed motion in a joint.

            Tremor Involuntary cyclic movement associated with alternating activation bursts in the agonist and antagonist muscles. Parkinsonian tremor is typically postural; it may be alleviated by voluntary movement.

    Indirect loop A loop from the cortex through the striatum, internal part of the globus pallidus, subthalamic nucleus, external part of the globus pallidus, thalamus, and back to cortex.

Bell-shape (velocity) profile A symmetric velocity-time (or velocity-distance) curve with a single maximum.

Bernstein's problem The problem of finding a way to perform a motor action when the number of degrees of freedom (elemental variables) is larger than the number of constraints. Also known as the *problem of motor redundancy*.

Bernstein levels of movement construction Five levels: The paleokinetic level (A), the level of synergies (B), the level of the spatial field (C), the level of action (D), and the level of symbolic, highly coordinated action (E).

Body

> **Homogeneous body** A body in which the masses contained in any two equal volumes are equal.

> **Rigid body** A body for which the distance between any two points within the body remains constant.

Body link A representation of a body segment in which a central straight line extends longitudinally through the body segment and terminates at both ends in axes about which the adjacent segments rotate.

Body segment A part of the body considered separately from the adjacent parts.

Bowstringing Tendency of a curved tendon or muscle to straighten.

Body segment inertia parameters Inertial characteristics of a human body segment, such as mass, tensors of inertia, and location of the center of mass.

C

Catapult action The reversible muscle action in which the elastic energy is accumulated over a long time period and then released over a short period of time.

Catchlike property (of a muscle) Increase of muscle tension in response to an extra-pulse of high-frequency stimulation applied over a constant frequency stimulation.

CC See *Contractile components*.

Center of gravity The point of application of the resultant gravity force. The center of gravity coincides with the center of mass.

Center of mass (CoM) A point where all of the mass of the system could be considered to be located. Coincides with the *Center of gravity*.

Center of mass model (of total body energy) A model in which the kinetic energy of a body is represented as the sum of the kinetic energy of the center of mass and the kinetic energy of the body links in their motion relative to the center of mass.

Center of pressure The point of application of the resultant force normal to the surface.

Central commands (finger mode) Hypothetical efferent commands to the muscles serving a finger that range from 0 if the finger is not intended to produce force (no voluntary activation) to 1 if the finger is intended to produce maximal force (maximal voluntary activation). In some studies the central commands are expressed in units of force, for example, newtons.

Central nervous system The brain and the spinal cord.

Central pattern generator (CPG) A structure in the central nervous system able to generate patterned activity without being subjected to a patterned input.

> **Half-center model** A simple CPG model with two groups of neurons inhibiting each other.

Centrifugal Directed away from a center.

Centripetal Directed toward a center.

Centroid Center of a volume.

Centroidal axis A line through the center of mass.

Cerebellum The "small brain." A brain structure located behind the brain stem with more neurons than the rest of the central nervous system.

> **Cerebellar disorders** Disorders associated with an injury to the cerebellum or to input pathways into the cerebellum.
>
>> **Dysmetria** Movements over a wrong distance; can be hypometria (movements over a smaller distance) or hypermetria (movements over a larger distance).
>>
>> **Tremor** Involuntary cyclic movement associated with alternating activation bursts in the agonist and antagonist muscles. Cerebellar tremor has postural, kinetic, and intentional components.
>>
>> **Cerebellar nuclei** Three pairs of nuclei—fastigious, dentate, and interpositus—that mediate the output of the cerebellum.
>>
>> **Inputs** Two pathways, the mossy fibers and the climbing fibers.

Cervical Pertaining to the neck.

Chain effects (in the control of prehension) Sequences of local adjustments necessitated by the task mechanics leading to nontrivial correlations among variables. The word "sequence" in the present context does not imply a chronological order; it only refers to the cause–effect relations.

Chain inertia matrix A square matrix in which the elements represent the inertial resistance felt at one joint to the angular acceleration at another joint.

Clasp-knife phenomenon A phenomenon of increased resistance of a joint to external motion (via stretch reflex) followed by a quick collapse of the joint (possibly due to reflex effects from Golgi tendon organs).

Closed-loop problem Indeterminacy in solving of inverse problem of dynamics when a kinematic chain is closed via the environment.

Closure A grasp that immobilizes the held object completely. Closure imposes six constraints on the object.

> **Force closure** A grasp that can be broken under external forces without changing the finger positions.
>
> **Form closure** A grasp that cannot be broken without changing the finger configuration.

CNS See *Central nervous system*.

Co-activation Simultaneous activation.

> **Alpha–gamma** Parallel excitation of alpha- and gamma-motoneuronal pools by descending signals associated with an action.
>
> **Agonist–antagonist** Parallel changes (usually, a parallel increase) in the levels of activation of two muscles with opposing actions (agonist–antagonist pair).
>
> **Co-activation (c-) command** One of the basic two commands to a joint within the equilibrium-point hypothesis. It leads to an increase in the joint angle range, within which both agonist and antagonist muscles are activated without a change in the midpoint of the range.

Coherence A property of waves. In movement studies, commonly, a measure of stability of the phase shift between pairs of signals, which is computed for each frequency separately.

Collagen Main protein of connective tissue; see *Center of mass model*.

CoM model See *Center of mass model*.

Compliance The amount of deformation per unit of force. Compliance is the inverse of stiffness.

> **Compliance matrix** The inverse of the *Stiffness matrix*.

Components (of a vector) Elements into which a vector quantity can be resolved.

Concentric muscle action Exerting forces while shortening. Same as *Miometric action*.

Conservation of energy (in human movement) Transformation of potential energy into kinetic energy and back.

Constitutive equations Mathematical relations between stresses and strains.

Constraint (a) In mechanics: Any restriction to free movement. (b) In mathematics: A restriction on values of variables.

    Actual constraints Tangible physical obstacles to movement.

    Anatomical constraints The constraints imposed by the structure of the musculoskeletal system.

    Geometrical constraint The constraint in transferring joint angular velocity into linear velocity of the end effector in a given direction. For instance, the more the knee is extended the smaller the contribution of the angular knee extension velocity into the linear velocity of the leg extension.

    Mechanical constraints The constraints necessitated by mechanics (e.g., balance requirements).

    Motor task constraints The constraints imposed in order to execute the planned motor task.

        Instructional constraints The motor task constraints defined by an instructor, experimenter, competition rules, etc.

        Intentional constraints The motor task constraints imposed by the performer.

Contact (mechanical) A collection of adjacent points where two surfaces touch each other over a contiguous area.

    Hard-finger contact (**model**) The contact takes place over an area but the fingertip deformation and rolling are neglected.

    Point contact (**model**) The digit tip force is assumed to act at a point. In the point model with friction, the digit can exert a force but not a torque (moment of force) on the sensor.

    Soft-finger contact (**model**) The contact takes place over a certain area, the finger deforms, and the point of force application can change during performance.

Contractile components (in muscle models) The force generators in the muscle (actin-myosin cross-bridges).

Controlled-release method A technique of muscle research: an active muscle is allowed to shorten and develop force at a new length.

Control theory An area of mathematics (engineering) that considers a system as composed of a controller and a controlled object. The controller manipulates the inputs to the controlled object to obtain a desired effect on its output.

    Control trajectory A time profile of a control variable.

    Control variable A time varying variable sent by a controller to a controlled object.

Convergence A pattern of distribution of neural signals from multiple sources to a single target.

Coordinates A set of numbers that locates the position of a body in a reference system.

Coriolis coupling coefficients Coefficients in the dynamic equations of kinematic chains that characterize the inertial resistance to Coriolis force.

Cortex of large hemispheres The upper layer of the cerebrum.

    Anterior commissure One of the two major neural pathways that connect the two hemispheres directly, not via subcortical structures.

    Brodmann areas Common nomenclature of cortical areas.

    Corpus callosum One of the two major neural pathways that connect the two hemispheres directly, not via subcortical structures.

    Layers Neuroanatomical thin slices of cortical tissue parallel to its surface.

    Lobes The cortex consists of five lobes: frontal, parietal, temporal, occipital, and insular.

    Premotor area Lies anterior to the primary motor area (Brodmann area 6); implicated in sensory guidance of movements.

Primary motor area, M1 An area in the posterior portion of the frontal lobe (Brodmann area 4); a source of the corticospinal tract.

Pyramidal cells Cortical neurons that make projections both within the cortex and to subcortical targets.

Stellate cells Interneurons within the cerebral cortex; their axons do not leave the cortex.

Supplementary motor area Located just in front of the primary motor cortex, in Brodmann area 6. Implicated in movement planning.

Cosine tuning (of muscle force) Dependence of the muscle activation level on the cosine of the angle between the direction of the exerted endpoint force and direction of the endpoint force at which muscle demonstrates the highest activity (the *preferred direction*). The term is also used to describe the tuning of cortical neuronal activity during movements.

Cost function A real-valued function whose value is minimized or maximized in an optimization problem.

Coulomb's law The independence of the coefficient of friction magnitude on the normal force. The law is not valid for all materials.

Couple Two equal, opposite, and parallel forces acting concurrently at a distance apart. The couple creates a nonzero moment of force.

Couple vector A free vector that represents a couple. The couple vector is normal to the plane of the couple; the sense of the couple vector is determined by the right-hand rule.

Coupling coefficients (in the dynamic equations of the multilink kinematic chains)

Centripetal coupling coefficients Coefficients representing the effects of the angular velocity of one joint on the torques at other joints.

Coupling inertia coefficients Coefficients representing the inertial effects of the accelerated motion at one joint on the torque at another joint. The coefficients are the off-diagonal elements of the inertia matrix of a kinematic chain.

Creep Increase in length under a constant load, a retarded deformation.

Cross-bridges Molecular links between the myosin and actin filaments in the muscle cell; the basic element of force production.

Cross product (of vectors P and Q) The vector of magnitude $P \times Q \sin \alpha$ (α is the angle formed by **P** and **Q**), with direction perpendicular to the plane containing **P** and **Q**, according to the right-hand rule.

Cross-sectional area (of a muscle)

Anatomical cross-sectional area Area of the muscle slices cut perpendicularly to the longitudinal muscle direction.

Physiological cross-sectional area (PCSA) Cross-section area orthogonal to the long direction of muscle fibers.

Projected physiological cross-sectional area Projection of the PCSA on the muscle's line of action.

Cross-talk (of EMG signals) Contamination of EMG signals by electric activity of neighboring muscles.

D

Damping (coefficient) Coefficient of proportionality between the external force and velocity of deformation.

Dashpot (in mechanical models) An element, which resists the deformation in proportion to the loading velocity.

Deafferentation Removal of sensory (afferent) signals into the central nervous system from parts of the body.

Decerebration Surgical procedure (typically performed in animals for research purposes) involving transection of the midbrain between the inferior and superior colliculi.

Deformable body A body whose shape is changed under external forces.

Degree of freedom (DoF) An independent coordinate used to specify a body, system, or position.

Degree-of-freedom problem See *Motor redundancy*.

Density The amount of mass per unit volume.

Descending limb (of a force–length curve) A section of the force–length relation where the muscle generates smaller active forces at larger length values.

Direct perception A hypothesis within the ecological psychology approach that sensory signals could be coupled to motor commands directly without updating the picture of the world and one's own body.

Displacement The difference between the position coordinates of a body in its final and initial positions. Displacement is a vector; it does not depend on the route traveled.

Distal Away from the center of the human body or origin of a reference system; the opposite of *proximal*.

Distance Magnitude of a traveled path, a scalar.

Distribution problem Selecting specific muscle forces to generate necessary joint moments.

Divergence Distribution of neural signals from a single source to multiple targets.

Dualism (mind-body) A philosophical view positing that the mind is a nonphysical substance independent of the body.

Dual-strategy hypothesis A hypothesis on the origin of muscle activation patterns. It assumes that two classes of movement exist, those with intentional control over movement time and those without such control.

> *Excitation pulse* A hypothetical input into an alpha-motoneuronal pool modeled as a rectangular function with modifiable height and duration.

> *Speed-insensitive strategy* The subject is not trying to control movement time explicitly— the height of the excitation pulse is assumed constant, while its duration can be modified.

> *Speed-sensitive strategy* The subject is trying to control movement time explicitly—both the height and duration of the excitation pulse are modified.

Dynamic force enhancement Muscle force increase immediately after muscle stretch.

Dynamic optimization Optimization problems that consider time-dependent variables.

Dynamic points Via points of a muscle path, which coordinates with respect to the skeleton change during motion.

Dynamical system Formalization for any system that follows a deterministic rule that describes the time transition of a system from a given state to another state.

> *Approach* An approach to biological systems that views them as dynamical systems that can potentially show loss of stability and qualitative changes in the patterns of variables that they produce.

Dynamics The study of relations between forces and the resulting motion; a branch of kinetics. The term is also used with the meaning "time varying."

E

Eccentric muscle action Exerting muscle force while lengthening. Same as *Plyometric action*.

Ecological psychology An area of psychology exploring interactions of a person (animal) with the environment, particularly those that occur under the action of the laws of nature without obligatory cognitive involvement.

Effective insertion (of a muscle at a joint) A via point of a muscle at the distal bone forming a joint that determines the muscle moment arm at this joint.

Effective origin (of a muscle at a joint) A via point of a muscle at the proximal bone forming a joint that determines the muscle moment arm at this joint.

Effects (of perturbation to a finger)

Local effects Effects of a perturbation applied to a finger on this finger's behavior, for example, effect of changing friction beneath a finger on the finger force.

Synergic effects Effects of a perturbation applied to a finger on behavior of other fingers, for example, effect of changing friction beneath a finger on the forces exerted by other fingers.

Efferent (efference) copy (a) A copy of signals from motoneurons to muscles; (b) A copy of neural signals associated with the generation of a motor action. Commonly used for sensory purposes.

Efferent fibers Neural fibers from motoneurons to muscles. *More general*: Neural fibers from a more central part of a system to its more peripheral part.

Efficiency See *Mechanical efficiency*.

Efficiency

Delta efficiency The ratio: Change in work/Change in metabolic energy expenditure.

Gross efficiency The ratio: Mechanical work/Metabolic energy expenditure.

Net efficiency The ratio: Mechanical work/Metabolic energy expenditure above the rest level.

Work efficiency The ratio: Mechanical work/Metabolic energy expenditure above that in unloaded conditions (e.g., during pedaling without resistance).

Eigenvector (of a square matrix) A nonzero vector that, after being multiplied by the matrix, produces a vector parallel to the original vector.

Elastic body A deformable body that returns to its original size and shape when external force is removed.

Elastic modulus The slope of the stress–strain curve.

Elastic resistance (in joints) The resistance that depends on joint angle and joint displacement.

Elastic response (of human muscles to stretch with constant moderate speed) The first phase of the response when the force rises quickly as a linear function of length change.

Elasticity Capacity of objects to resist external forces and resume their normal shape after force removal.

Electroencephalography (EEG) A method of recording changes in the electrical field produced by neurons in the brain with electrodes placed on the skull. Characterized by poor spatial resolution and excellent temporal resolution.

Brain waves 2–4 Hz—delta rhythm; 4–8 Hz—theta rhythm; 8–13 Hz—alpha rhythm; 13–30 Hz—beta rhythm; >30 Hz—gamma rhythm.

Electromechanical delay The time period between the first signs of electrical activity of a muscle and the first signs of its mechanical action.

Electromyography (EMG) A method of recording muscle action potentials.

Filtering A procedure that narrows the range of frequencies represented in the signal.

Intramuscular EMG A method using thin electrodes placed inside the muscle of interest.

Normalization A method of making signals comparable across persons, muscles, conditions, and trials. Frequently uses signals recorded in a standard task or in a maximal output task.

Rectification A procedure that eliminates negative values in the signal.

Surface EMG A method using pairs of electrodes placed on the skin over the muscle of interest.

Elemental variable A variable at a selected method of analysis that can potentially be changed without changes in other variables at that level.

Energy (in mechanics) The capacity for doing work.

> **Kinetic energy** Energy due to motion at a nonzero velocity.
>
>> **Rotational kinetic energy** Kinetic energy associated with rotation of a body around its center of mass.
>>
>> **Translational kinetic energy** Kinetic energy associated with movement of the center of mass of a body.
>
> **Mechanical energy (of a rigid body)** The sum of the kinetic and potential energy of the body.
>
> **Potential energy** Energy stored in the system in latent form.
>
>> **Elastic potential energy** Potential energy due to deformation.
>>
>> **Gravitational potential energy** Potential energy due to the location of the body in a gravity field.
>
> **Total mechanical energy (of a rigid body)** See *Mechanical energy*.

Energy compensation during time Transformation of kinetic energy into potential energy and its subsequent use.

Energy conservation coefficient Proportion of the mechanical energy conserved in a given movement.

Energy recuperation See *Energy compensation during time*.

Energy transfer Redistribution of kinetic energy among body links, also redistribution of mechanical energy among muscles: When one muscle shortens extending another muscle the first muscle loses energy while the second muscle absorbs energy. The absorbed energy can be stored as elastic potential energy and/or dissipated into heat.

End effector The last link of an open kinematic chain.

Engram A term introduced by N. Bernstein meaning a pattern of a neural variable stored in memory and related to topological (nonmetrical) properties of a learned movement.

Enslaving See *Finger enslaving*.

Enthesis Tendon-bone junction.

Equations

> **Closed-form equations** Equations with all the variables explicitly presented.

Equifinality Reaching the same final position despite possible transient changes in the external forces.

> **Violations of equifinality** Consistent effects of transient forces on the final position.

Equilibrium A state, at which the resultant force and moment of force acting on the mechanical system are zero.

> **Stable** The system returns to the original state after a transient perturbation.
>
> **Unstable** The system moves away from the original state after a transient perturbation.

Equilibrium (of a material particle) A particle is said to be at equilibrium when it either is at rest or moves along a straight line with constant velocity.

Equilibrium length (of a muscle) The maximal length at which elastic force is zero; also, the length of an isolated passive muscle.

Equilibrium-point hypothesis Voluntary movements are produced by neurophysiological signals that change the equilibrium states of the system "body plus external forces."

> **Alpha-model** The neurophysiological signal used by the central nervous system to produce movements is assumed to represent the level of activation of alpha-motoneuronal pools.
>
> **Lambda-model** The neurophysiological variable used by the central nervous system to produce movements is assumed to be the threshold of the tonic stretch reflex.

Equilibrium potential Electrical potential inside the membrane measured with respect to the extracellular space, at which there is no net charge movement across the membrane.

Error compensation If, within a multielement system, one element introduces an error into the common output of the system, other elements adjust their outputs to reduce the error of the common output.

Evoked potentials Changes in a neurophysiological electrical signal associated with a sensory stimulus or an action that is time locked to the stimulus or action.

Excitation-contraction coupling Processes involved in the production of mechanical muscle fiber contraction following the generation of an action potential on the muscle membrane.

Exponential Changing with the power of e, the base of natural logarithms, for example, e^x.

Expressed section (of a force–length curve) The part of the force–length curve seen in vivo.

Extrafusal fibers Ordinary, power-producing muscle fibers that make up the main part of the muscle.

Extrinsic coordinates Coordinates of the body parts with respect to the environment.

F

Fascia Fibrous connective tissue that binds adjacent muscles together.

Fascial compartments Groups of muscles separated by connective tissue.

Fascicles Bundles of muscle fibers.

Feedback control A mode of control when the controller changes the control variables based on their effects on the state of the object.

> **Delay** The time interval between a sensory signal reflecting an error and the time when the correction takes place.
>
> **Gain** The ratio between the magnitude of a corrective vector and the magnitude of the original error vector.

Feedforward control A mode of control when the controller does not change the control variables based on their effects on the state of the object.

Field A function that assigns a scalar or vector to each point in some region or space.

Filament See *Myofilament*.

Finger interaction Dependence of finger force/movement on forces/movements of other finger(s).

> **Enslaving (enslavement, lack of individuation)** Unintentional force production by or movement of a finger when another finger of the hand produces force or moves intentionally.
>
> **Force deficit** A drop in the maximal force produced by a finger when it acts in a group of fingers as compared to a single-finger action.
>
> **Sharing** A pattern of percentage of force produced by each finger in multi-finger tasks.

Finger

> **Virtual finger (VF)** An imagined digit that generates the same mechanical effect as a set of actual fingers.

Finger mode A hypothetical neural variable reflecting a person's intent to change force by one digit only.

Finite element method A method to analyze stress–strain relations in complex mechanical objects. In the method, the entire structure is divided into smaller, more manageable (finite) elements.

Fitts's law A logarithmic dependence of movement time on the ratio of movement distance to target size.

Fixed points Points of a muscle path, whose coordinates with respect to the skeleton do not change during motion, for example, the points of muscle origin and insertion.

Flexibility Maximal range of joint motion.

Force (in mechanics) A measure of the action of one body on another. Force is a vector quantity that is defined by its magnitude, direction, and a point of application (or the line of force action).

Absolute force See *Specific force.*

Active force Rise in muscle force in excess of the passive force due to activation of the muscle.

Centripetal force A force directed toward a center. The force induces centripetal acceleration.

Centrifugal force The force of reaction to the centripetal force. The centripetal and centrifugal forces are applied to different bodies.

Coriolis force The force associated with Coriolis acceleration.

Coupling forces The forces of action and reaction acting according to Newton's third law between adjacent segments in a kinematic chain.

Direct forces Forces exerted by a finger in response to a command sent to this finger; direct forces neglect enslaving effects.

Effective force An imaginary force that produces the same effect on the center of mass (but not on the entire body) as all the external forces combined.

Gravity force Force between two objects proportional to the product of the masses of the objects and the gravity constant and inversely proportional to the distance between the objects squared. On Earth, it is equal to the product of the object mass and acceleration of gravity, *g.*

Inertial force Force proportional to the acceleration and mass of the object (for rotational movement, the moment of force is proportional to the angular acceleration and moment of inertia).

Interaction force Forces acting on a body segment due to movement of another segment or to forces acting on the other segment.

Internal force **(in human body)** Force acting between body parts.

Internal forces **(in grasping)** The forces that are in their entirety cancel each other and do not affect the resultant (manipulation) force, for example, the grip force.

Manipulation force The resultant force and moment acting on the grasped object.

Master force See *Direct force.*

Motion-dependent forces Forces that depend on the kinematics of the object.

Passive force The force of resistance of a relaxed muscle to stretch.

Reaction force Forces acting on an object from another object due to Newton's third law.

Resultant force **(or simply *Resultant*)** Force that produces the same effect on a rigid body as two or more forces.

Specific force Muscle force per unit of the physiological cross-sectional area.

Slip force The minimal grip force required to prevent slippage of a handheld object.

Shear force Forces tangential to the surface of an external object.

Total muscle force The sum of the passive and active muscle forces.

Force control An approach based on an assumption that there are neural structures that model interactions within the different structures of the body and between the body and the environments and use these models to precompute requisite force time profiles to fit specific motor tasks.

Force couple See *Couple.*

Force depression Force decrease because of an immediately preceding muscle action, for instance the muscle shortening.

Force-length curve Muscle force versus muscle length relation.

Force platform A device for measuring the ground reaction forces.

Force–velocity curve Relation between the muscle force and its speed of shortening. Such curves are commonly obtained by interpolating the force and velocity data recorded in several trials while one of the parameters of the task, for example, the load, is systematically changed.

Forward dynamics problem Computations of system's motion given known (or assumed) muscle activations and the system's initial conditions.

Fraction approach (of computing the mechanical energy expenditure) A method of computing mechanical energy expenditure from the changes in the fractions of the mechanical energy of the body.

Fractions (of mechanical energy of a body) Potential energy and two fractions of kinetic energy, associated with translation and rotation, respectively.

Fractions (of the grasp force)

Dynamic fraction The fraction of the grasp force depending on the inertial load of the handheld object (object acceleration in the direction tangential with respect to the grasping surfaces). The fraction represents the changes in the grip force that are solely due to the forces of inertia.

Static fraction The fraction depending on the static weight of the handheld object.

Statodynamic fraction The difference between the grasp forces in statics and during vertical oscillations of the object at the instances of zero acceleration.

Friction Force due to molecular interaction between two objects; maximal friction force is commonly assumed to be proportional to the force normal to the surface of interaction (see *Coulomb's law*).

Coefficient of static friction The ratio of shear friction force to normal force that corresponds to the transition from rest to motion.

Friction cone A geometric representation of coefficient of friction in space: A cone showing magnitudes of shear force that can be applied without slippage per unit of normal force.

Static friction The friction that must be overcome to set a body in motion.

Free moment A moment that is preserved under parallel displacements. *Moment of a couple* is a free moment.

Functional (in mathematics) A scalar-valued function of a function: An argument of a *functional* is a function and its output is a number.

G

Gamma motoneurons Small motoneurons innervating intrafusal muscle fibers of muscle spindles.

Gamma-scanner method A method of determining body-segment parameters by measuring the intensity of a gamma-radiation beam before and after it passes through a human body.

Generalized motor program (a hypothesis) The central nervous system stores a neural representation of essential features of a movement. During movement execution, the GMP is scaled in magnitude and duration to produce required forces.

Give See *Yield effects.*

Golgi tendon organs Force-sensitive receptors located at muscle–tendon junctions and innervated by group Ib afferents.

Golgi tendon organ reflexes (GTO reflexes) GTO reflexes produce disynaptic inhibition of alpha-motoneurons innervating the muscle of their origin and trisynaptic disinhibition of antagonist alpha-motoneurons.

Goniometer A device that produces an electrical signal proportional to the joint angle.

Grasp (or Grip) Finger forces acting on the grasped object that produce zero resultant force. Grasp is also formally defined as a system of unit wrenches and wrench intensities exerted on a grasped object.

Aperture Distance between the tips of the digits opposing each other while grasping an object.

Feedforward control of grip force Changes in the grip force in anticipation of or simultaneously with self-produced or predictable changes in inertial or other forces acting on the object.

Grasp force See *Grip force.*

Prismatic grasp Grasp in which the thumb opposes the fingers and the contact surfaces are parallel.

Safety margin Proportion of normal force acting on the grasped object above the minimum required to prevent the object slipping.

Grasp matrix A $6 \times n$ matrix, also known as the *matrix of moment arms* (n corresponds to the number of digits). The rows of the matrix correspond to the resultant forces and moments acting on the object, and the columns correspond to the unit digit forces and moments (wrenches).

Grasp moment A moment of force exerted on the grasped object.

Grip force (in prismatic grasps) Internal normal forces in the null space of the grasp matrix. Grip force is mathematically independent on the manipulation force, that is, it does not affect the resultant force acting on the grasped object.

Gravity line The vertical line thought to be the center of gravity.

Ground reaction force The reaction force exerted on the performer by the supporting surface.

H

H-reflex A monosynaptic reflex induced by an electrical stimulation of a peripheral muscle nerve.

Halen-Kelso-Bunz equation An equation that describes properties of relative phase (f) between trajectories of two effectors, $df/dt = -\sin f - 2k \sin 2f$.

Hamstrings The three posterior thigh muscles: The semitendinosus, semimembranosus, and biceps femoris.

Hand muscles Extrinsic muscles (with the belly in the forearm) produce forces transmitted to all four fingers via tendons attached at the intermediate and distal phalanges; intrinsic muscles (with the belly in the hand) produce finger-specific forces at the proximal phalanges.

Extensor mechanism A connective tissue structure on the back of the digits that transmits forces produced by intrinsic muscles to more distal phalanges.

Hill's force–velocity curve See *Force–velocity curve.*

Hill's model A lumped-parameter muscle model that includes a *contractile component* (CC), *series elastic component* (SEC), and *parallel elastic component* (PEC).

Homonymy Using the same word with different meanings. In science, homonymy should be avoided.

Homunculus (a) A "small man" in the brain making decisions; (b) A distorted human-like figure drawn on the surface of an anatomical brain structure and representing sensory or motor projections.

Hooke's law Linear relation between the force and deformation.

Human body

Axes of the human body

Anterioposterior axis Interception of the sagittal and transverse planes.

Lateromedial (frontal) axis Interception of the frontal and transverse planes.

Longitudinal axis Interception of the sagittal and frontal planes.

Planes of the human body
  Cardinal (principal) plane Any plane that passes through the body's center of mass.
  Frontal (coronal) plane A plane dividing the body into anterior and posterior sections.
  Sagittal plane A plane dividing the human body into the left and right sections.
  Transverse plane For bodies in an upright posture, a horizontal plane passing through the body.
Position of the human body Human body's location, attitude (orientation), and posture.
Hypertonia (muscle hypertonia) Abnormally high muscle tone.
Hypotonia (muscle hypotonia) Abnormally low muscle tone.
Hysteresis (muscle) Difference between the loading and unloading curves in the stress–strain cycle.

I

In situ In the original biological location but with partial isolation.
In vitro Isolated from a living body and artificially maintained.
In vivo Within a living body, as it is.
Impedance (in mechanics) For a system undergoing simple harmonic motion, impedance is the ratio of the force applied at a point to the resulting velocity at that point.
Impedance (in biomechanics of human motion) Total resistance of a body to motion that depends on the displacement (*apparent stiffness*), velocity (*apparent viscosity*), and acceleration (*inertia*). Such an understanding of the term is relatively new and is not commonly accepted.
  Impedance control A view that the central nervous system modifies impedance of effectors to produce movements.
Index of difficulty (in human movements) A characteristic of a movement task commonly used in studies of Fitts's Law: $ID = \log_2(2D/W)$, where ID is the index of difficulty, D is the distance to the target, and W is the target width along the movement direction.
Index of softness of landing (ISL) The ratio: (Negative work of joint moments)/(Decrease of total energy of the body).
Inferior olives Paired structures in the medulla, the origin of the climbing fibers, one of the two major inputs into the cerebellum.
Insertion (of a muscle) The distal attachment of a muscle to a bone or another tissue.
Inter-finger connection matrix A matrix of coefficients that relate central commands (finger modes) to individual fingers with actual finger forces.
Interaction torque Component of joint torque that depends on movement in another joint(s).
Internal model (a) Any neurophysiological structure that participates in predictive behavior (synonymous with neural representation); (b) A neural structure that computes/predicts effects of the interactions among parts of the body and between the body and the environment to produce desired muscle forces.
  Direct A neural structure that computes/predicts effects of the current efferent command on the body motion and associated sensory signals. Commonly used within the force-control approach.
  Inverse A neural structure that computes/predicts a requisite neural command based on a desired motor outcome. Commonly used within the force-control approach.
Interneurons Neurons within the central nervous system that project on other neurons.
  Ia-interneurons Interneurons that receive excitatory projections from Ia-afferents and make inhibitory projections on alpha-motoneurons innervating the antagonist muscle (reciprocal inhibition).

Ib-interneurons Interneurons that receive excitatory projections from Ib-afferents and make inhibitory projection on alpha-motoneurons innervating the muscle of origin; in addition, Ib-interneurons contribute to facilitation of alpha-motoneurons innervating the antagonist muscle.

Renshaw cells Interneurons that receive excitatory projections from axons of alpha-motoneurons and make inhibitory projections on alpha-motoneurons of the same pool (recurrent inhibition) and on gamma-motoneurons innervating muscle spindles in the same muscle.

Intrafusal fibers The fibers inside the muscle spindles; the intrafusal fibers are innervated by gamma motoneurons.

Intrinsic coordinates Coordinates of body part with respect to each other.

Invariant characteristics (of a movement) Movement characteristics and muscle activation patterns that are similar across skilled performers, for example, straight-line endpoint trajectory, bell-shaped velocity profile, and triphasic muscle activation pattern.

Inverse optimization Finding an unknown objective function (or functions) from experimentally recorded optimal solutions.

Inverse problem A problem of defining a group of variables based on known values of another group of variables, which are a function of the first group.

Inverse problem of dynamics **(in biomechanics)** Determining joint moments during human movements from the known kinematics, especially limb accelerations. To solve the inverse problem, mass-inertial properties of body segments must be known.

Inverse problem of kinematics **(in biomechanics)** Finding a joint configuration from the known position of the endpoint of a kinematic chain.

Inverted pendulum model A model of postural control, which considers the body as an inverted pendulum swinging about the ankle joints.

Isokinetic Maintaining a constant joint angular velocity.

Isometric Maintaining a constant length of the "muscle plus tendon" complex.

J

Jacobian A matrix of partial derivatives between two sets of variables.

Chain Jacobian Jacobian relating the differential endpoint displacement with the differential angular displacement at the joints.

Muscle Jacobian Jacobian relating the differential changes in muscle length with the differential angular displacements at the joints.

Jerk The rate of change of acceleration.

Joints

Ideal joint Joint without energy dissipation due to friction or deformation.

Revolute joint Joint allowing only rotation of the adjacent segments.

Joint angles

Anatomical joint angle (External joint angle) The angle between a segment's anatomical position and the position of interest.

Included joint angle (Internal joint angle) The angle between the longitudinal axes of two segments defining a joint (the angle is $<180°$).

Joint force The force at a joint that represents the effects of gravity and inertia. Because the force does not account for the muscle forces, it does not represent the actual forces acting on the articular surfaces.

Joint moment See *Torque—Joint torque.*

Joint strength curve A plot of maximal voluntary force versus joint angle.

K

Kimocyclography A method invented by Bernstein that allows recording trajectories of markers placed on the body.

Kinematics Description of motion disregarding forces.

Kinematic chain A linkage of rigid bodies.

Branched (complex) kinematic chain A chain containing at least one link entering more than two kinematic pairs.

Closed kinematic chain The chain constrained on both its ends.

Open kinematic chain The chain with one end free to move.

Parallel kinematic chains (also *Parallel manipulators*) Several kinematic chains working in parallel, for example, fingers of a hand. The word *parallel* in the present context means that the position of the endpoint of each linkage is independent of the position of the other linkages. Parallel chains are *branched kinematic chains*.

Redundant kinematic chain An open kinematic chain in three-dimensional space with more than six degrees of freedom.

Serial kinematic chain A chain in which each of the links enters no more than two kinematic pairs.

Kinematic pair The chain consisting of two adjacent links connected by a joint.

Kinematic problems

Direct kinematic problem The joint coordinates are known; the end-effector position is sought.

Inverse kinematic problem The position of the end effector is known; the joint coordinates are sought.

Kinematic redundancy The number of kinematic degrees of freedom is larger than strictly necessary for motion.

Kinesthetic illusion An erroneous perception of body location, configuration, or forces acting on or within the body.

Kinetics Study of the relations between the forces and their effects on bodies at rest (*statics*) and bodies in motion (*dynamics*).

L

Large fiber peripheral neuropathy A rare disorder characterized by selective loss of signal transmission along main groups of peripheral afferent fibers.

Leading-joint hypothesis A hypothesis from the force-control group that assumes that one joint plays a leading role and generates motion in all other joints with interaction torques; other joints are assumed to implement small torque adjustments.

Length

Anatomical (**of a body segment**) The distance between predetermined anatomical landmarks.

Biomechanical (**of a body segment**) The distance between joint axes.

Linear oscillator A system with a mass on a spring and a damping element.

Damping ratio A dimensionless parameter characterizing transitions of a linear oscillator between equilibrium states.

Natural frequency A parameter expressed in radians per second that characterizes the frequency at which the oscillator oscillates in the absence of external forces.

Overdamped An oscillator with the damping ratio over unity. In the absence of external force shows no oscillations but a smooth transition to the resting length of the spring.

Underdamped An oscillator with the damping ratio under unity. In the absence of external force shows oscillations with an exponential drop in their amplitude.

Linear phase The first phase of muscle force decay during muscle relaxation.

Linear spring (in mechanical modeling) An object whose length changes instantaneously in proportion to the load change.

Link A body in a kinematic chain.

Locomotion A motor activity associated with displacement of the whole body in the environment.

Corrective stumbling reaction A phase-dependent, complex preprogrammed reaction that can be seen in leg muscles at a delay of 50–70 m in response to foot stimulation, mechanical or electrical.

Gait A pattern of leg motion characterized by a particular stable relative phase of homologous joint trajectories across limbs; for example, walking, trotting, and galloping.

Fictive locomotion Phasic patterns of activity of alpha-motoneuronal pools that can be observed in a paralyzed animal.

Midbrain locomotor area An area in the midbrain that, when electrically stimulated, leads to locomotion of the decerebrated animal.

Passive Walking-like patterns demonstrated by inanimate objects, commonly under the action of gravity.

Lombard's paradox Under certain conditions, a two-joint muscle can cause extension of a joint where it usually has a flexion action.

Long-term depression (LTD) A long-lasting decrease in efficacy of a synaptic connection following a vigorous activation through that synapse or a particular combination of inputs into the target neuron.

Long-term potentiation (LTP) A long-lasting increase in efficacy of a synaptic connection following a vigorous activation through that synapse or a particular combination of inputs into the target neuron.

Lumbar enlargement Part of the spinal cord at the level of vertebrae T12-L1 where many spinal segments are tightly packed.

Lumped parameters (in mechanical modeling) Mechanical properties assigned to individual parts of the model, for example, elasticity—to a linear spring, damping—to a dashpot, etc.

M

Magnetic resonance imaging (MRI) A method of imaging internal body organs that uses a brief pulse of the electromagnetic field at a radio frequency to perturb the orientation of the spin axes of hydrogen atoms. Characteristics of the electromagnetic waves released while the spin axes return to their original orientation are analyzed.

Functional MRI A method that uses two MRI examinations, before and after an action, and quantifies differences between the results of the two scans.

Magnetoencephalography A method of noninvasive recording and analysis of the magnetic fields produced by electrical currents in brain cells.

Mass A property of matter to resist a change in velocity (inertial mass). A property of matter to create gravitational attraction between bodies (gravitational mass).

Mass-inertial characteristics (of the human body segments), See *Body segment parameters* Mass, location of the centers of mass, and moments of inertia.

Mass-spring models Simplified models that view muscles and moving body segments as inertial objects on springs with or without additional force-generating elements.

Matrix (in composite materials) Material that surrounds the reinforcement materials, for example, fibers.

Matrix (in mathematics) A rectangular array of numbers or other elements.

    Inter-finger connection matrix Matrix of weight coefficients that relates central commands to individual fingers (finger modes) with finger forces.

Maximum maximorum Highest among maximal values.

Mechanical advantage The ratio of mechanical output (e.g., muscle moment of force) to mechanical input (e.g., muscle force). In this particular example, mechanical advantage equals muscle moment arm.

Mechanical efficiency The ratio of work done to total energy expenditure.

Mechanical efficiency of positive work The amount of positive work done per unit of total available energy, where the total available energy is the sum of metabolic energy spent and the absolute value of muscle negative work.

Mechanical energy expenditure The total amount of mechanical energy expended for a given motion.

Membrane A biological structure built primarily from lipids that separates the cell from the environment.

    Channel Membrane sites specialized for transport of specific molecules across the membrane.

    Depolarization A decrease in the magnitude of the negative equilibrium potential inside the membrane that brings it closer to the threshold for action potential generation.

    Hyperpolarization An increase in the magnitude of the negative equilibrium potential inside the membrane that brings it further away from the threshold for action potential generation.

    Potential Electric potential inside the membrane as compared to the potential of the extra-cellular space.

    Threshold for activation Transmembrane potential, at which the cell generates an action potential.

Metabolic Related to metabolism, the biochemical processes occurring within a living organism.

Miometric action See *Concentric muscle action.*

Mirror neurons Cortical neurons that increase their activation level when an animal performs a particular action and when the animal observes another animal or a human performing this action.

Mode (in production of forces by fingers) A set of forces exerted by all four fingers due to a command sent to one of the fingers.

    Muscle mode A group of muscles with activation levels scaled in parallel.

Moment arm (of a force) A coefficient relating a force and the moment produced by this force about a certain center or axis.

    Effective moment arm A moment arm determined with functional methods; such a moment arm fits the equation "*moment of force = force × moment arm.*"

Moment arm vector The moment per unit of force expressed as a vector.

Moment axis The axis along the moment arm vector.

Moment of a couple A measure of the turning effect of a couple.

Moment of force A measure of the turning effect of a force.

    Primary moment An intended moment of force at a joint produced by a given muscle.

    Secondary moment A moment of force at a joint produced by a given muscle in an undesired direction. The moment should be negated by other muscles.

    Moments of force **(in object manipulation)**

        Agonist moments Moments resisting the moment of external force.

        Antagonist moments Moments assisting the moment of external force.

Moment of inertia A measure of the body's inertial resistance to changes in angular motion.

Monosynaptic A neural circuit involving only one neuro-neural synapse.

Motion analysis system A system that converts kinematic variables, in particular coordinates of special points on the body or on other objects, into electrical signals.

> *With active markers* Systems that use markers (light-emitting diodes) connected to the main computer.

> *With passive markers* Systems that use infrared emitters, reflective markers, and cameras that record marker positions in space.

Motoneuron A neuron that innervates (sends the axon to) muscle fibers.

> *Alpha* A motoneuron that innervates force producing, extrafusal fibers.

> *Gamma* A motoneuron that innervates intrafusal fibers inside muscle spindles.

Motor abundance The availability of more elements that are strictly necessary to solve a motor task viewed as a useful feature of the design. See also *Motor redundancy*.

Motor control Area of science exploring natural laws that define how the nervous system interacts with other body parts and the environment to produce purposeful, coordinated movements.

Motor evoked potential (a) Muscle activation (EMG) signals time locked to a stimulus. (b) Neural activation signals time locked to the initiation of an action.

Motor primitives (a) Hypothetical neural structures producing relatively simple blocks for complex actions; (b) Relatively simple patterns of mechanical variables that are used to construct complex actions.

Motor program An input into a subsystem within the neuromotor hierarchy expressed in neurophysiological variables and leading to actions produced by all the hierarchically lower subsystems.

> *Generalized motor program* A concept assuming that functions of control variables, directly related to forces and torques, are stored in the central nervous system and can be scaled in time and magnitude.

Motor redundancy The availability of more elements that are strictly necessary to solve a motor task viewed as a computational problem. Redundant motor systems are described mathematically by underdetermined sets of equations, that is, by the equation sets where the number of unknowns exceeds the number of equations.

> *Kinematic* Availability of more joint-related kinematic variables (rotations and translations) as compared to the number of constraints associated with typical motor tasks.

> *Kinetic* Availability of more force and moment of force variables at the level of effectors (e.g., muscles, digits) as compared to the number of constraints associated with typical motor tasks.

> *Muscle* Availability of more muscles as compared to the number of constraints associated with typical motor tasks.

> *Problem of motor redundancy* The problem of finding solutions for a system with more elemental variables than constraints.

Motor unit A collection of muscle fibers innervated by the same motoneuron.

> *Fast, fatigable (Type IIB)* Large motoneurons innervating large groups of muscle fibers characterized by fast conduction velocities along the axon, large forces, high rates of force development, and a quick drop in force during maintained force production.

> *Fast, fatigue-resistant (Type IIA)* Medium-size motoneurons innervating medium-size groups of muscle fibers characterized by medium-level forces, intermediate rates of force development, and resistance to fatigue.

> *Slow, fatigue-resistant (Type I)* Small motoneurons innervating small groups of muscle fibers characterized by slow conduction velocities along the axon, low forces, low rates of force development, and no signs of fatigue.

Movement economy Ability to move spending small amounts of energy.

Muscle action Production of muscle force.

Concentric muscle action The muscle produces force while shortening.

Eccentric muscle action The muscle produces force while lengthening.

Isometric muscle action The muscle produces force without change in the length of the muscle–tendon complex.

Muscle architecture Internal anatomy of the muscles.

Muscle belly A muscle without the tendons.

Muscle centroid Locus of the geometric centers of the muscle transverse cross-sections.

Muscle compartment A group of muscle fibers with similar mechanical actions within a muscle. Each compartment is surrounded by a sheath of connective tissue called *perimysium*.

Muscle coordination Purposeful activation of many muscles to produce a given motor task.

Muscle fiber A muscle cell.

Muscle matrix Intramuscular connective tissue that surrounds muscle fibers.

Muscle morphometry Anatomical characteristics of a muscle in the body, for example, location of its origin and insertion, moment arms.

Muscle path Line of muscle force action from one attachment site to another.

Muscle redundancy The number of muscles exceeds the number of kinematic degrees of freedom that the muscles serve.

Muscle spindles Complex anatomical structures housing sensory endings sensitive to muscle length and velocity.

Muscle–tendon complex See *Muscle–tendon unit*.

Muscle–tendon junction The connection between a muscle and its tendon.

Muscle–tendon plane The plane that contains all fixed and moving points of a muscle path.

Muscle–tendon unit (MTU) A muscle together with the attached tendon(s).

Muscle tone A poorly defined notion reflecting a feeling of resistance to an imposed motion in comparison to what is expected in a typical healthy person (normal tone).

Muscles

Biarticular muscles See *Two-joint muscles*.

Bi-pennate muscles Pennate muscles with the fibers converging to both sides of a central tendon.

Compartment A group of muscle fibers that shows anatomical and functional separation from other fibers within the same muscle.

Convergent muscles Muscles that are broad at the origin and converge at one attachment point at the insertion.

Extrafusal fibers Muscle fibers that lead to tendon force production during contraction.

Extrinsic hand muscles Muscles with the belly in the forearm that have several tendons attached to different fingers of the hand.

Homogeneous muscles Muscles with fibers of similar length.

Intrafusal fibers Muscle fibers inside muscle spindles. Have no direct effect on tendon forces.

Intrinsic hand muscles Muscles with the belly in the hand that produce finger-specific flexion action and contribute to the action of the extensor mechanism.

Invariant characteristic The dependence of active muscle force on muscle length; the term introduced within the equilibrium-point hypothesis.

Muscles with uniform architecture See *Homogeneous muscles*.

Poli-articular muscles Muscles that serve more than one joint.

Series-fibered muscles Muscles with nonspanning fibers.

Tetanic contraction Lasting contraction in response to a sequence of action potentials to muscle fibers.

Tone A poorly defined notion reflecting a feeling of resistance to an imposed motion in comparison to what is expected in a typical healthy person ("normal tone").

Triangular muscles See *Convergent muscles*.

Twitch contraction Short-lasting contraction in response to a single volley of action potentials to muscle fibers.

Two-joint muscles Muscles that cross and serve two joints.

Unipennate muscles Pennate muscles with the fibers to one side of a tendon.

Myelin A substance that covers neural fibers and increases the effective distance of local currents. This leads to higher speeds of action potential transmission.

Myelinated fibers Fibers covered with myelin. Characterized by high speeds of action potential transmission.

Myofibril A muscle fibril, a slender striated thread within a muscle fiber, composed of *Myofilaments*.

Myosin A contractile protein, the second is *Actin*.

Myotendinous junction Site where muscle fibers connect with the tendon.

N

Neural maps Topographic patterns of neural activation over relatively large areas in brain structures associated with motor or sensory events.

Neural pathways Groups of axons traveling together within the central nervous system.

Neuromuscular transmission Propagation of excitation from the axon of a motoneuron to muscle fibers.

Nonnegative matrix factorization (NNMF) A method of representing behavior of large groups of elemental variables that cannot have negative values with relatively few independent variables selected in an optimal way.

Nonparametric force–velocity relation Relation between the maximal maximorum force in a muscular strength test (F_{mm}) and the maximal speed V_m against a constant resistance, for example, body weight, for instance, the relation between the F_{mm} in a leg extension and the takeoff velocity in squat vertical jumps.

Nonspanning fibers Muscle fibers that end in the middle of the fascicle.

Normalized maximum shortening velocity Inverse of *Time-scaling parameter*.

Normalized moment vector See *Moment arm vector*.

Normalized shortening velocity Velocity of muscle shortening in units of muscle length per second.

Null space of a matrix The null space of matrix A is the set of all solutions to the equation $Ax = 0$, where **x** is a vector.

Null space (of a grasp matrix) A space of internal digit forces and moments; such forces and moments cancel each other and do not contribute to the manipulation (resultant) force.

Number of actions (of a muscle) Number of independent motions that can be imparted by the muscle on an unconstrained bone.

O

Objective function See *Cost function*.

Occam razor A simpler explanation of a phenomenon should be preferred over a more complex one.

Operant conditioning Development of a behavior when the animal is allowed to explore the environment and gets a reward for certain types of action.

Optimal control A branch of mathematics developed to find ways to control a system, which changes in time, such that certain criteria of optimality are satisfied.

Optimal feedback control (in movement studies) An optimization method based on a cost function that combines a measure of internal effort spent on control and a measure of accuracy of performance.

Optimal length (of a muscle) The muscle length at which maximal isometric force is produced.

Optimization An approach to controlling a redundant system by optimizing (typically, minimizing or maximizing) magnitude of a cost function.

Cost function A function of elemental variables that is minimized or maximized by the controller.

Minimum fatigue The cost function is related to fatigue associated with actions.

Minimum jerk The cost function represents the integral of jerk squared over movement time.

Minimal (second-order) norm Minimization of the euclidean distance between the initial and final states in the multidimensional space of elemental variables.

Minimum time Performing the action within the shortest time.

Minimum torque-change The cost function is the integral of torque derivative squared over movement time.

Direct Assuming a cost function and computing values of elemental variables based on task constraints.

Inverse Using observations to compute a cost function.

Origin (of a muscle) The proximal attachment of a muscle to a bone or another tissue.

P

Parallel elastic component, PEC (in muscle models) Muscle elements responsible for the passive resistance of a nonstimulated muscle to stretch.

Parallel manipulator A manipulator with two or more end effectors, for example, fingers grasping a rigid object.

Passive elements (of the musculoskeletal system) Ligaments, tendons, inactive muscles, and inactive joints. Passive elements cannot change their length (angle) without external forces.

Pathways Groups of axons that travel together within the central nervous system.

Ascending Pathways that carry sensory information in a rostral direction, from sensory neurons to structures within the central nervous system.

Descending Pathways that carry information in a caudal direction, related to a planned motor action.

Corticobulbar tract A neural tract originating in the cortex and innervating the nuclei of the cranial nerves.

Corticospinal tract A neural tract originating in the cortex and projecting on various spinal neurons.

Pyramidal tract A neural tract from the cortical frontal motor areas with a contribution from neurons in the parietal somatosensory areas; it splits into the corticobulbar and corticospinal tracts.

Relay nuclei Neuroanatomical formations that receive inputs from neural pathways and serve as outputs of other pathways.

Topographic organization Preservation of topography: Neighboring original neurons project onto neighboring neurons in relay nuclei, which in turn project onto neighboring neurons in the target.

PCSA See *Physiological cross-section area.*

PEC See *Parallel elastic component.*

Pendular motion Motion of an ideal pendulum, in which the total mechanical energy is constant over an oscillation cycle.

Pennate muscles Muscles with the fascicles attached to the tendon at an angle.

Pennation angle Angle between the direction of muscle fibers and either (a) the line of muscle force action (external portion of the tendon) or (b) the aponeurosis (internal tendon).

Pennation plane Plane formed by the tangent vector of the fascicle and the tangent vector of the aponeurosis at the point of contact; the pennation angle is measured in this plane.

Perception-action coupling Coupling of sensory signals to motor commands directly without first using these signals to update the picture of the world and one's own body.

Phase portrait The dependence between object's velocity and coordinates.

Plane of the grasp (in the prismatic grasps) A plane that contains all the points of digit contacts with the object.

Plastic body A deformable body that does not return to its original size and shape when external force is removed.

Plasticity An ability of neural structures to change gains of synaptic connections and to establish new connections with practice or injury.

Plateau region (of a sarcomere's force–length curve) Length at which the sarcomere generates maximal forces.

Plyometric action See *Eccentric muscle action.*

Population vector A vector (typically, of a mechanical variable) representing activity within a large group of neurons.

Positron emission tomography (PET) A method of imaging internal body organs using a radioactive isotope incorporated into a molecule, which participates in biological processes within the body, injected into the blood stream.

Posture A combination of the relative positions of body segments and/or of the whole body with respect to a reference frame.

> *Inverted pendulum model* Analysis of vertical posture as the problem of balancing an inverted pendulum.
>
> *Light touch effects* Reduction of postural sway and other effects on posture by a light finger touch to an external object.
>
> *Posture-movement paradox* Posture-stabilizing mechanisms resist deviations produced by external forces but not those produced by voluntary movements.
>
> *Posture-stabilizing mechanisms* All the mechanisms that resist mechanical perturbations of posture, peripheral, reflex, preprogrammed, and voluntary.
>
> > *Ankle strategy* A pattern of muscle activation that starts with major changes in the activation of muscles crossing the ankle joint and involves body motion mostly about the ankle joints.
> >
> > *Hip strategy* A pattern of muscle activation that starts with major changes in the activation of muscles crossing the hip joint and involves body motion mostly about the hip joints.
>
> *Sway* Involuntary changes in the coordinates of the center of pressure or the center of mass.
>
> > *Rambling* A sway component representing an interpolation of the coordinates of instantaneous equilibrium points.
> >
> > *Trembling* A sway component representing the difference between the sway and rambling.

Power The rate of doing work.

> *Muscle power* Power of muscle forces.

Power law Dependence of the endpoint velocity on the trajectory curvature.

> **2/3-power law** The relation between angular velocity $\mathcal{A}$ and curvature C of the movement path segment $\mathcal{A} = k \cdot C^{2/3}$, where k is a constant.

> **1/3-power law** The relation between the limb endpoint tangential velocity v and the curvature C of the movement path segment $v = k_1 \cdot r \cdot C^{1/3}$, where r is the radius of curvature C and k_1 is a coefficient.

Preferred direction (of a neuron) Direction of motion corresponding to the largest activity level of a neuron. See *Cosine tuning*.

Preflex The instantaneous reaction of a muscle to an external perturbation that depends on the muscle activation level and, hence, can be changed in advance.

Prehension The combined sensory-motor function of the grasping hand.

> *Principle of superposition* Elemental variables form relatively independent subgroups, which are used for the control of different performance variables such as the total force and the total moment of force.

> *Synergy* Covaried adjustments of elemental variables produced by the digits that stabilize the mechanical action of the hand.

Preprogrammed reactions (M2-3, triggered reactions) Reactions to perturbations that come at an intermediate latency, between spinal reflexes and voluntary reactions, and depend on the instruction to (intention of) the person.

Pressure The amount of force per unit of area in a direction perpendicular to the surface of an object.

Principal component analysis (PCA) A correlation or covariation matrix factorization method that allows representing a multidimensional set of data with a smaller number of orthogonal principal components.

> *Factor analysis* An extension of PCA. It usually involves optimal rotation of the principal components.

Principle of minimal action A general physical principle according to which an object moves in such a way as to minimize a certain physical variable (e.g., time, distance, and work). In accordance with this principle, the control system chooses movement trajectory and endpoint velocity such that the mechanical work done on the endpoint and the arm is minimal.

Principle of mechanical advantage A principle according to which effectors with larger moment arms produce larger shares of the resultant moment of force.

Principle of superposition (a) Decomposing complex actions into elemental actions that are controlled independently by separate controllers. (b) Viewing a process as the sum of two or more elemental processes that proceed independently of each other.

Principle of virtual work The sum of works of all forces and moments done during virtual displacements in a system with workless constraints is zero.

Problem of dynamics

> *Direct problem of dynamics* Determining the changes in motion caused by the given forces.

> *Inverse problem of dynamics* Determining the forces that produce a given motion.

Proprioception Perception of position of and load on body segments.

Proximal Closer to the center of the human body or to the origin of a referent system, the opposite of *distal*.

Pseudo-inverse A particular solution for the problem of redundancy that minimizes the euclidean distance between the initial and final state in the space of elemental variables.

Pulse-step model A model that assumes separate control over the movement trajectory (pulse) and over the final position (step). Typically, it assumes control with patterns of muscle activation.

Purkinje cells The largest cells in the cerebellum that produce output to the cerebellar nuclei.

Q

Quick-release method A technique of muscle research: a muscle exerts force isometrically and then it is allowed to shorten against a smaller load.

R

Range of motion The difference between the two extremes of joint movement.

Rambling (in the rambling–trembling hypothesis) Migration of the reference point with respect to which the balance during standing is instantly maintained.

Rate coding Force control through changes in the firing rate of motoneurons.

Readiness potential (Bereitschaftspotential) A slow, gradual change of the brain potential, typically into negative values, seen in preparation to a voluntary action 1 s or more prior to movement initiation.

Reafference principle A principle of comparing afferent signals with a copy of efferent signals. As a result, afferent signals are interpreted as reporting deviations of the body from the referent coordinate, not from an absolute posture.

Receptor (a) A cell or subcellular structure (sensory ending) that converts physical variables into changes in the membrane potential; (b) A molecular structure specialized for making chemical bonds with specific molecules.

>*Exteroceptors* Receptors sensitive to physical variables that carry information about events in the external world.

>*Interoceptors* Receptors sensitive to physical variables that carry information about events within the body.

>*Proprioceptors* Receptors sensitive to physical variables that carry information about contacts between the body and the external world, forces between body parts, and about the body configuration and its changes.

>>*Articular receptors* Sensory endings located in the joint capsule. They generate action potentials reflecting joint angle and joint capsule tension.

>>*Cutaneous receptors* Various receptors sensitive to skin deformation (pressure). They differ in size of the receptive field and rate of adaptation to maintained deformation.

>>*Golgi tendon organs* Sensory endings located at the junction between the muscle fibers and the tendon, sensitive to tendon force.

>>*Muscle spindle* A complex structure in the muscle that contains small muscle fibers (intrafusal fibers) and two types of sensory endings, primary and secondary. Connected to power-producing muscle fibers by strands of connective tissue.

>>>*Primary endings* Sensory endings sensitive to muscle length and velocity, innervated by Ia-afferents.

>>>*Secondary endings* Sensory endings sensitive to muscle length, innervated by group II afferents.

Reciprocal inhibition An inhibitory reflex connection involving two synapses within the central nervous system from primary spindle endings mediated by Ia-interneurons to alpha-motoneurons innervating the antagonist muscle.

Reciprocal (r-) command One of the two basic commands within the equilibrium-point hypothesis. It leads to shifts of the midpoint of the joint angle range, within which both agonist and antagonist muscles are activated, without a change in the size of the range.

Reciprocal activation Decrease of activity of an anatomic antagonist when activity of an agonist increases.

Recruitment An increase in the number of active neurons (e.g., motoneurons).

Recuperation (of energy) See *Energy compensation during time.*

Recurrent inhibition Inhibition of alpha-motoneurons of a pool produced by small interneurons (Renshaw cells) that are excited by branches on the axons of the alpha-motoneurons of the same pool.

Red nucleus A small brain structure, the source of the rubrospinal tract. It receives an input from cerebellar nuclei.

Redundancy Availability of numerous (an infinite number of) solutions to motor tasks associated with a smaller number of constraints as compared to the number of degrees of freedom (elemental variables).

  Redundancy **(of a kinematic chain)** The same endpoint position can be assumed by various joint configurations.

Reference system (reference frame) System of coordinates.

Referent configuration (a) A body configuration, at which all the muscles are at the threshold of activation via the tonic stretch reflex; (b) A set of salient spatial variables, for which referent values are set at a high level of the control hierarchy.

Reflex A poorly defined notion. Typically, a quick reaction to a stimulus that is relatively stereotypical and does not depend on the instruction to the subject. It is frequently assumed that reflexes are mediated by spinal structures.

  Autogenic Reflexes seen in a muscle produced by sensory signals from receptors in the same muscle.

  Conditioned Reflexes developed by repetitive simultaneous presentation of two stimuli with gradual substitution of a natural stimulus with a different one.

  Crossed extensor Reflex contractions seen in many major extensor muscles of an extremity in response to stimulation of flexor reflex afferents in the contralateral extremity.

  Flexor Reflex contractions seen in many major flexor muscles of an extremity in response to stimulation of flexor reflex afferents within the same extremity.

  H-reflex A monosynaptic reflex produced by stimulation of Ia-afferents that make excitatory projections on alpha-motoneurons of the muscle from which the afferents originate.

  Heterogenic Reflexes seen in a muscle produced by sensory signals from receptors located in another muscle.

  Inborn Reflexes that are seen in newborn animals.

  Latency Time delay between a stimulus and the initiation of the earliest response.

  Long-loop A reflex response with a relatively long latency (on the order of 50–90 m) that shows complex patterns of muscle activation that can change with action and instruction.

  Monosynaptic Reflex response mediated by only one synapse within the central nervous system.

  Oligosynaptic Reflex response mediated by a few (two to three) synapses within the central nervous system.

  Phasic Transient, short-lasting, reacting to a change in the respective physical variable.

  Polysynaptic Reflex response mediated by many (usually, an unknown number of) synapses within the central nervous system.

  Reversal Switching of reflex responses from a muscle to its antagonist with changes in the conditions such as body posture and/or contact forces with the environment.

  Tonic Steady state, reacting to magnitude of the respective physical variable.

  Tonic stretch reflex A polysynaptic reflex that defines the dependence of active muscle force on muscle length.

    Characteristic The dependence of active muscle force on muscle length.

    Threshold The shortest muscle length, at which muscle activation is seen during a very slow stretch of the muscle.

Tonic vibration reflex Tonic muscle contraction seen in response to high-frequency, low-amplitude vibration applied to the muscle or its tendon (sometimes seen in other muscles of the body).

Unloading reflex A short-latency change (usually, a decrease) in the muscle activation level in response to an unloading of the muscle leading to its quick shortening.

Vestibulo-ocular reflex Short-latency reflex eye rotation produced by sensory signals in the vestibular apparatus that helps keep the projection of an object of interest on the fovea.

Wiping reflex In some animals, a reflex action by an extremity wiping the irritating stimulus off the body.

Refractory period A period of decreased or lacking response to stimulation.

Absolute The membrane does not generate an action potential even to a very strong stimulus.

Relative The membrane can generate an action potential but requires a stronger stimulus than usual.

Relay nuclei Neuroanatomical formations that receive inputs from neural pathways and serve as outputs of other pathways.

Renshaw cells Small interneurons located in the anterior horns of the spinal cord. They are excited by branches of the axons of alpha-motoneurons and inhibit alpha-motoneurons of the same pool as well as gamma-motoneurons innervating spindles within the same muscle.

Repetition without repetition A term coined by N. Bernstein that implies solving the same motor task with different neural and motor patterns.

Residual force enhancement Long-lasting muscle force increase after stretch.

Rest length (of a muscle) The natural muscle length in situ.

Reversible muscle action See *Stretch-shortening cycle*.

Rigid body A body in which the distance between any two points within that body is constant.

Rigidity A clinical symptom involving increased velocity-independent resistance to muscle stretch.

Rotation Movement of a body about an axis or center during which all particles of the body travel in the same direction through the same angle.

S

Safety margin Extra gripping force above the minimal level necessary to avoid slippage of the handheld object.

Sarcolemma A membrane enclosing a muscle fiber.

Sarcomere A contractile unit of skeletal muscle.

Sarcomere nonuniformity Uneven distribution of muscle strain across sarcomeres.

Scalar A physical variable characterized by magnitude but not by direction (e.g., mass, distance, temperature, and speed).

Screw axis A line in space. At any given instant, the translation and rotation of a body occur along and around a screw axis.

SEC See *Series elastic component*.

Secondary moment Moment of force that is not necessary for task execution.

Secondary moment hypothesis The CNS tries to minimize secondary moment to avoid needless muscle activity.

Sensory reweighting The importance of signals from different groups of receptors may be task specific and may also show changes with practice.

Series elastic component (SEC) Muscle elements that behave as a spring connected in series to the force generator in the muscle (the contractile component). The SEC is a functional element of muscle mechanics, not a precisely defined anatomical structure.

Servo-control A particular version of a feedback control system that acts to keep a variable at a level specified by a hierarchically higher system despite possible changes in the environment.

Servo-hypothesis A hypothesis on muscle control based on sending a signal to gamma-motoneurons to the muscle. This signal is viewed as setting a magnitude of muscle length that is kept constant with the help of the tonic stretch reflex viewed as a perfect servo.

Sharing problem See *Distribution problem.*

Shortening factor Actual fiber length divided by its length at rest.

Shortening heat The heat proportional to the shortening distance of the muscle and independent of load.

Shortening-induced force depression Reduction of muscle force after muscle shortening and redevelopment of isometric force compared to isometric force developed at the same length without preliminary shortening.

Singularity (in kinematics) A specific position for which a given mathematical operation is not defined or its result equals zero.

SISO model Single input–single output model.

Size principle (Henneman principle) The principle of orderly recruitment of motor units within a muscle, from the smaller motoneurons to the larger ones. Derecruitment follows the opposite order.

Slack length (of a tendon) The length at which a tendon starts resisting extension.

Slip See *Yield effects.*

Sliding filament theory Muscle force is generated by the cross-bridges causing relative sliding of the actin and myosin filaments.

Source approach (of determining mechanical energy expenditure) A method of computing mechanical energy expenditure as the time integral of joint power.

Sources of mechanical energy Forces and moments of force.

Intercompensated sources Sources that compensate for each other. Two-joint muscles are intercompensated sources, they can absorb energy at one joint and generate energy at the other.

Nonintercompensated sources of energy The energy expended by one source is not compensated by the energy absorbed by another source. For instance, joint torques on joints served by single-joint muscles are not intercompensated.

Sources compensated over time (Recuperative sources) Sources that can absorb mechanical energy for subsequent use.

Spasticity A clinical condition characterized by excessive involuntary muscle activation, spasms, and exaggerated reflex responses to peripheral stimuli.

Negative signs Signs that are absent in a person with spasticity as compared to a healthy person (such as weakness, discoordination, and quick fatigability).

Positive signs Signs that are present in a person with spasticity but not in a healthy person.

Clasp-knife phenomenon A specific pattern of joint reaction to an externally imposed motion: First, the joint resists the motion strongly, but at some point it collapses like a pocketknife.

Clonus Involuntary alternating activation of muscles within an agonist–antagonist muscle pair, typically at 6–8 Hz.

Muscle tone A misnomer reflecting subjective impression that a clinician feels when trying to move a body segment while the subject tries to be relaxed.

Spasms Involuntary brisk or long-lasting episodes of strong muscle contractions. Spasms can be local or involve large muscle groups (flexor spasms and extensor spasms).

Specific tension See *Specific force*.

Spectral analysis A method of signal processing that represents a signal as a superposition of sine waves at different frequencies and with different amplitudes.

Speed The time rate of covering distance. Speed is a scalar.

Speed-accuracy trade-off When a person moves at different speeds to a visual target, movement time varies as a linear function of the ratio of distance to the standard deviation of the final position.

Speed-difficulty trade-off (Fitts's Law) Under the instruction to be both fast and accurate, movement time shows a logarithmic dependence on the ratio of movement distance to target size.

Spinal cord The elongated structure within the central nervous system, from the medulla to the lumbar spinal vertebrae, which plays a major role in the sensorimotor function. It contains pathways that deliver sensory information to and neural commands from the brain, and interneurons that mediate reflexes and play a role in the control of certain movements.

　　Roots Sites of entrance and exit of axons.

　　　　Dorsal Sites of entrance of axons that carry sensory information into the spinal structures.

　　　　Ventral Sites of exit of axons that carry signals from the spinal cord to peripheral structures such as muscles.

　　Segments Each segment receives sensory information from a particular area of the body and innervates muscles in more or less the same area of the body. There are 8 cervical segments, 12 thoracic segments, 5 lumbar segments, and 5 sacral segments.

Spinal ganglia Groups of neurons outside the spinal cord but in close proximity to it. Spinal ganglia contain bodies of proprioceptive neurons.

Spring constant See *Stiffness*.

Spring-like actions (of muscles and extremities) Elastic energy storage and recoil.

Stability A physical property of a system reflecting its ability to keep the current state or trajectory when subjected to a transient perturbation.

　　Dynamic An ability to move toward the original trajectory after a transient perturbation.

　　Postural An ability to maintain posture despite possible external perturbations.

　　Static An ability to return to the same steady state after a transient perturbation.

Stages of motor learning Three stages suggested by N. Bernstein: freezing redundant degrees of freedom, releasing those degrees of freedom, and utilizing external forces to one's advantage.

Static analysis (of kinematic chains)

　　Direct static analysis Computing the end-effector force from known joint torques.

　　Inverse static analysis Computing the joint torques from known external forces.

Static optimization Optimization that neglects time.

Stiffness The amount of force per unit of deformation. Stiffness is the inverse of compliance. Application of this term to "active objects," such as muscles or human extremities, is questionable because their resistance to deformation, for example, to stretch, is time dependent and is under neural control. See also *Apparent stiffness*.

　　Apparent stiffness A stiffness-like measure obtained from an active object, for example, an active muscle. The word *apparent* underscores that this measure differs from the analogous parameters of passive objects.

　　Joint stiffness A misnomer trying to represent joint mechanical behavior as that of a spring. "Apparent joint stiffness" is recommended.

　　Limb stiffness A misnomer trying to represent limb mechanical behavior as that of a spring.

Short-range stiffness Muscle resistance to stretch during the initial phase of muscle elongation.

Stiffness ellipse (at the endpoint) The ellipse obtained by connecting the tips of the restoring forces in response to a unit deflection in all directions.

   Ellipse orientation The angle between the major axis of the ellipse and the X-axis of a fixed reference system.

   Ellipse shape (Eccentricity) The ratio of the major to minor axes of an ellipse.

   Ellipse size The ellipse area.

Stiffness matrix A matrix relating the differential force or torque increments with the differential deflection or deformation.

Strain A relative elongation $\Delta l/l$ where l is the initial length of the object and Δl is its elongation.

   ***Apparent strain* (of a tendon)** Amount of elongation per unit of tendon length.

   Eulerian strain The ratio of the change in length of a line segment to its final length after the deformation.

   Lagrangian strain The ratio of the change in length of a line segment to its initial length before the deformation.

   ***Macroscopic strain* (of a tendon)** See *Apparent strain*.

   ***Material strain* (of a tendon)** The tendon fiber length change per unit of fiber length.

   Microscopic strain See *Material strain*.

   Transverse strain Strain normal to the applied load.

Stress The amount of force per unit of area.

   Normal stress Stress in a plane orthogonal to force direction.

   Shear stress Stress in a plane parallel to the force direction.

Stress relaxation Decrease of stress over time under constant deformation.

Stretch reflex Change in muscle activation induced by muscle stretch.

Stretch-shortening cycle The sequence of muscle stretch and shortening.

Synapse A structure consisting of a presynaptic membrane, a synaptic cleft, and a postsynaptic membrane that uses physical and chemical processes to enable transmission of excitation and inhibition between excitable cells.

   Excitatory A synapse that causes depolarization of the postsynaptic membrane, that is, brings its membrane potential closer to the threshold for action potential generation.

   Inhibitory A synapse that causes hyperpolarization of the postsynaptic membrane, that is, brings its membrane potential further away from the threshold for action potential generation.

   Neuromuscular A synapse between a terminal branch of the axon of a motoneuron and a muscle fiber.

   Non-obligatory A synapse that, when it acts alone, is unable to bring the postsynaptic membrane to the threshold for action potential generation. Neuro-neural synapses are typically nonobligatory.

   Obligatory A synapse that always brings the postsynaptic membrane to the threshold for action potential generation. Neuromuscular synapses are typically obligatory.

Synergy (a) Stereotypical patterns of activation of major flexor or extensor muscles typical of certain movement disorders, for example, those seen after stroke; (b) A set of variables that show correlated changes in time or with changes in task parameters; or (c) A neural organization that shares performance among a redundant set of variables and ensures stability of performance variables by using flexible combinations of elemental variables.

   Covariation A feature of a synergy characterized by covariation of elemental variables that helps maintain a desired value of a performance variable.

Linear torque synergy Linear scaling of joint torques during voluntary movement of a multi-joint chain.

Multi-muscle (a) A group of muscles that show correlated changes in activation levels over time or with changes in task parameters; (b) A neural organization ensuring covaried adjustments in muscle activation levels that keep a desired value or a desired time profile of a performance variable.

Sharing A feature of a synergy characterized by a pattern of relative involvement of elemental variables.

T

Tangent modulus of elasticity See *Young's modulus*.

Tendon A band of fibrous tissue that connects a muscle to a bone.

Tendon action (of muscles) Muscles' behavior as nonextensible struts.

Tetanic Pertaining to *Tetanus*.

Tetanus A sustained muscle contraction caused by high-frequency firing of motoneurons.

Thalamus A brain structure that consists of a number of nuclei; it is part of several major loops that link the cortex of the large hemispheres, the basal ganglia, and the cerebellum, participating in the production of movement.

Thixotropy Property of a substance to decrease its viscosity when it is shaken or stirred.

Three-element muscle model See *Hill's model*.

Toe region A region of the stress–deformation curve of a tendon at low levels of the stress.

Toe-region modulus The average slope of the stress–deformation curve in the toe region.

Tonic stretch reflex A reflex mechanism with an unknown neural path that links muscle length to active muscle force production.

Torque See *Moment of a couple*.

Active joint torques Joint torques generated by torque actuators.

Joint torque Two moments of force about the joint rotation axis acting on the adjacent segments. The moments are equal in magnitude and opposite in direction.

Interaction torques (forces) The joint torques and forces induced by motion in other joints.

Net joint torque The total torque produced at a joint; the sum of the active and passive torques.

Passive joint torques Joint torque resisted by passive structures, in particular by the skeleton and ligaments.

Reactive torques The torques arising due to Newton's third law; see also *Interaction torques*.

Torque agonists (in grasping) The fingers resisting external torque and hence helping to keep rotational equilibrium.

Torque antagonists (in grasping) The fingers assisting external torque.

Trajectory In kinematics, a path of the moving point. Generally, a continuous time function of a physical variable.

Control The time function of a control variable.

Equilibrium The time function of variables characterizing equilibrium states of the system.

Virtual An imprecise term, used sometimes as a synonym of equilibrium trajectory and sometimes as a trajectory that the system would have shown if the external load were zero.

Transcranial magnetic stimulation A method of noninvasive stimulation of neural structures using eddy currents produced by a quickly changing magnetic field.

Translation (in kinematics) Movement of a body so that any line fixed with the body remains parallel to its original position.

Trembling (in the rambling-trembling hypothesis) Oscillation of the body around the rambling trajectory while standing.

Tremor Involuntary cyclic movement associated with alternating activation bursts in the agonist and antagonist muscles.

Parkinsonian tremor Is typically postural; it may be alleviated by voluntary movement.

Cerebellar tremor Has postural, kinetic (increases during movement), and intentional (increases when the extremity approaches the target) components.

Triphasic activation pattern A muscle activation pattern consisting of an initial agonist burst accompanied by a low level of antagonist co-contraction, followed by an antagonist burst (during which the agonist is relatively quiescent), and ending with one more agonist burst.

Twitch A brief muscle contraction in response to a single stimulus.

Two-joint muscles Muscles that cross and serve two joints.

U

Ultrasound Sound pressure with a frequency exceeding the upper limit of human hearing, approximately 20 kHz.

Uncontrolled manifold A subspace in the space of elemental variables corresponding to a fixed value of a performance variable; the controller does not have to exert control as long as the elemental variables stay within the uncontrolled manifold.

Hypothesis A hypothesis that the central nervous system organizes an uncontrolled manifold (UCM) corresponding to a desired value of a potentially important performance variable and then acts to keep the elemental variables within the UCM.

V

Variability (of human motion) A feature of motor performance reflecting different solutions for the same motor problems that are seen over time or during repetitive trials.

Variance A measure of variability equal to standard deviation squared.

"Good" variance A component of variance that does not affect a specific performance variable.

"Bad" variance A component of variance that affects a specific performance variable.

Vector A physical or mathematical variable having magnitude and direction; a unidimensional array of numbers.

Binormal vector Vector perpendicular to the tangent and normal vectors.

Couple vector A free vector that represents a force couple. The couple vector is normal to the plane of the couple.

Curvature vector See *Normal vector*.

Normal vector Vector in the direction of normal acceleration, also called the *Curvature vector*.

Tangent vector Vector in the direction of the derivative at the point of interest on a curve. If the curve were representing a trajectory of a point in time, the tangent vector would be a unit vector in the direction of the instantaneous velocity of the point.

Vector product See *Cross product*.

Velocity The time rate of change of position. Velocity is a vector.

Angular velocity The rate of movement in rotation.

Joint velocity Angular velocity at a joint.

Segment velocity Velocity of a body segment as viewed by an external observer, in an absolute reference frame. Segment velocity can be both linear and angular.

Velocity-dependent response (of human muscles to stretch with constant moderate speed) The second phase of the response, during which force continues to increase but with progressively slower rate until the force reaches its peak at the end of stretch.

Vestibular apparatus Structures in the inner ear that contain vestibular receptors; these receptors convert linear and angular accelerations into sequences of action potentials.

Vestibular nuclei Four paired nuclei located in the brain stem: the superior, lateral, medial, and inferior nuclei. They receive projections from the vestibular receptors and from other structures including the cerebellum.

Vestibulospinal tracts The lateral vestibulospinal tract innervates antigravity muscles; the medial vestibulospinal tract innervates neck muscles that are responsible for stabilizing the head.

Vibration-induced falling (VIF) Violations of vertical posture produced by high-frequency, low-amplitude vibration applied to postural muscles (VIFs can also be induced by vibration applied to some of the other muscles).

Virtual displacement A hypothetical small displacement of a body or a system from an equilibrium position.

Virtual finger An imaginary digit with the mechanical action equal to that of a set of fingers (commonly, the four fingers of a hand) combined.

Viscoelasticity Combination of viscous and elastic properties in materials.

Viscosity Internal friction between molecules.

Viscosity hypothesis Under standard stimulation a muscle generates similar internal forces but these forces are not manifested externally due to the viscous resistance within the muscle. The hypothesis was abandoned after A.V. Hill (1938) showed that the heat production is sharply different in concentric and eccentric muscle actions.

W

Weakest-link principle Force development rate is determined by the slowest muscle fibers.

Weight (of an object) The gravity force exerted on the object.

Work (of a force) Scalar product of the vectors of force and displacement.

External work (**in biomechanics**) Work done by the performer on the environment.

Internal work (**in biomechanics**) Work done by the performer to change the mechanical energy of own body or body segments.

Negative muscle work Work done by a muscle to resist its elongation. The absolute value of this work equals to the value of work done on the muscle by an external force to extend the muscle.

Positive work (**of a force or a moment of force**) Work done over a linear or angular displacement in the direction of the force or moment.

Quasi-mechanical work An imagined work that would be done on a body segment by the resultant force and couple if there were no transformation between the gravitational potential energy and kinetic energy.

Virtual work The work done by a force over a virtual displacement.

Work-energy principle (for a rigid body) The total work done on the body by several forces equals the change in the body's mechanical energy.

Working envelope The set of boundary points that can be reached by the end effector.

Wrench A force and a couple with the vectors along the same line.

X

X-rays Electromagnetic radiation with the frequencies outside the visible range, 10^{16}–10^{19} Hz.

Y

Yield effect Sudden reduction of muscle force during a fast stretch.

Young's modulus The ratio of the uniaxial stress over the uniaxial strain in the range of the linear stress–strain relation.

Z

Z-membrane A membrane separating two sarcomeres.

Zero-work paradox For movements beginning and ending at rest at the same location the energy is expended but the total work is zero.

Index